Vegetable Production and Marketing in Africa

Socio-economic Research

MIX
Paper from responsible sources
FSC® C013604
FSC www.fsc.org

Vegetable Production and Marketing in Africa

Socio-economic Research

Edited by

Dagmar Mithöfer

World Agroforestry Centre, Kenya

and

Hermann Waibel

Leibniz University of Hannover, Germany

www.cabi.org

CABI is a trading name of CAB International

CABI Head Office
Nosworthy Way
Wallingford
Oxfordshire OX10 8DE
UK

Tel: +44 (0)1491 832111
Fax: +44 (0)1491 833508
E-mail: cabi@cabi.org
Website: www.cabi.org

CABI North American Office
875 Massachusetts Avenue
7th Floor
Cambridge, MA 02139
USA

Tel: +1 617 395 4056
Fax: +1 617 354 6875
E-mail: cabi-nao@cabi.org

A catalogue record for this book is available from the British Library, London, UK.

Library of Congress Cataloging-in-Publication Data

Vegetable production and marketing in Africa : socio-economic research/ edited by Dagmar Mithöfer, Hermann Waibel.
p. cm.
Includes bibliographical references and index.
ISBN 978-1-84593-649-5 (alk. paper)
1. Vegetable trade–Africa. 2. Vegetables–Africa–Marketing. 3. Agriculture–Economic aspects–Africa. I. Mithöfer, Dagmar. II. Waibel, Hermann, 1951- III. Title.

HD9220.A352V44 2011
338.1'75096–dc22

2011003308

ISBN-13: 978 1 84593 649 5

Commissioning editor: Sarah Hulbert
Production editors: Tracy Head and Simon Hill

Typeset by AMA DataSet, Preston, UK.
Printed and bound in the UK by CPI Antony Rowe.

Contents

Contributors

Mary O. Abukutsa-Onyango, Associate Professor of Horticulture, Maseno University, Kisumu-Busia Road, Maseno, PO Box 333, 40105-Maseno, Kenya. Present address: Professor of Horticulture, Jomo Kenyatta University of Agriculture and Technology, Thika Road, Juja PO Box 62000-00200, Nairobi, Kenya. mabukutsa@yahoo.com

David Amudavi, Consultant, International Centre of Insect Physiology and Ecology (*icipe*), PO Box 30772-00100, Nairobi, Kenya, and Lecturer, Egerton University, Department of Agricultural Education and Extension, PO Box 536-20155, Egerton, Njoro, Kenya. damudavi@africa-online.co.ke

Anna Asfaw, Research Assistant, Leibniz University of Hannover, Faculty of Economics and Management, Königsworther Platz 1, 30167 Hannover, Germany. Present address: Economic Adviser, Deutsche Gesellschaft für Technische Zusammenarbeit (GTZ), Postfach 5180, 65726 Eschborn, Germany. anna.asfaw@gmx.de

Solomon Asfaw, Research Associate, Leibniz University of Hannover, Faculty of Economics and Business Administration, Königsworther Platz 1, 30167 Hannover, Germany. Present address: Economist, Agricultural and Development Economics Division, Food and Agriculture Organization of the United Nations, Rome, Italy. Solomon.Asfaw@fao.org

Nigat Bekele, Postgraduate Research Fellow, International Centre of Insect Physiology and Ecology (*icipe*), PO Box 30772-00100, Nairobi, Kenya, and PhD Candidate, Egerton University, Department of Agricultural Economics and Business Management, PO Box 536-20155, Egerton, Njoro, Kenya. nbekele@icipe.org

Emma Blackmore, Researcher, International Institute for Environment and Development, 3 Endsleigh Street, London, UK. emma.blackmore@iied.org

Jantje Blatt, Landscape Ecologist, Centre for Advanced Training in Rural Development (SLE), Humboldt-University, Hessische Str. 1–2, 10115 Berlin, Germany. www.berlinerseminar.de. sabine.doerr@agrar.hu-berlin.de

Kristian Brakel, MA in Islamic Studies and Education, Centre for Advanced Training in Rural Development (SLE), Humboldt-University, Hessische Str. 1–2, 10115 Berlin, Germany. www.berlinerseminar.de.sabine.doerr@agrar.hu-berlin.de

Liesbeth Colen, Junior Researcher and PhD Student, Centre for Institutions and Economic Performance (LICOS), Katholieke Universiteit Leuven, Belgium. Liesbeth.Colen@econ.kuleuven.be

Christoph Ehlert, Research Assistant, Leibniz University of Hannover, Faculty of Economics and Management, Königsworther Platz 1, 30167 Hannover, Germany. Present address: Research Associate, Rheinisch-Westfälisches Institut für Wirtschaftsforschung e.V., Hohenzollernstraße 1–3, 45128 Essen, Germany. christoph.ehlert@rwi-essen.de

Munguzwe Hichaambwa, Research Associate, Food Security Research Project, Lusaka, Zambia. munguzwe.hichaambwa@iconnect.zm

Kees Jansen, University Lecturer, Technology and Agrarian Development Group, Wageningen University, the Netherlands. kees.jansen@wur.nl

Phyllis G. Kasambula, Senior Technician, National Agricultural Research Organization Kawanda Agricultural Research Institute Horticulture Research Programme, PO Box 7065, Kampala, Uganda. Present position: Agronomist, National Agricultural Research Organization National Crops Resources Research Institute (NaCRRI) Horticulture Research Programme – Namulonge, PO Box 7084, Kampala, Uganda. mirembegaliwango6@gmail.com

Kristina Kloss, Business Administration, Centre for Advanced Training in Rural Development (SLE), Humboldt University, Hessische Str. 1–2, 10115 Berlin, Germany. www.berlinerseminar.de.sabine.doerr@agrar.hu-berlin.de

Thomas König, Agricultural Economist, Centre for Advanced Training in Rural Development (SLE), Humboldt University, Hessische Str. 1–2, 10115 Berlin, Germany. (www.berlinerseminar.de). Present position: Senior Adviser for Sustainable Development/Agricultural Economist, Office of the Mountain-River-Lake Development Committee of Jiangxi Province (MRLDO), North One Road, 14, Provincial Governmental Complex, Nanchang City, 330046 China. thomas.koenig@cimonline.de

Jillian M. Lenné, Consultant, North Oldmoss Croft, Fyvie, Turriff, Aberdeenshire AB53 8NA, UK. jillian.lenne@btopenworld.com

Bernhard Löhr, Coordinator, Development of biocontrol-based IPM for the diamondback moth, Plutella xylostella L. in Eastern and Southern Africa, International Centre for Research on Insect Physiology and Ecology (*icipe*), PO Box 30772-00100, Nairobi, Kenya. Present address: Consultant, zum Breitbusch 7, 76855 Annweiler, Germany. blohr55@gmail.com

James MacGregor, Senior Researcher, International Institute for Environment and Development, 3 Endsleigh Street, London, UK. Present address: Regional EcoNomics™ Director, Europe & Africa WorleyParsons, Parkview, Great West Road, Brentford, Middlesex, UK. James.MacGregor@WorleyParsons.com

Ibrahim Macharia, Research Associate, Leibniz University of Hannover, Faculty of Economics and Business Administration, Königsworther Platz 1, 30167 Hannover, Germany. mach4ibra@yahoo.com

Miet Maertens, Professor, Division of Agricultural and Food Economics, Department of Earth and Environmental Sciences, Katholieke Universiteit Leuven, Belgium. Miet.Maertens@ees.kuleuven.be

Kai Mausch, Research Assistant, Agriculture and Development Economics, Faculty of Economics and Management, Leibniz University of Hannover, Hannover, Germany. Present Address: Associate Professional Officer (Economics), International Crops Research Institute for the Semi-Arid Tropics (ICRISAT), Chitedze Agricultural Research Station, PO Box 1096, Lilongwe, Malawi. K.Mausch@cgiar.org

Dagmar Mithöfer, Scientist/ Economist, International Centre for Research on Insect Physiology and Ecology (*icipe*), PO Box 30772-00100, Nairobi, Kenya. Present address: Senior Marketing Specialist/Economist, World Agroforestry Centre (ICRAF), PO Box 30677-00100, Nairobi, Kenya. d.mithoefer@cgiar.org.

Mukwiti N. Mwiinga, Lecturer and Researcher, University of Zambia, PO Box 32379, Lusaka, Zambia. mwiingam@gmail.com or mwiingam@msu.edu

Thorsten Nilges, Political Scientist, Centre for Advanced Training in Rural Development (SLE), Humboldt-University, Hessische Str. 1–2, 10115 Berlin, Germany. www.berlinerseminar.de.sabine.doerr@agrar.hu-berlin.de

Gideon Obare, Associate Professor, Department of Agricultural Economics and Business Management, Egerton University, PO Box 536-20155, Egerton, Njoro, Kenya, and Visiting Professor, Food Security Centre, University of Hohenheim, Wollgrasweg 43, 70599 Stuttgart, Germany. obarega@egerton.ac.ke

Janneke Paalhaar, Research Assistant, International Development Studies, Department of Technology and Agrarian Development, Wageningen University, the Netherlands. Present address: Adviser Eelerwoude, Goor, the Netherlands. jannekepaalhaar@hotmail.com

Margaret W. Pasquini, Research Officer, CAZS Natural Resources, Bangor University, Deiniol Road, Bangor, LL57 2UW, Wales, UK. Present address: Assistant Professor, Centro Interdisciplinario de Estudios sobre Desarrollo, Universidad de los Andes, Calle 18ª No. 0–03 este, Bogotá, Colombia. mpasquin@uniandes.edu.co

Jo Swinnen, Professor and Director, Centre for Institutions and Economic Performance (LICOS), Katholieke Universiteit Leuven, Belgium. Jo.Swinnen@econ.kuleuven.be

David Tschirley, Professor, International Development, Department of Agricultural, Food, and Resource Economics. Michigan State University, East Lansing, Michigan, USA. tschirle@msu.edu

Hermann Waibel, Professor, Institute of Development and Agricultural Economics Faculty of Economics and Management, Leibniz University of Hannover, Königsworther Platz 1, 30167 Hannover, Germany. waibel@ifgb.uni-hannover.de

Andrew F. Ward, Senior Adviser, DFID Research into Use Programme, PO Box 9200, Umoja House, Dar es Salaam, Tanzania. Present address: PO Box 50050, 15101 Ridgeway, Lusaka, Zambia. dr_a_f_ward@hotmail.com

Katinka Weinberger, Global Theme Leader Marketing, The World Vegetable Center (AVRDC), PO Box 42, 74199 Shanhua, Taiwan. Present Address: Director, Centre for Alleviation of Poverty through Secondary Crops' Development in Asia and the Pacific (UN-CAPSA), Jalan Merdeka 145, 16111 Bogor, Indonesia. k.weinberger@uncapsa.org

Franziska Woellert, Human Geographer, Centre for Advanced Training in Rural Development (SLE), Humboldt-University, Hessische Str. 1–2, 10115 Berlin, Germany. www.berlinerseminar.de.sabine.doerr@agrar.hu-berlin.de

Foreword

The demand for fresh and processed vegetables is increasing worldwide, putting pressure on domestic and international markets. Vegetables are high in vitamins and essential micronutrients that are otherwise lacking in the diets of many poor rural and urban consumers. Vegetables can be sustainable and affordable sources of these micronutrients.

Vegetable production provides high returns to both land and labour, thereby creating employment opportunities and incomes for rural small-scale farmers, and it contributes to the income diversification of rural smallholders. Market access and participation contribute to economic development and poverty alleviation. Vegetables lend themselves to value addition and employment further along the value chains. While horticulture – including vegetables – is an important emerging cash crop in many African countries, its production-to-marketing system faces many challenges. At the production level, access to high-quality inputs and information on pest management is essential. Consumer concerns for food safety and the adherence to good agricultural practices increasingly drive changes in vegetable production-to-marketing systems. Perishability poses particular challenges to marketing and necessitates good access to markets and up-to-date market information, a challenge given the dynamic nature of horticultural markets. The World Development Report 2008 refers to the 'horticulture revolution', but much more needs to be done to live up to this expectation.

This book is a major contribution towards making the African horticultural revolution happen. The book provides a well-chosen collection of conceptual and methodological chapters on the largely neglected socio-economic aspects of vegetable production-to-marketing systems in Africa. The diverse topics covered in this book include the conceptual challenges in economic research on vegetable production systems, the implications of good agricultural practice standards, the challenges and opportunities of meeting the growing market demand and issues in pest management. The book will provide important insights for students, researchers, development practitioners and policy makers into the current issues in vegetable production and marketing. This work will stimulate further studies through its contribution to, and advancement of, the methods and applied research in this area, and will serve as a guideline for policy makers and development partners.

Christian Borgemeister
Director-General
International Centre of Insect Physiology and Ecology (*icipe*)

Acknowledgements

We acknowledge financial support by the Federal Ministry for Economic Cooperation and Development (BMZ, Germany) granted for two research projects, namely the 'Economic impact assessment as a decision-making tool for resource allocation in horticultural research in East Africa' and 'Economic impact assessment of biological control of the diamondback moth in crucifers in East Africa'. In the course of these two projects, several empirical and conceptual research studies were conducted that provided interesting material that stimulated the production of this book. Most of the chapters are outputs from joint PhD and MSc research projects between the International Centre of Insect Physiology and Ecology (*icipe*) and the University of Hannover, Germany, Wageningen University, The Netherlands, as well as Egerton University, Kenya. Many of the research studies included in this book were presented at the workshop 'Socio-economic research in vegetable production and marketing in Africa', Utalii Hotel, Nairobi, 5–6 March 2009, from where the idea to publish a book on this topic was created. As editors, we would especially like to thank all participants of this workshop for the constructive feedback on the presentations that gave us an excellent basis for writing this book.

We wish to thank *icipe* management for technical support in the organization needed to produce the book. We would also like to thank the World Agroforestry Centre, which very generously allowed the first editor to complete the book after having joined its Global Research Program 3, 'Improving smallholder tree product marketing' in March 2009. Many colleagues at *icipe* – as well as at the World Agroforestry Centre – contributed much support at various stages of the research itself, and during the editorial process of the book, and we would like to thank them.

We want to thank all contributors, authors and co-authors of the book who submitted their ideas and chapters, shaping the book into its present form. We are grateful for their patience with the lengthy editorial processes and multiple rounds of reviewing and editing. In addition to submitting their own contributions, many authors and co-authors thoroughly reviewed other chapters and we gratefully acknowledge this input. All chapters were reviewed by colleagues outside the authors' consortium and we would like to thank them for contributing their time and thoughts, bringing in highly constructive fresh perspectives on the topics covered. As many of the studies were based on extensive fieldwork, we would like to acknowledge all the input from vegetable producers, other stakeholders along the value chain, field assistants and experts consulted who made these studies possible.

At CABI, we would like to thank Meredith Caroll, Sarah Mellor and Sarah Hulbert for support. Kathleen Sinclair of the Nottawasaga Institute did a tremendous job in the language editing of all chapters, and Lucy Kiilu supported the administrative side of the preparation of the manuscript. We thank them both.

Dagmar Mithöfer
Hermann Waibel

External Reviewers

Hippolyte Affognon, International Livestock Research Institute, Nairobi, Kenya.
Mica Bennet, Committee on Sustainability Assessment and World Agroforesry Centre, Nairobi, Kenya.
Hugo de Groote, International Maize and Wheat Improvement Centre, Nairobi, Kenya.
Luis Flores, Institute of International Agriculture, Michigan State University, East Lansing, USA.
Steve Franzel, World Agroforesry Centre, Nairobi, Kenya.
Heike Hoeffler, Institute of African Studies, University Leipzig, Leipzig, Germany.
Moti Jaleta, International Livestock Research Institute, Addis Ababa, Ethiopia.
Simeon Kaitibie, Department of Finance and Economics, College of Business and Economics, Qatar University.
Thanaporn Krasuaythong, Department of Agricultural Economics, Faculty of Agriculture, Khon Kaen University, Thailand.
George Owuor, Department of Agricultural Economics, Egerton University, Njoro, Kenya.
Diemuth Pemsl, World Fish Centre, Penang, Malaysia.
Frank Place, World Agroforesry Centre, Nairobi, Kenya.
Suwanna Praneetvatakul, Department of Agricultural and Resource Economics, Faculty of Economics, Kasetsart University, Bangkok, Thailand.
Karl Rich, Department of International Economics, Norwegian Institute of International Affairs, Oslo, Norway.
Erich Schmidt, Institute for Environmental Economics and World Trade, Faculty of Economics and Management, Leibniz University of Hannover, Hannover, Germany.
Franklin Simtowe, International Crops Research Institute for the Semi-Arid Tropics, Nairobi, Kenya.
David Spielman, International Food Policy Research Institute, Addis Ababa, Ethiopia.
Aad van Tilburg, Marketing and Consumer Behaviour Group, Wageningen University, Wageningen, the Netherlands.
Detlef Virchow, Food Security Center, University of Hohenheim, Stuttgart, Germany.
Rudolf Witt, International Food Policy Research Institute and Centre for World Food Studies, Vrije Universiteit, Amsterdam, the Netherlands.
Meike Wollni, Department of Agricultural Economics and Rural Development, Georg-August University, Göttingen, Germany.

1 An Overview

Hermann Waibel and Dagmar Mithöfer

The eradication of extreme hunger and poverty, as stated in the first Millennium Development Goal, requires not only increasing agricultural productivity but also its diversification. In a more integrated world economy, agricultural growth in developing countries crucially depends on the expansion of market opportunities, and improving competitiveness in international, regional and domestic markets for crops which traditionally were undervalued in the crop portfolio of small-scale farmers. Vegetables are an example of such crops that were not given much attention in the past. To date, vegetables offer many opportunities for developing-country agriculture due to advances in production and transportation technology. The increasing demand, due to changing consumer preferences and positive income elasticity, is expanding market opportunities for vegetables. Vegetables have multiple features. They can be commercial crops for profit and they are important from a food security and nutritional point of view. They are produced in many different systems and locations, i.e. on specialized farms as crop rotations or mixed cropping, as field crops, as house garden crops, in rural, peri-urban and even in urban areas. Vegetables may even be a risk management strategy in times of crisis (Nath *et al.*, 1999).

The governments of developing countries and development organizations have promoted horticultural value chains providing better access to lucrative export markets (e.g. Labaste, 2005). Increasing volumes of international trade for both fresh and processed vegetables over the past two decades have been well documented (Aksoy and Beghin, 2005). Developing countries have become the main exporters of fresh and processed vegetables. Although developing countries have expanded vegetable production continuously during the recent past, this growth shows large regional disparities. While the growth in vegetable production in terms of production volumes as well as per capita consumption has concentrated on Asia and Latin America, in sub-Saharan Africa vegetable production is still lagging behind (Weinberger and Lumpkin, 2007). However, vegetable exports from sub-Saharan Africa have increased tremendously over the past two decades. Some sub-Saharan African countries, such as Kenya for French beans, Côte d'Ivoire for green onions (Weinberger and Lumpkin, 2007) and Senegal for French beans and tomatoes (Maertens and Swinnen, 2009), have become important vegetable exporting countries; if North Africa is included, the continent has become the main external supplier of vegetables to the EU (Weinberger and Lumpkin, 2005). This development emphasizes that vegetable production must receive much more attention from development agencies and the research community than has been the case in the past, a fact recognized in the *World Development Report 2008*, which discusses the 'horticulture revolution' (World Bank, 2007).

The reasons for the increasing attention given to vegetables are manifold. Among these are the fact that they are high-value crops offering market and income opportunities, especially for small-scale producers; they are labour intensive and, therefore, have positive employment effects in both the production and the postharvest sector (McCulloch and Ota, 2002; Maertens and Swinnen, 2009); and vegetables have positive nutritional traits (Mubarik and Tsou, 1997) that are essential in improving health conditions, especially of children. Research and development organizations, such as the Global Forum for Agricultural Research (GFAR) and the Consultative Group of International Agricultural Research (CGIAR), have pointed out the need for more research on vegetables, in particular in the context of African farming systems (GFAR/CGIAR, 2005).

Vegetable research may not have reached the intensity that would be necessary to harness its potential for development, because it is a challenging subject. Vegetables include a large number of different crop species grown under different production and marketing conditions and, therefore, require carefully chosen priorities. This is especially true in the field of socio-economic research, where more scientifically rigorous studies are needed in order to improve the understanding of production-to-marketing systems. Understanding the behaviour of the different actors in the vegetable value chains is a precondition for the design of effective and efficient policy interventions for sustainable development. Many of the socio-economic studies on vegetables available to date are rather detailed and descriptive. However, more often than not such studies are data driven and sometimes lack a good theoretical basis. To conduct reliable socio-economic research requires practical, but at the same time scientifically rigorous, methods of research.

There are, of course, many research topics that deserve attention. Many of these have been addressed in this book; others require more research and, it is to be hoped, will be taken up by succeeding research. One major area of research addressed herein is in the field of environment and human health aspects related to vegetable production. Producing vegetables that are safe for humans and the environment is a requirement for successful establishment in the export market. Publicly and privately set food safety and environmental standards (see, for example, Augier *et al.*, 2005) driven by the preferences of developed-country consumers raise questions for developing-country stakeholders, such as what the impact of high standards on the welfare position of smallholder producers and farm workers will be. It is important to contribute to the understanding of issues, such as whether standards constitute a barrier to trade or whether they are a market-driven governance tool contributing to the modernization of a sector, so that advantage can be taken of such institutional changes to promote development. The need for taking stock, collating and synthesizing existing socio-economic studies and good practices for such research is obvious.

Rapidly growing urban populations and growth in per capita incomes in sub-Saharan Africa are creating major opportunities for local farmers by driving domestic and regional market demand for food. Due to their perishable nature, vegetables depend on well-functioning supply chains and current knowledge of producers on markets and market access (University of California Davis, 2005). The formal retail revolution for fresh vegetables is expected to be much slower in Africa than earlier predicted (Traill, 2006); thus, for the near future, growing urban and peri-urban populations will depend on traditional supply chains mainly characterized by spot-market trading and open wet markets. Therefore, understanding vegetable marketing and supply chains in domestic and regional markets is key to promoting efficient market structures for rural poverty alleviation and feeding growing cities.

Next to market linkages, pest and disease control is a major constraint to realizing the full development potential of vegetable production-to-marketing systems (University of California Davis, 2005). Vegetable production usually relies on high levels of (sometimes potentially harmful) external inputs and is not always practised with safe and clean technologies. The often heavy use of chemical pesticides and the low level of adoption of available technical alternatives, such as integrated pest management and biological control, are major research questions.

In this book, some of these pertinent research questions and policy challenges are addressed through conceptual, as well as empirical, research. Policy recommendations and directions for further research are derived based on the findings in each chapter. The book aims to inform researchers, development partners and policy makers on the opportunities and constraints of vegetable production-to-marketing systems for development. It aims to start addressing some of the open questions highlighted in the *Global Horticulture Assessment* (University of California Davis, 2005).

The book is structured as follows: in Chapter 2, theoretical concepts for socio-economic research of vegetables in Africa are introduced. To some extent, these concepts are reflected in several other chapters of this book, but they also report on research experiences from other continents that might serve as a guide for similar research in Africa.

In Chapter 3, a conceptual framework for impact assessment of production standards is derived, structured around a value chain approach. The chapter outlines potential impact pathways of production standards, which, due to traceability requirements and close linkages to the export market, go beyond the producer level. It synthesizes current thinking in impact assessment, addressing attribution and selection bias. The chapter draws parallels between an intervention aiming at good agricultural practices, as done through private standards, and other interventions, such as those promoting integrated pest management and other natural resource management projects. The chapter further synthesizes empirical evidence on impact pathways, as well as the impact of production standards on various stakeholders in developing-country horticultural production-to-marketing systems, paying particular attention to rigorous attribution. At the conceptual level, the chapter concludes that impact assessment methods focusing on attribution need to be developed to capture the impact at the meso- and macro-levels of a sector. At the empirical level, observations on the medium- to long-term financial impact, as well as the environmental impact, are not available, so it is not yet clear whether production standards are a sustainable tool for sustainable resource management.

Chapter 4 deals with food safety standards and rural household welfare. It looks at the impact of the proliferation and enhanced stringency of food safety standards on small-scale farmers and econometrically analyses to what extent these represent potential barriers to farmers seeking to expand their trade in higher-value food or to what extent they are a catalyst for upgrading production operations, and for improved collaboration between the public and private sectors. The chapter points out clearly that it is necessary to understand the current status and likely trajectory of agrifood standards, the feasible range of commercial, administrative and technical options available to African farmers, firms and governments, and the underlying economics of such responses. The chapter concludes that consideration needs to be given to policies that shift small-scale producers away from the most demanding global markets. It is important for smallholders to diversify their product categories, invest in better postharvest qualities and partake in domestic and south–south trade, the market that has the potential to grow quickly in the next two decades. From the standard-setting point of view, it is crucial that the emerging private standards be smallholder friendly, be acceptable to both buyers and producers and be capable of implementation without significant donor support.

Chapter 5 investigates the differences in the impact of compliance with GlobalGAP standards on small and large Kenyan export vegetable-producing farms. It presents empirical findings on the effects of vegetable export standards across farm sizes, namely smallholder, large-scale contracted and large-scale exporter-owned farms. The chapter assesses the economic performance of the three different certified farm types. Performance indicators applied are the financial cost of compliance with the GlobalGAP standard, which sheds light on the risk that arises from the certification investment, as well as transaction cost indicators that can be attributed to the three different farm types from the buyers' perspective. The chapter addresses the research gap as defined by Jaffee (2003), by providing some indicators on the costs of running a monitoring, standard-compliant smallholder production system.

Chapter 6 discusses food production standards and farm worker welfare in Kenya. The chapter applies Sen's functionings and capabilities approach to investigate the influence of private food production standards on farm worker welfare. A comparison is made between certified and non-certified small farms using four welfare criteria: (i) being trained; (ii) earning a decent income; (iii) being physically healthy; and (iv) being mentally healthy. The chapter finds that GlobalGAP certification has a positive impact on welfare, since it positively influences the amount of training a worker receives. At the same time, certification does not translate into higher wages or better mental or physical health.

Chapter 7 deals with group culture and its importance for the successful participation of smallholders in the export market, using Kenyan French bean value chains as an example. The chapter shows that, although associations in Kenya are widespread among small-scale horticultural farmers, mainly with the purpose to gain access to the export market, considerable variation in success among associations persists, in terms of access to the international market and the level of integration into the supply chain. The chapter uses the cultural theory to analyse group-related factors, such as structure and social cohesion, and grid-related factors, such as rules and penalties, and their role in successful market linkages. It shows that both are important but that grid factors still carry greater weight. Such knowledge is valuable for guiding collective action among the producers of French beans for the stringent export market.

Chapter 8 presents a synthesis of two interesting studies of the export vegetable supply chains and rural households in Senegal, which is the second largest exporter of green beans and the largest exporter of tomatoes in sub-Saharan Africa. The chapter provides evidence on the welfare implications of rural households' participation in green bean and tomato export markets. Both case studies show that the growth in vegetable exports has contributed importantly to poverty reduction. In addition, the authors show that, while the inclusion of smallholder suppliers in high-value vegetable export chains certainly creates benefits for these farmers, the main poverty-reducing effects are attributable to the labour markets through the creation of employment accessible to the poorest rural households.

In Chapter 9, a comparative assessment of the marketing structure and price behaviour of three major vegetables (tomatoes, onions and rape) in the capital of Zambia is undertaken. The lessons learned from this research are relevant to other countries in sub-Saharan Africa. One is that the growing importance of regional trade in fresh vegetables contributes to the harmonization of trade regulations, reduces the likelihood of arbitrary border closings and improves regional market information sharing and reduces and stabilizes consumer prices. The second lesson is that supply chains for these three staple vegetables, and most farmers, continue to rely on the existing trading system. The third conclusion is that production in the urban setting is marginal for tomatoes and onions, but could be important for rape and other leafy green vegetables. More information is needed on this issue. Fourthly, the authors find extremely high price variability driven by the unpredictable fluctuation in quantities.

Chapter 10 analyses value chains in domestic markets and regional trade between Kenya and Tanzania. The chapter's approach combines the concept of value chain analyses with that of livelihood analyses to identify the poorest actors in these value chains and, from that, to derive recommendations on further poverty alleviation opportunities. The study finds that porters and handcart drivers in urban centres are amongst the poorest actors of domestic and regional value chains. For them, a legalization of their status would improve their situation enormously. They could further benefit from empowerment through the establishment of (in)formal groups, as well as by having better access to market information.

Chapter 11 brings a gender perspective to supply chain analysis and applies this to the supply chains of indigenous vegetables in urban and peri-urban areas of Uganda and Kenya. The authors find that indigenous vegetables support a large number of small businesses along the supply chain in urban and peri-urban areas. The authors describe the actors involved and provide an overview on the value and size of the market for indigenous vegetables in both capital cities, as well as in a smaller city in both Uganda and Kenya. It is clearly pointed out that women participate in all segments of the chain, but are more prominent in the retail sector. Interestingly, the authors find that the income of women along the supply chain is generally lower than that of their male counterparts.

Chapter 12 assesses the potential of private voluntary standards in horticultural exports for providing incentives to the private sector investing in small-scale growers, to the mutual benefit of growers as well as exporters. The chapter then traces characteristics of such co-investment and its role for inclusive business models through a number of case studies. It concludes that investment in private voluntary compliance cost is not a subsidy for small-scale growers, but rather a cost-sharing arrangement characterizing a sustainable trading relationship. The chapter further concludes that the role of cooperatives and marketing partners in providing services – particularly credit and technical expertise – has been invaluable in overcoming the barriers to working with small-scale growers and filling the gaps typically left by the public sector in the developing world. Inclusive business models in the export market can provide lessons for trading relationships in domestic markets, as well as for the development of viable alternatives to traditional outgrower schemes.

Chapter 13 assesses the possibilities of strengthening the vegetable value chains in East and southern Africa through capitalizing on potential spillovers from export to domestic markets by synthesizing current evidence. This comparison contributes to identifying the major constraints of domestic vegetable value chains. The authors emphasize the important role of the export vegetable sub-sector as a source of knowledge, technology and experience for the domestic sub-sector. Further, institutions and organizations in the export sector can share valuable lessons and provide a platform for similarly useful institutions and organizations tailored to the needs of the domestic and regional market.

Chapter 14 presents a case of economic impact assessment of classical biological control and pesticide use in Kenya and Tanzania, targeting the diamondback moth, one of the most destructive pests of cabbage worldwide. The effect of the release of an exotic parasitoid by the International Centre of Insect Physiology and Ecology in 2001 to control the pest in East Africa is thoroughly analysed, using econometric methods based on data from a large random survey of cabbage producers in Kenya and Tanzania with and without the biological control. The authors show that farmers producing cabbage in areas where the parasitoid is present use significantly fewer pesticides. Methodologically the damage control function method was applied, which shows that use of pesticides or the presence of the parasitoid increases cabbage output. Interestingly, the chapter indicates incompatibility between current pesticide use practices and the requirements of classical biological control.

Chapter 15 deals with the externalities of pesticide use in vegetable production in Kenya. Through secondary data analysis, individual farmer interviews, pesticide residue analysis, expert interviews and workshops, the health and environmental costs of farmers' pesticide use are estimated. This study fills a major gap for many countries where production of intensive, high-value crops is increasing. The chapter distinguishes between costs, which are actually borne by farmers but are often not attributed to pesticide use, and non-market effects, which are paid by society at large. The former include, for example, damage to crops and livestock and the health effects to the farmer applying the pesticides – or to his family members who might suffer from drift or contaminated water. Some of the external costs, such as the destruction of beneficial organisms, could be verified qualitatively only through valid farmer observations. These provide a good starting point for more research in this field. The study carries a strong message to policy makers who are involved in decision making in agriculture, environmental and health policies. The overall costs of pesticides are at least twice the farmer's pesticide bill, a fact that demands more attention at all levels.

Chapter 16 contributes to understanding the uptake of less external input-intensive vegetable production systems, such as integrated pest management (IPM) practices. It assesses the factors linked to the acquisition of IPM knowledge and sharing. It overlays this analysis with two different training approaches that are currently being promoted in Kenya: the farmer field school (FFS) approach and the common interest group (CIG) approach. The study finds that, compared with farmers who do participate in such group-based extension approaches, farmers participating in FFS and CIG are more likely to acquire knowledge of IPM, but not to share the acquired knowledge with other farmers. The study finds that the distance to extension services was a major constraint to information flow. Furthermore, group membership beyond the two group-based extension approaches increases the likelihood of acquiring IPM information.

It is hoped that this overview has stimulated enough appetite for readers to explore further detailed information in the different chapters of the book and that the rich blend of empirical and methodological chapters will be used as a point of departure for more socio-economic research on vegetables in Africa.

References

Aksoy, A. and Beghin, J.C. (2005) *Global Agricultural Trade and Developing Countries*. World Bank, Washington, DC.

Augier, P., Gasiorek, M. and Lai Tong, C. (2005) The impact of rules of origin on trade flows. *Economic Policy* 20, 567–623.

GFAR/CGIAR (2005) How can the poor benefit from growing markets for big value products? Workshop held at the International Center for Tropical Agriculture (CIAT), Cali, Colombia, 3–5 October.

Jaffee, S. (2003) From challenge to opportunity: The transformation of the Kenyan fresh vegetable trade in the context of emerging food safety and other standards. Agriculture and Rural Development Discussion Paper 1, World Bank, Washington, DC.

Labaste, P. (ed.) (2005) The European horticulture market: Opportunities for sub-Saharan African exporters. Working Paper 63, The World Bank, Washington, DC.

Maertens, M. and Swinnen, J.F.M. (2009) Trade, standards and poverty: Evidence from Senegal. *World Development* 37(1), 161–178.

McCulloch, N. and Ota, M. (2002) Export horticulture and poverty in Kenya, IDS Working Paper 174, Institute for Development Studies, Brighton, UK.

Mubarik, A. and Tsou, S.C.S (1997) Combating micronutrient deficiencies through vegetables – A neglected food frontier in Asia. *Food Policy* 22(1), 17–38.

Nath, P.M., Piluek, P.K. and Herath, E.M. (1999) *The Vegetable Sector in Thailand: A Review*. FAO Regional Office for Asia and the Pacific, Bangkok, Thailand.

Traill, B. (2006) The rapid rise of supermarkets? *Development Policy Review* 24(2), 163–174.

University of California Davis (2005) *Global Horticulture Assessment*. University of California Davis, Davis, California.

Weinberger, K. and Lumpkin, T.A. (2005) Horticulture for poverty alleviation: The unfunded revolution. AVRDC Working Paper 15, The World Vegetable Center, Shanhua, Taiwan.

Weinberger, K. and Lumpkin, T.A. (2007) Diversification into horticulture and poverty reduction: A research agenda. *World Development* 35(8), 1464–1480.

World Bank (2007) *World Development Report 2008: Agriculture for Development*. World Bank, Washington, DC.

2 Theoretical Concepts for Socio-economic Research of Vegetables in Africa

Hermann Waibel

Introduction

In developing countries, vegetable production has increased steadily over the past four decades (Weinberger and Lumpkin, 2005). This production increase is a response to the growing demand for vegetables caused by growing populations and rising incomes, since vegetable demand is income elastic. The increasing volumes of international trade of both fresh and processed vegetables (Aksoy and Beghin, 2005) are part of this development. Nevertheless, in spite of impressive growth rates in production in some parts of the world, supply has been falling behind domestic demand in some countries leading to an increase in the real price of vegetables. While the growth in vegetable production has concentrated on Asia and Latin America, in sub-Saharan Africa the 'vegetable revolution' has not fully 'exploded' yet. However, there have been some exceptions, such as Kenya and Senegal, where foreign investment and a favourable policy environment have upgraded the role of vegetables to an important non-traditional export commodity.

As shown in a wide body of literature, vegetables have multiple features. They play a role as a commercial activity and are important from a food security point of view. They are produced in many different systems and locations, i.e. specialized farms as crop rotation or mixed cropping, as field crops and as house garden crops and in rural, peri-urban or even urban areas. Hence, vegetable production touches upon many problems related to the environments in which they are grown. For example, it has been reported from Thailand that, in times of economic crisis when migrant household members lose their jobs, they tend to engage in vegetable production (Isvilanonda, 1992; Nath *et al.*, 1999).

In many respects vegetables represent a challenging subject of research, mainly due to the large number of crop species with different production and marketing conditions. This is especially true for socio-economic issues where researchers have focused on cereal crops. There is a great need for scientifically rigorous socio-economic studies of vegetable systems in order to improve the understanding of vegetable producers' decisions and behaviour, and for the design of effective and efficient incentive conditions for the promotion of sustainable vegetable supply chains.

Many of the socio-economic studies on vegetables available to date contain comprehensive and detailed information, but such studies are data driven and are not always based on good theory. To conduct reliable socio-economic research requires practical, but at the same time scientifically rigorous, methods of research. One reason why economic studies of vegetables in developing countries are so limited is the complexity of vegetable cropping systems, with their large number of crops, varieties and production systems. This chapter, therefore, offers a contribution towards a framework for the analysis of supply chains as deemed adequate for African conditions. In sub-Saharan Africa, vegetable production systems are still less sophisticated in terms of technology, and highly commercialized or export production is still limited. This chapter maps out some important methodologies that are believed to be useful for the analysis of vegetable production systems and supply chains. Examples of economic research in various parts of the world conducted under the supervision of the author at Leibniz University of Hannover, Germany, are added to illustrate some of the concepts. However, the framework presented here must remain incomplete because of the wide range of methods existing for socio-economic analysis of vegetables. Nevertheless, it is believed that this chapter can be a useful guide for conducting future socio-economic research into vegetable production systems in Africa.

The chapter is structured as follows: in the next section the criteria for a theory-driven socio-economic research framework are described. In the third section (Methodologies of the Analysis of Vegetable Production) a series of methodologies that have been successfully applied in the socio-economic analysis of vegetable systems are presented. In the fourth section (Selected Examples of Socio-economic Research in Vegetables), two illustrative examples of economic studies on vegetables from Asia and Central America are briefly presented. These are intended to be a stimulant for the initiation of similar studies in the African context. The chapter ends with conclusions and suggestions for further socio-economic research of vegetable production in Africa.

Criteria for a Socio-economic Research Framework of Vegetable Production Systems

Socio-economic studies of vegetable production have many facets, including, for example, the field of natural resources, the use of technologies and the uncertainties that demand careful consideration. In addition, it is well recognized that socio-economic research on vegetables cannot ignore marketing aspects. However, it is beyond the scope of this chapter to discuss the rich blend of marketing research methods. Since the supply side of vegetables is already a highly complex issue, this chapter concentrates on the economic methods of production, albeit not ignoring market issues. While the methods proposed here are not only suitable for the socio-economic research of vegetables in Africa, they might also be those least frequently applied, to date, in the studies available.

To begin, any theoretically sound production economic framework for vegetables has to fulfil a few requirements in order to produce results that allow solid conclusions to be drawn and facilitation of policy recommendations. As a minimum, these requirements must include the following:

- critical analysis of available statistical data complemented by expert assessments;
- analysis based on a good understanding of the existing production, farm and household systems;
- analysis that captures the heterogeneity of vegetable production systems;
- analytical methods that capture production cycles and the seasonality of vegetables; and
- an assessment of the relative merits of normative versus positive methods versus the background of data requirements.

In the following sub-section, the components of the analytical framework are outlined.

Situation analysis

Foremost, a more sophisticated economic analysis of vegetable production systems requires baseline information. At a minimum, the productivity of the particular vegetable, as well as alternative cropping activities, must be known. Some information on the resource endowment of the producers, farmers or households, depending on the type of system, is needed. Especially important are labour profiles, as labour is a major input in vegetable production. Marketing options need to be recognized as well, because marketing channels largely determine the economics of the system. Information on the efficiency of the production methods and the existing knowledge gaps of production managers and labourers is necessary to assess the feasibility and relative attractiveness of new technologies. In addition, information on the environmental and health implications of vegetable production is increasingly important. This has been demonstrated in many studies on the impacts of production, on social and health standards, on productivity and on farmer well-being (e.g. Humphrey and Schmitz, 2002; Asfaw *et al.*, 2009). Due to the multitude of vegetable crops grown, detailed data on inputs and outputs are sparse. However, if even only indicative information on material costs, labour costs, yields and prices is available, the gross margins of the production activities can be established. In many cases, the exact material inputs may be unavailable, not only because of the lack of records, but also as a result of overly high variation. It is often not possible to establish special data collection efforts that can generate these detailed accounts. An effective way to overcome this deficiency is to establish expert panels where such data can be generated through a participatory process.

Heterogeneity production functions and multiple cropping

Analytical methods for the socio-economic research of vegetable supply systems must capture the multi-product nature of vegetable production systems. The conventional production function analysis as widely applied in the analysis of cereal crops is not well suited to vegetable crops. The hypothesis of declining marginal yield, and consequently of the question of optimal input use, is less important. Instead, the main question is that of optimal crop combinations defined in the context of the principle of 'equi-marginal' profits. While input allocation decisions are also important, production functions in vegetable production are often Leontief-type 'plateau functions':

$$Y(x) = \begin{cases} A + bx \text{ for } x \leq x' \\ y' \quad \text{otherwise} \end{cases} \tag{2.1}$$

where A = intercept and b = slope. The function $Y(x)$ in Eqn 2.1 shows that Y will increase linearly up to a certain level (plateau), after which further inputs x' will not increase output y'. If technical relationships as portrayed in Eqn 2.1 exist, the focus of economic analysis is no longer the optimization of variable input factors; the focus of the analysis shifts to other issues. For example, the large heterogeneity in vegetable crops with numerous production choices and often highly volatile prices makes the identification of cropping plans, which maximize producers' total profits (Π) at acceptable levels of risk, a difficult decision. To account for risk the variance (δ) of total profit has to be considered, and a risk parameter (r) needs to be introduced to reflect the decision maker's risk preferences. The covariance (δ_1 δ_2) must be accounted for to identify an optimal portfolio selection.

Seasonality and sequential cropping

Physiological and economic factors determine the seasonality of vegetable production. Vegetable crops vary in their characteristics for growth duration and the demand for natural production conditions, especially in terms of temperature and rainfall. Information on growth duration and biologically optimal planting periods is generally known. In addition, demand-side patterns (e.g. festivities) augment seasonal price changes, making vegetable production highly dynamic. The perishable nature of vegetables makes the timing of production an important component of crop management. Thus, analytical methods must capture the dynamic nature of production cycles. This again makes the requirements different from the experience gained in the analysis of food staples such as cereals.

The short growth duration of vegetable crops facilitates sequential cropping where the same or different vegetable crops are planted sequentially on the same piece of land. In the simplest case of two crops, the profit of the second crop will depend on the interaction with the first crop. The profits of the first and second crop will assume the following form:

$$\Pi_1 = f_1\{y_1, p_1, c_1\} \tag{2.2}$$

$$\Pi_2 = f_2\{y_2, p_2, c_2, q\} \tag{2.3}$$

where Π_1 and Π_2 are the profit of the first and second crops, respectively; y_1, y_2 are the yields; p_1, p_2 are the output prices and c_1, c_2 the production costs of the two sequential crops; and q is a quality parameter indicating the degree of crop interaction. Interaction is preferably positive, but negative interactions are possible in the case of pest and disease build-up when subsequent crops are host plants for the same pests. The notion of risk in the context of a portfolio analysis, as described in the previous section, can be added by comparing alternative cropping plans.

Externalities

In vegetable production quality and appearance are important criteria, and hence pesticide use is often high. Pesticides are known to generate negative externalities, even if their use follows the regulation requirements. External effects of pesticides appear in several areas, such as human occupational health, the loss of domestic animals, the loss of biodiversity, the loss of pollinators and other wildlife, the destruction of natural enemies and the development of pesticide resistance (Crissman *et al.*, 1998; Waibel *et al.*, 1999; Pretty *et al.*, 2000; Maumbe and Swinton, 2003; Pretty and Waibel, 2004; Pimentel, 2005). To account for externalities in socio-economic studies of vegetables, the profit function needs to be complemented by a pollution function:

$$\Pi_n = f_n\{(y_n\, p_n) - c_n(x_n q_n), z_n\} \tag{2.4}$$

$$Z_n = z_n\{y_n, p_n, x_n\} \tag{2.5}$$

where Π_n represents the profits of a vegetable cropping system, p_n the product prices, x_n the production inputs, e.g. pesticides, and q_n the input prices. The pollution function Z_n is related not only to production inputs x_n, but also to output (y_n) and quality as reflected in prices. Various approaches exist to establish the pollution or damage function, but in practice an accounting approach that uses average values is often the only feasible way, especially in developing countries (see Macharia, 2010).

Normative versus positive methods

The choice of the type of method is often data driven. Generally, positive or normative methods are applied in economic analysis. Positive methods use empirical data to test different hypotheses on technological and economic issues without postulating any specific behaviour. In this volume several examples for African vegetable production are presented that apply positive methods (see, for example Chapters 4, 6 and 9). Most economic analysis uses some types of positive methods. Normative methods are generally used when information is more sparse. These studies draw upon existing information often derived from expert knowledge, rapid rural appraisals or purposive sampling surveys; an example can be found in Chapter 5. Generally a typical set of production technologies, resource endowments and socio-economic conditions are formulated and an objective function is maximized or minimized. The limitations are that such models in principle are 'non-behavioural', i.e. assumed, rather than observed, behaviour and decision rules are reflected in the results.

For both positive and normative models, primal and dual methods can be used. Primal methods depart from a set of physical and technological conditions specified in production and demand functions. The corresponding dual approaches do not use quantities of products and factor inputs, but solely rely on prices as a reflection of their marginal product or utility. Whole-farm and sector-wide mathematical programming models provide primal and dual solutions with optimal portfolios, shadow prices of scarce resources and opportunity costs of unused production activities. Dual approaches, such as the profit function approach, can be a useful tool for testing the implications of sector-wide policy interventions. However, the level of aggregation inherent in these methods limits the applicability for generating practical field level information. In many cases, the point of departure in vegetable research will be a partial budgeting type of analysis that provides a good entry point to the understanding of the economics of the systems. Given the sparse data situation in vegetable production, dual approaches through mathematical programming are a useful complement, if it is indeed possible to generate the necessary typical farm or household model that can serve as a representation of the majority of the production conditions.

Methodologies of the Analysis of Vegetable Production

At a minimum, socio-economic analysis of vegetable supply systems should include farm-level economic analysis and a sector analysis of the vegetable supply sector, including the spatial arrangement of production areas. To capture the natural resource implications of vegetable production systems, the analysis of externalities is increasingly needed. The impact assessment of technological or institutional innovations, as well as policy interventions, is required to provide accountability for public funds, to generate lessons learned for better project designs and to guide decision makers in project resource allocations. While impact assessment consists of several components, in many cases adoption studies are the starting point. In the following sub-sections these methods are outlined, paving the way for some illustrative examples presented in the fourth section (Selected Examples of Socio-economic Research in Vegetables).

Farm management analysis

Important baseline information needed to conduct farm management analysis is the establishment of annual net revenue functions (Mausch *et al.*, 2009). The annual net revenue function (Φ) is given by the sum of the individual net revenues of crops grown by the farm as a function of each crop's respective area. The net revenue of each crop per unit area is calculated as revenue less variable cost, including the opportunity cost of family labour in production (Eqn 2.6):

$$\Phi = \sum_{k}^{n} \begin{bmatrix} ((y_k p_k) a_k) - \\ \left(\begin{array}{l} (s_k a_k) p_{s_k} + (c_k a_k) p_c \\ +(f_k a_k) p_f \\ +(i_k a_k) + (l_k a_k) \end{array} \right) \end{bmatrix} \tag{2.6}$$

where y is output in quantity per unit of land and time (year), p represents prices of output and inputs; c, f, s, i and l are the input quantities of chemical pesticides, fertilizer, seeds, irrigation and labour, respectively, while a is the area planted per crop.

The net revenue function varies depending on the efficiency of production and the level of technology, which can result in differences in fixed costs. In many cases, fixed costs are difficult to obtain, and thus profits cannot be calculated. Instead, annual net revenues (or gross margins if enterprises use mainly family labour) are a good basis for assessment of the financial performance of vegetable cropping portfolios. If variance can be established – ideally through time series data or cross-sections of similar systems – risk can be incorporated in the analysis by establishing the mean and variance of the annual net revenue across locations or over different vegetable production systems.

Annual net revenue analysis is a cornerstone for whole-farm budgeting by simple spreadsheet analysis or farm programming using mathematical programming. The concept of establishing typical farms can be a suitable methodology for assessment of technologies in vegetable production systems. However, the definition of typical vegetable producers has to be based on a range of data sources because the statistics usually do not allow the derivation of meaningful parameters of distributions (e.g. average farms) (Hemme, 2000). However, the data sources used for constructing a typical vegetable producer model can be based on either a sample survey or a census of producers in a region or watershed. Usually, a two-step procedure is applied, the process starting with an expert consultation to discuss prior assumptions about a farming system established by the researcher, with the purpose of identifying and quantifying the major constraints and opportunities of a production system through key indicators. In the second step, the indicators are validated by triangulation with information from available sources, including sector statistics, case study reports, additional communication with experts and primary data obtained via farm surveys.

Models based on typical farms are different from those based on statistical averages, as the former are open to expert assessment. As a result, these exhibit a consistent set of resource constraints and production activities. Another advantage is that expert panels are cost effective compared with comprehensive data collection efforts, because up-to-date information can be incorporated. The complexity of information can be adapted to the requirements of the modelling purpose. Typical farm models can be formulated with more realistic functional relationships, including multidimensional objectives (Hemme *et al.*, 1997; Häring, 2000; Michel, 2001).

While the maximization of gross margins, net revenues or profit in most cases is the appropriate objective function, approaches that consider risk in coefficients or resource constraints and multiple objectives can be formulated in a similar fashion (Hazell and Norton, 1986; Buysse *et al.*, 2007). Especially in smallholder systems, a multi-objective choice criterion may more adequately capture the decision process.

Spatial and sector-level analysis

Location factors play an important role in vegetable production. Production can be concentrated in either close neighbourhood-to-consumer markets, export facilities/processing units or in distant areas – for example, in tropical highland areas with favourable climatic conditions for the production of specific vegetable crops. Location factors related to climatic factors affect vegetable crops in different ways. Leafy vegetables tend to grow well under tropical lowland conditions, but are more susceptible to transport losses, while root and bulb vegetables require a cooler production environment and are better suited for transport. The analysis of production locations, therefore, is a major issue in socio-economic studies of vegetables. The classic approach in handling this problem is the Von Thünen location theory, which is basically a theory of transportation over space. In its simplest form the Von Thünen location model considers only a single crop, with transportation cost as a linear function of distance subject to weight and perishability (Bissett, 2004). In this case, the farm gate price is determined by the commodity price in the marketplace less transportation costs. Vegetable production land, therefore, generates a rent accruing to an operator or owner of that land. The land rent is defined as revenue less production and transportation costs. Under homogeneous production conditions, differential rents are solely a function of distance from a market, as described by the following equation:

$$R_V(r) = y(p-c) - ytr \qquad (2.7)$$

where

$R_V(r)$ = location rent per unit area of land used for vegetable production$_V$ at distance r from the centre of demand;

y = output per unit area;

p = market price per unit of output at market location ($r = 0$);

c = average cost of production;

t = average rate of transportation cost per unit of commodity and distance; and

r = distance of production location to market location.

Land rent for vegetable production areas R_V at each production location is a linear function, decreasing with the distance from the marketplace. At some critical distance, land rent becomes zero when $r = r_m$, i.e. the marginal production locations where revenues are fully offset by production and transportation costs. Producers enjoy a positive land rent only if $r < r_m$. This is also the zone where competition for vegetable-growing land exists. Beyond the margin non-vegetable crops (R_{NV}) are economically more attractive, i.e. the land rent $R_V < R_{NV}$. Location theories are the conceptual basis for analysis of vegetable production at the sector level, taking into account the spatial distribution of production locations.

A useful methodological tool for modelling vegetable supply chains is a regionally disaggregated mathematical programming model (Hardeweg and Waibel, 2009). The perishable nature of vegetables, multiple and sequential cropping schemes, as well as the seasonality of production as a consequence of climatic conditions, can be depicted by static or comparative–static equilibrium models. As the inter-regional trade of vegetable commodities takes place at the wholesale level, the inclusion of the transport system is necessary. As vegetable production represents only a minor part of agricultural resource use and demand for external inputs and labour, supply equations can be assumed as fully elastic. To be practical and cost effective against the background of data limitations, a spatial model must be limited to the major vegetable crops or crop aggregates and cropping technologies. At the very least, a significant percentage of vegetable crops covering the major production locations of a country, together with regional time-specific (monthly) supply and demand balances, must be incorporated. A major problem is model calibration. In some cases, data can be derived from agricultural census and crop production statistics. Often, however, primary data on vegetable production technologies must be collected in topically focused surveys with mainly purposive sampling, which can be discussed in expert workshops.

As mentioned in the preceding section, typical farm models can be established representing a significant group of producers. To be of use in the assessment of the impact of a change in exogenous factors and interventions such as economic and environmental policies, such models must incorporate the main components of production and marketing. For each of the main production locations, a typical production system (vegetable farm) must be formulated and a reasonably precise depiction of the wholesale and retail market interactions has to be incorporated. One limitation of such a normative model is the presentation of the demand side. Own-price and cross-price elasticities for individual vegetable commodities are rarely available, so that a welfare analysis with the maximization of the sum of consumer and producer surplus is not possible. Hence, the objective function of a sector analysis of vegetables may have to be formulated as cost minimization that yields the marginal costs of supply for specified demand quantities. Other limitations are that dynamic processes, such as labour migration or structural change, cannot be reflected in a static sector model. On the other hand, a spatial resolution of the results and the combination with geographic information systems can facilitate the presentation of the results and make such models useful for policy recommendations.

An example of a sector model will be given in the fourth section of this chapter (Selected Examples of Socio-economic Research in Vegetables).

Impact assessment of technological and institutional innovations

Impact assessment must be a crucial part of socio-economic research in vegetable production in Africa. Funding agencies and research managers, accountable for their allocation decisions, need a notion of the rate of return of the research investment. In addition, impact assessment results often stimulate improvements in project designs and there is a general information gap for vegetables, unlike for cereals where ample evidence has been collected for crop genetic improvement (CGI) research (Evenson and Gollin, 2003) and in natural resource management research in the Consultative Group on International Agricultural Research (CGIAR) (Zilberman and Waibel, 2008).

In vegetable production systems, impact assessment has to take into account the special features of this commodity. First, in the case of new input technologies, one problem is small market size as a result of the multitude of vegetable species. For example, a pesticide or a new biocontrol agent requires a certain market size to recover high development costs.

A similar problem can apply for output markets with small market volumes or specialized contractual arrangements with agents, which can make the assessment of productivity effects complex; therefore, it is unlikely that a new technology related to vegetable production will reach the high rates of return to agricultural research that are found in research on cereals. Secondly, vegetable technologies are knowledge intensive. The level that needs to be spent on research and development for innovations depends on prior knowledge. For many vegetable crops, prior research may be either limited or not publicly accessible.

Thirdly, the benefit of technological and institutional innovations in vegetable production may be found in the field of environment and human health, e.g. by reducing chemical pollution or health hazards. However, producers may not adopt such technologies, especially when such benefits are external and mechanisms to internalize them are absent. This will lead to under-adoption of such technologies, unless regulation is changed or appropriate incentive schemes are established. On the other hand, new mechanisms have emerged through the requirements to access international value chains. Such specific topics of impact assessment that deal with vegetables (production, environmental health and social and human health standards) will be specifically discussed in Chapter 3 of this volume.

While economic theory provides well-defined principles for impact assessment, the complexity and specific features of projects focusing on vegetable production may require more complex approaches than those suggested by much of the literature on the productivity supply shift (e.g. Griliches, 1957; Evenson and Gollin, 2003). Improvements in vegetable production may affect consumers' and farmers' surplus beyond the traditional supply-shifting effects associated with increased yields or reduced costs. These often touch upon externalities, risk reduction, stabilization of ecosystems or the quality enhancement of products. Hence, while the impact assessment of projects related to the improvement of vegetable production should provide a notion of the internal rate of return of the R&D investments, such analysis must be complemented by additional indicators. Specific aspects of project design and management need to be considered in order to generate some lessons learned. This can provide additional value to impact assessment studies.

The proper implementation of impact assessment studies requires the application of at least the basic toolbox of information and methods. These generally include: (i) adoption; (ii) income effects; (iii) risk aspects; (iv) economic surplus and rate of return; and (v) externalities.

The most crucial aspect in impact assessment is adoption. As pointed out by Asfaw *et al.* (2009) for the case of production and health standards in Kenyan horticulture, a time lag exists between the introduction of a technology and the time of adoption. Hence, the diffusion process needs to be taken into account and the shape of the diffusion curve needs to be established. For the classic S-curve, statistical tools such as logit and probit models exist to analyse discrete choices by producers, such as early and late adopters. However, in vegetables, the assumptions underlying the imitation model often do not hold, and the homogeneity assumption is hardly met in reality. In vegetables, different market segments with rather distinct features exist and, under these conditions, the likelihood of technology adoption may vary greatly and therefore a different likelihood of technology adoption exists. This calls for the threshold model of diffusion. David (1969) has developed a threshold model of adoption that considers an explicit economic micro-level decision-making mechanism, recognizes the heterogeneity of the adopters and incorporates dynamic processes and diffusion drivers, such as policies and institutions that move the diffusion process forward over time.

Assessment of the externalities of vegetable production

While this topic could also be included in Chapter 3 of this book (Framework for Economic Impact Assessment of Production Standards and Empirical Evidence), a separate section will be devoted to it here due to its importance. Research on the externalities of vegetable production is a necessary component of a socio-economic research portfolio for vegetable research. Often, the perception exists that vegetable production systems in Africa are of low input intensity and, thus, the problem of pesticide use is almost ubiquitous. However, vegetable production in Africa tends to be highly intensive, especially when targeting export markets, but only a few attempts have been made to describe and quantify the external costs of pesticides in vegetable production in Africa. Studies on externalities can be grouped into three broad categories:

1. Accounting approach: this begins with establishing scientific evidence of the externality and uses actual market prices to value the physical effects. For example, the treatment costs of cleaning pesticide-contaminated water can be valued.
2. Economic approach: in the absence of market prices, hypothetical and surrogate markets are used to value the negative effect of pesticides.
3. Mixed approach: combining both approaches.

The accounting approach includes three steps: identification, quantification and monetization. A clear identification of externalities means the collection of all available scientific evidence from reliable sources in order to establish the causality between an action and the effect. For example, the health symptoms of a farmer must be linked to his pesticide application practices. Quantification is the measurement of the physical or biological effect of the pesticide externality. Finally, monetization means attaching a suitable price to the quantities identified for the different externalities. A total is established by aggregating different externalities, e.g. human health and environment and others (see Waibel *et al.*, 1999). The accounting approach is theoretically questionable as it does not consider the choices made by economic agents. However, it is a first attempt that may stimulate useful discussion leading to further studies.

The economic approach is based on decisions made by economic agents. However, since by definition externalities are non-market effects, pesticide externalities can be derived only from surrogate or hypothetical markets (Zilberman and Katti, 1997; Crissman *et al.*, 1998; Cuyno *et al.*, 2001; Hanley *et al.*, 2003).

Indirect methods rely on observations in related markets (surrogate market). The following methods have become especially popular: hedonic prices, travel cost approach and avoidance costs. The hedonic method postulates that the price of a good is a function of its attributes whose influence can be quantified. For example, the health hazards from pesticides can be reflected in the wage paid to labourers hired to spray fruit orchards regularly. The precondition for the successful application of this approach is that the markets for the surrogate good are competitive. The travel cost method draws upon the price of visiting outdoor recreation sites as a proxy for environmental amenities. For pesticide externalities, this method may be relevant only in the case of recreational activities, such as fishing, where more distant locations must be used as a result of pesticide pollution of nearby water bodies. The avoidance cost method involves the purchase of goods in markets that can be used to mitigate adverse effects, e.g. the price of protective clothing as a proxy and minimum value for the expected health damage.

Direct methods are based on stated preferences in hypothetical markets for a good or a bad to be assessed and include the two approaches, namely contingent valuation (CV) and choice experiments (CE). In contingent valuation respondents are asked to reveal their 'willingness to pay' (WTP) for improvements in non-market goods and services or their 'willingness to accept' (WTA) compensation for the loss of well-being associated with a change. For example, respondents can be asked about their WTP for a pesticide that is less toxic to their health. In the following section (Selected Examples of Socio-economic Research in Vegetables), an example for pesticide use in vegetable production in Nicaragua is given.

In choice experiments, respondents are asked to judge their most preferred choice among several attributes of a bundle of goods, one of which involves an environmental good. The precondition for a successful application of this method is that respondents are knowledgeable about the item to be judged. Furthermore, preferences for the market goods in one bundle and non-market goods in the other bundle must exist. Random utility functions are used to relate external effects, and the economic costs attributed to the choice sets.

In most studies to date, a combination of the economic and the accounting approach has been applied (see Waibel *et al.*, 1999; Pretty *et al.*, 2000). For example, in a study about pesticide externalities in Germany for estimating losses in various production sectors an accounting approach was used, while for the loss of species contingent valuation approaches were applied. In the case of Pretty *et al.* (2000), the cost of monitoring and remediation of damaged habitats, bee colony losses, treatment of pesticide-contaminated drinking water, as well as the treatment cost of acute pesticide poisonings, were carried out by relying on the accounting approach for the most part, but were partially complemented by the economic approach.

Selected Examples of Socio-economic Research in Vegetables

Sector modelling of Thai vegetable production

Applying the theory of location to agricultural production as outlined in the previous section, an example is presented of a spatial model of the vegetable sector in Thailand (Hardeweg, 2008). This model portrays the entire vegetable supply chain and allows identification of the factors that determine the spatial mobility of vegetable production. It shows to what extent the ongoing shift away from the traditional vegetable production locations (i.e. from peri-urban to more rural areas) can result in a reduction in the potentially polluting high external inputs such as pesticide and nitrogen fertilizer. As such, the model is a policy tool for the assessment of policies that can facilitate the adoption of environmentally friendly production technology in this process.

For the multi-period mathematical programming model, activities are included that reflect the multiple and sequential cropping nature of the vegetable production sector to mirror the seasonality of production.

Inter-regional trade is presented at the wholesale level, with transport activities reflecting different technologies. Completely elastic supply equations for production inputs and agricultural resources, such as land and unskilled hired labour, can be assumed since vegetable production represents only a minor part of agricultural resource use and demand for external inputs. The multitude of the cropping activities is covered by regionally and seasonally differentiated cropping technologies for 23 vegetable crops or crop aggregates that represent 90% of the vegetable supply in eight regions of the country. Monthly supply and demand balances at the regional level are formulated in the model. Data for model parameterization have been derived from the Thai Agricultural Census 2003 and crop production statistics. Primary data on current vegetable production technology were collected in expert workshops carried out in 2002. The model, which is based on the typical farm approach (see Methodologies of the Analysis of Vegetable Production; Farm management analysis), was calibrated to reproduce a 3-year average of monthly production data at the regional level by applying a positive mathematical programming approach.

The initial solution of the model shows the spatial distribution of production and the transport flows to local (in the production region) and external (Bangkok) markets (Fig. 2.1). More than half of the vegetable production is consumed outside the respective regions of origin (Fig. 2.1), and thus the supply of vegetables is highly transport intensive. Moreover, in Bangkok a central marketplace exists that receives the majority of produce from the nearby central region (CE). However, a significant proportion comes from the upper northern region (UN), which has good climatic conditions for the temperate vegetable types. On the other hand, inter-regional trade, other than with the central marketplace, is less pronounced and most regions are surplus regions. Additional model outputs are the total resource use (e.g. of land) of the sector and the input quantities supplied by factor markets, e.g. for pesticides and fertilizer. Land use for vegetable production over all regions at peak production time amounts to 148,000 ha, equivalent to 0.93% of the total arable land in Thailand. The labour profile for resources on vegetable farms shows that most of the labour is allocated to vegetable production, with averages between 53% in the lower northern region (LN) and 83% in the southern region (SO). Labour is a major constraint to the expansion of vegetable production, and the model can show that farm land suitable for vegetable production is underused because of the lack of labour.

In terms of resource use, the model shows, for example, that the total transport of vegetables in Thailand amounts to 3.6 million t per year over an average transport distance of 293 km. This requires 90 million litres of diesel, corresponding to about 0.61% of the national diesel consumption.

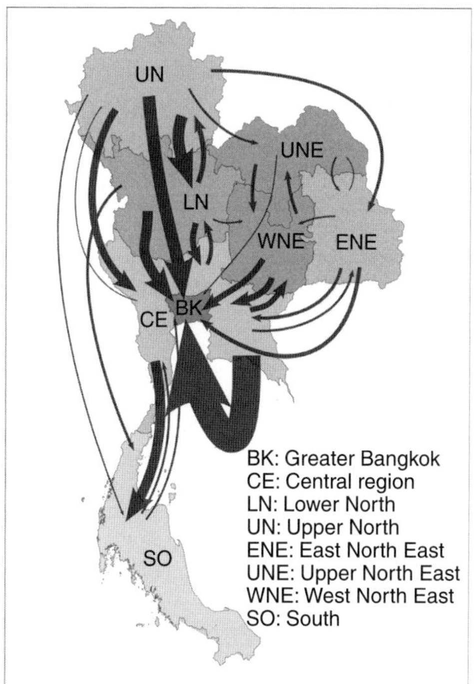

Fig. 2.1. Production locations and trade flows derived from Thai vegetable sector model (from Hardeweg, 2008). CE, central region; ENE, east-north-eastern region; LN, lower northern region; UN, upper northern region; WNE, west-north-eastern region.

One of the environmental consequences of fossil energy use for transportation is CO_2 emission, which amounts to 245,000 million t annually.

Total nitrogen (N) use in the vegetable sector in Thailand is 60,000 milion t, equivalent to about 6% of the national nitrogen fertilizer consumption, which is used on less than 1% of the arable land. The regional disparity in nitrogen intensity shows, however, that in the central region (CE) nitrogen use is higher than 600 kg/ha, much above the national average. Similarly, pesticides in the vegetable sectors account for 1528 million t of active ingredients, which corresponds to 3.9% of total pesticide use in Thai agriculture and vis-à-vis a land use of below 1%. The average pesticide use per hectare in vegetable production thus amounts to more than four times the sector average.

The model was applied to simulate the impact of various policy interventions, such as energy and environmental taxes or area zoning for intensive production. It can also simulate the impact of changes on external factors, such as demand or new technologies in production, transportation and storage.

One example of policy simulation, namely the introduction of a 'Pigouvian-type tax' is shown in Fig. 2.2. Through this model, it may be demonstrated that a tax on pesticides can be effective in reducing the environmental impact of pesticide use. The environmental impact quotient (EIQ), which is a composite measure of occupational, consumer and environmental hazards caused by pesticides, is reduced by nearly 25% compared with the baseline value. Figure 2.2 also shows the effects of a pesticide tax on other variables such as transportation (measured in 'ton-kilometres') and on nitrogen use.

One of the interesting results is that Greater Bangkok in particular benefits from pesticide reduction but, on the other hand, farm gate prices remain constant and the return on family labour is reducing.

This example shows that, even when data are scarce, intelligent usage of statistical data, survey data, expert knowledge and intuition allows the development of a model

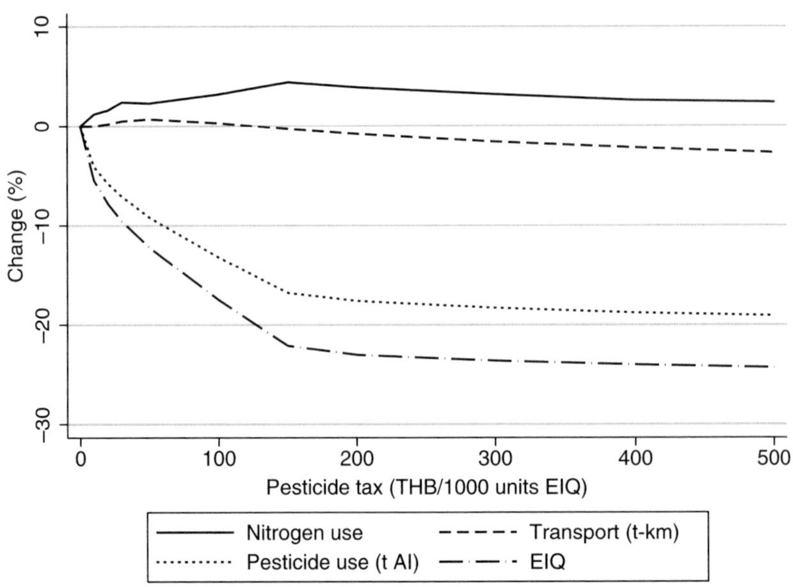

Fig. 2.2. Impact of introducing a pesticide tax on the vegetable sector in Thailand – modelling results. EIQ, environmental impact quotient; THB, Thai baht (currency); t AI, metric tons of active ingredient.

of the vegetable sector of a country that is capable of assessing the implications of policy interventions and can, thus, be a good entry point for guided policy discussion.

Vegetable farmers' willingness to pay for improved health in Nicaragua

A major issue in vegetable production and consumption is health. On the one hand, vegetables have nutritional benefits for consumers due to their high level of micronutrient contents (Bellin and Leitzmann, 1995). On the other hand, vegetable producers are at risk from pesticide use, which has potentially negative consequences for the environment. Hence, research on the health aspects of vegetable production systems is another area that needs to be given attention in the African context.

In Nicaragua a study was conducted looking at the health implications of pesticide use in vegetable farming, and at established farmers' willingness to embrace health improvements (Garming and Waibel, 2008). For the empirical basis, a total of 433 farmers in four vegetable-producing regions were interviewed. A contingent valuation approach to assess the health effects of chemical pesticides was applied. The value of health was measured by farmers' willingness to pay (WTP) for low-toxicity pesticides. Results show that farmers are willing to spend an additional amount of about 28% of their current pesticide expenditure to avoid health risks. The validity of the results was established in scope tests and by means of a two-step regression model. WTP depends on the farmers' experiences with poisoning, income variables and their current exposure to pesticides. The results provided inputs to the design of rural health policies and in the formulation of programmes aiming to reduce the negative effects of pesticides.

The survey instrument was a specially defined experiment combined with a questionnaire, where the respondents' understanding of pesticide-related health problems was updated by asking them to recall pesticide use in the previous growing period and experiences with poisoning and poisoning symptoms. Information was then given about the possible health effects of pesticides, using a list identifying the most commonly applied pesticides as either high, medium or low risk based on their respective World Health Organization (WHO) classification. The distinction between acute and chronic health risks was explained. Subsequently, WTP was established for two scenarios: (i) a pesticide avoiding chronic risks; and (ii) a pesticide avoiding both chronic and acute risks. The comparison of WTP in these scenarios was used for a scope test, indicating whether the respondents understood and valued the differences in the extent of health benefits. The elicitation of WTP was designed as an open-ended bidding game, starting with a 100% price premium, then lowering or increasing the price depending on the farmer's response. After two bidding rounds, respondents were asked to rethink their decision and the WTP question was repeated. Total WTP was calculated as the product of price premium and the purchased amount of the pesticide (Table 2.1).

The positive WTP for avoiding risks from pesticides demonstrates that Nicaraguan vegetable farmers are well aware of pesticide health risks. Table 2.2 shows that WTP represents a significant fraction of a family's health expenditure and pesticide use although the distribution is rather skewed, with more extreme values in the upper quartile.

The study showed that the health costs from pesticide hazards in vegetable production were underestimated, and that was for only the directly measurable and observable costs. In fact, farmers value their health more highly than what is expressed in productivity loss, protective clothing and medicine. The WTP estimations of the health costs of pesticides are important information for policy makers. For example, they can be used to assess the benefits of IPM (integrated pest management) programmes more realistically. Information about pesticide health costs can serve as a basis for government decision making regarding investments in rural

Table 2.1. Willingness of Nicaraguan vegetable farmers to pay (in US$) for health improvements in two scenarios (2004) (Source: Garming and Waibel, 2008).

Scenario	Mean (SE)	Median	1st Quartile	3rd Quartile	Skewness
Chronic health effects	25.8 (3.7)	6.0	0	20.3	6.2
Chronic and acute health effects	61.6 (9.6)	20.75	6.0	50.0	7.8

SE, standard error.

Table 2.2. Willingness to pay (WTP; US$) as a proportion of pesticide expenditure and income[a] (Source: Garming and Waibel, 2008).

Parameter	Mean (SE)	Median	1st Quartile	3rd Quartile
Pesticide expenditure/year	609.0 (61.7)	222.5	95.2	618.8
WTP 'chronic and acute'/pesticide expenditure	23.0 (2.5)	5.9	1.0	20.0
Agricultural net income/year	1846.5 (228.4)	666.7	143.3	1851.7
Household income/year	2096.0 (235.6)	904.7	265.0	2257.3
WTP 'chronic and acute'/ household income	3.1 (1.6)	1.2	0.07	3.8
Family expenditure on health care	97.8 (14.3)	30.0	0.0	66.7

[a] The displayed values are the means of the ratios calculated on an individual basis over the whole sample. SE, standard error.

health infrastructure, especially in intensively farmed areas.

Summary and Conclusions

The main message of this chapter is that more emphasis needs to be placed on a theory-based approach in future socio-economic analysis of vegetable production systems in Africa. In contrast, most of the studies to date have been driven by available data. The two examples presented in section four of this chapter (Selected Examples of Socio-economic Research in Vegetables) demonstrate that this is possible even with limited resources. While the technical aspects of vegetable production are important, the multitude of vegetable crops and varieties requires alternative ways of data collection. For example, expert opinions and workshop arrangements can be used. There is a need for innovative approaches of aggregation and incorporating the spatial dimension of production within and beyond the national level. Furthermore, addressing more openly the factors not related to productivity, but to health and the environment, deserves more attention in socio-economic studies.

This chapter has introduced concepts and examples of socio-economic research in vegetable systems from various parts of the world. These were drawn from research conducted under the supervision of the author at the Leibniz University of Hannover. Methodologies and examples were introduced that may not be used frequently in current socio-economic studies about vegetables in Africa. Therefore, they can be used as a guide for the design, planning and implementation of such issues concerning vegetable systems within the African context. Considering the complexity of the technical and socio-economic problems of vegetable systems in Africa, the framework presented here necessarily must remain partial.

However, the framework has addressed the main features of vegetable production, namely baseline information, seasonality, heterogeneity, externalities and research methodology. These were linked to varying degrees to the methods briefly described. Thereafter, two examples from other continents – one each from Asia and Central America – were presented, which are of particular value for Africa if such studies could be implemented.

Acknowledgements

The author greatly appreciates the useful comments and suggestions of Dr Diemuth Pemsl and Prof. Erich Schmidt on this chapter. He sincerely thanks all those PhD students who carried out their thesis work on the socio-economic aspects of vegetables in developing countries.

References

Aksoy, A. and Beghin, J.C. (2005) *Global Agricultural Trade and Developing Countries*. World Bank, Washington, DC.

Asfaw, S., Mithöfer, D. and Waibel, H. (2009) Food-safety standards, pesticide use and farm level productivity: The case of high-value crops in Kenya. *Journal of Agricultural Economics* 60(3), 645–667.

Bellin, F. and Leitzmann, C. (1995) Die Bedeutung der Mikronährstoffe für die menschliche Entwicklung – ein Plädoyer für Gemüse. *Entwicklung und Ländlicher Raum* 4, 7–9.

Bissett, L. (2004) Location rent and the spatial distribution of economic activities: A nonlinear reformulation of the Von Thünen paradigm. PhD Thesis, Regional Science, University of Pennsylvania, Philadelphia, Pennsylvania, pp. 109.

Buysse, J., Van Huylenbroech, G. and Lauwers, L. (2007) Normative, positive and econometric mathematical programming as tools for incorporation of multifunctionality in agricultural policy modelling. *Agriculture, Ecosystems and Environment* 120(1), 70–81.

Crissman, C., Antle, C., Capalbo, J.M. and Capalbo, S.M. (1998) *Economic, Environmental, and Health Tradeoffs in Agriculture: Pesticides and the Sustainability of Andean Potato Production*. Kluwer Academic Publishers, Dordrecht, Netherlands and Boston, Massachusetts.

Cuyno, L.C.M., Norton, G.W. and Rola, A. (2001) Economic analysis of environmental benefits of integrated pest management: A Philippine case study. *Agricultural Economics* 25, 227–233.

David, P.A. (1969) A contribution to the theory of diffusion. Memorandum No. 71, Stanford Center for Research in Economic Growth, Stanford University, California.

Evenson, R.E. and Gollin, D. (2003) *Crop Variety Improvement and its Effect on Productivity: The Impact of International Agricultural Research*. CAB International, Wallingford, UK.

Garming, H. and Waibel, H. (2008) Willingness to pay for health risks from pesticides in developing countries: A case study from Nicaragua. *European Journal of Health Economics* 88(2), 499–511.

Griliches, Z. (1957) Hybrid corn: An exploration in the economics of technological change. *Econometrica* 25(4), 501–522.

Hanley, N., Ryan, M. and Wright, R. (2003) Estimating the monetary value of health care: Lessons from environmental economics. *Health Economics* 12, 3–16.

Hardeweg, B. (2008) The spatial distribution and inter-regional dynamics of vegetable production in Thailand. Wissenschaftsverlag Vauk, Kiel, Germany, ISBN 978-3-8175-0413-8. Zugl. Dissertation, Leibniz University of Hannover, Germany.

Hardeweg, B. and Waibel, H. (2009) The spatial distribution and inter-regional dynamics of vegetable production in Thailand. In: *GEWISOLA Jahrestagung* 30 September–2 October 2009, Kiel, Germany. Gesellschaft für Wirtschafts- und Sozialwissenschaften des Landbaus e. V.

Häring, A. (2000) Economic simulation of organic farms guided by farmers – a participatory approach to agricultural policy research? In: *2nd SREN Workshop on Research Methodologies in Organic Farming – On-farm and Participatory Research*, Bari, Italy. FAO, Rome.

Hazell, P.B.R. and Norton, R.D. (eds) (1986) Mathematical programming for economic analysis in agriculture. In: *Biological Resource Management*, Macmillan, New York, p. xiv.

Hemme, T. (2000) Ein Konzept zur international vergleichenden Analyse von Politik- und Technikfolgen in der Landwirtschaft. PhD thesis, University of Göttingen, Germany.

Hemme, T., Isermeyer, F. and Deblitz, C. (1997) *Tipi-Cal Version 1.0: Ein Modell zur Politik- und Technikfolgenabschätzung für typische Betriebe im internationalen Vergleich.* Arbeitsbericht No. Arbeitsbericht 2/97. Institut für Betriebswirtschaft der FAL, Braunschweig, Germany.

Humphrey, J. and Schmitz, H. (2002) How does insertion in global value chains affect upgrading in industrial clusters? *Regional Studies* 36(9), 1017–1028.

Isvilanonda, S. (1992) *Vegetables and Fruits in Thailand: Supply and Demand.* Thailand Development Research Institute (TDRI), Bangkok, Thailand.

Macharia, I. (2010) Negative externalities of pesticide use in vegetable production in Kenya. PhD Thesis, Leibniz University of Hannover, Germany.

Maumbe, B.M. and Swinton, S.M. (2003) Hidden health costs of pesticide use in Zimbabwe's smallholder cotton growers. *Social Science and Medicine* 57(9), 1559–1572.

Mausch, K., Mithöfer, D., Asfaw, S. and Waibel, H. (2009) Export vegetable production in Kenya: Is large more beautiful than small? *Journal of Food Distribution Research* 40(3), 115–129.

Michel, K. (2001) Typische Betriebe als Instrument partizipativer betriebswirtschaftlicher Forschung - Eine Fallstudie zur Baumschulwirtschaft im Kreis Pinneberg. Diplomarbeit-Thesis, Fachbereich Gartenbau, University of Hannover, Hannover, Germany.

Nath, P.M., Piluek, P.K. and Herath, E.M. (1999) *The Vegetable Sector in Thailand: A Review.* FAO Regional Office for Asia and the Pacific, Bangkok, Thailand, pp. 71.

Pimentel, D. (2005) Environmental and economic costs of the application of pesticides in the United States. *Environment, Development and Sustainability* 7, 229–252.

Pretty, J. and Waibel, H. (2004) Paying the price: The full costs of pesticides. In: Pretty, J. *The Pesticide Detox: Solutions to Safe Agriculture.* Earthscan Publications Ltd., published in association with the United Nations Food and Agriculture Organization, London, UK, pp. 39–54.

Pretty, J., Brett, C., Gee, D., Hine, R., Mason, C.F., Morison, J.I.L. *et al.* (2000) An assessment of the total external costs of UK agriculture. *Agricultural Systems* 65, 113–136.

Waibel, H., Fleischer, G. and Becker, H. (1999) The economic benefits of pesticides: A case study from Germany. *Agrarwirtschaft* 48(6), 219–229.

Weinberger, K. and Lumpkin, T.A. (2005) Horticulture for poverty alleviation – the unfunded revolution. In: A.P.N. 05-613 AVRDC (ed.) AVRDC Working Paper Series, The World Vegetable Center, Shanhua, Taiwan, pp. 20.

Zilberman, D. and Katti, M. (1997) Pesticide use and regulation: Making economic sense out of an externality and regulation nightmare. *Journal of Agricultural and Resource Economics* 22, 321–332.

Zilberman, D. and Waibel, H. (2008) Productivity enhancement and natural resource management. In: Waibel, H. and Zilberman, D. (eds) *International Research in Natural Resource Management: Advances in Impact Assessment.* CAB International, Wallingford, UK. pp. 21–55.

3 Framework for Economic Impact Assessment of Production Standards and Empirical Evidence

Dagmar Mithöfer

Introduction

Economic development and global trade in products from developing to developed countries have steadily increased over time, while the institutions governing such trade have become more complex and intertwined. Driven by the increasing demand for products of high quality, production standards have proliferated (Swinnen and Maertens, 2007). Henson (2006) distinguishes mandatory standards, voluntary consensus standards and de facto mandatory standards. Mandatory standards are set by public institutions whereby compliance is obligatory in a legal sense. Voluntary consensus standards arise from a formally coordinated process involving market and non-market actors including the government, while de facto mandatory standards arise from an uncoordinated process of market-based competition between private actors (Henson, 2006). Freedom of response to these groups of standards increases from mandatory, de facto mandatory to voluntary consensus (Henson, 2006).

Standards can be differentiated by their point of intervention: concerning food safety: performance standards regulate the upper level of risk in food, while process standards specify risk control inputs, and often a combination of the two prevails (Antle, 2000). Often, performance standards are assumed to perform better economically when inefficient firms prevail in a sector, for situations of highly variable input use and if a regulator is more interested in the level of the standard than in the variability of compliance across a sector or industry (Cho and Hooker, 2009). However, according to Fulponi (2007), standards tend to shift from product (i.e. considering output only) to production process perspectives and take more of a management system approach, which governs the complete production system. Standards can act as business-to-business signals if the standard is not used to label the final product or, when used for the labelling of the final product, standards can serve for product differentiation (Henson and Humphrey, 2008). Standards have increased in scope, starting from food safety to social standards, environmental health (energy, water use, packing, distribution) and even to animal welfare (Fulponi, 2007).

Comprehensive production standards, such as GlobalGAP (a process standard in fresh fruit and vegetable production) or Rainforest Alliance (a process standard in coffee production), are often defined and enforced through private actors, while standards of the public domain often concentrate on a certain risk factor, such as maximum residue level. Both types of standards impact production systems. The impact of standards has been subject to a lively debate in recent years. On the one hand, concerns were raised that standards constitute technical barriers to trade potentially associated with negative distributional effects (Reardon *et al.*, 2003; Weatherspoon and Reardon, 2003; Jensen, 2004; Augier *et al.*, 2005; Okello, 2005, unpublished doctoral thesis), while on the other hand standards may constitute a catalyst for modernization (Jaffee and Henson, 2004; Maertens and Swinnen, 2008). So far, few studies have formally quantitatively assessed the impact of standards, although the number of such studies has increased (see, for example, case studies presented in Chapters 4, 5 and 6 of this book) and, for those conducted, different approaches and indicators have been used.

Standards define the 'dos and don'ts' in the production-to-marketing chain and, due to credibility requirements, they impact on enforcement structures. Standards are an institutional change, following North's definition of an institution as 'the rules of the game – both formal rules, informal norms and their enforcement characteristics' (North, 2005, p. 22). Different impact pathways have been suggested for standards and certification, varying from being a market-based mechanism providing market advantages, to a learning mechanism for knowledge transfer, to assurance mechanisms of non-observable characteristics and practices (Overdevest and Rickenbach, 2006). For the market-based mechanism, certification schemes are seen as a market-based approach for the conservation of natural resources and the provision of environmental services by collecting price premiums for environmentally sustainably produced products, whereby the premium can be considered as a reward (the price) for the environmental service provided. For the learning mechanism, new rules and the credibility of the label necessitate structures that provide stakeholders with knowledge on how to implement the rules and how to work in systems that are compliant with these rules. For the assurance mechanism, verification of the chain of custody is often seen as a necessary requirement to uphold trust in the certification schemes and to uphold consumer willingness to pay (Overdevest and Rickenbach, 2006). Another classification of standards differentiates between risk management and product differentiation standards. The former ensures that a product complies with the process and product requirements, while the latter is a marketing tool (Aragrande *et al.*, 2005; Henson and Humphrey, 2008).

The aim of this chapter is to outline a framework for the impact assessment of production standards[1] including an outline of potential impact pathways. Its main example is Kenyan horticulture and the private process standard GlobalGAP, but it also draws on other standards. The chapter links changes potentially triggered by standard introduction to methodological challenges associated with impact assessment and attribution in the context of a dynamic setting (changes over time), and a multidimensional (direct and indirect effects on a range of factors) setting. The chapter builds on recent research on the agrifood transformation in developing countries (Reardon *et al.*, 2009). The chapter expands the existing analysis by linking the existing standards on impact assessment work to general advances in methods for impact assessment, drawing on Shiferaw *et al.* (2005)[2] and by highlighting the ambiguity between achieving multiple goals, such as consumer needs, and development and environmental outcomes.

Impact Evaluation

The need for impact assessment (synonymously referred to as impact evaluation) and impact evaluation practices is not only of particular interest in the area of development and aid (see, for example, Network of Networks of Impact Evaluation (NONIE)[3]), but also in the area of agricultural research for development (see, for example, Consultative Group on International Agricultural Research (CGIAR)[4]).

Such impact evaluations in the field of development economics have increasingly been the focus of donors and development partners. NONIE and the Development Assistance Committee of the Organization for Economic Cooperation and Development define impact as 'the positive and negative, primary and secondary long-term effects produced by a development intervention directly or indirectly, intended or unintended' (OECD, 2002, p. 24). Impact evaluations are done before the start of an intervention as well as after. Much of this trend is driven by the need for accountability and by the wish to learn from past successes and failures in case of *ex post* evaluations. *Ex ante* impact evaluation can contribute to strategic guidance on intervention choice in the presence of alternative investments and limited resources.

Impact evaluations vary in their focus. Some emphasize an argument along a causal chain, such as a logical framework, while others place more emphasis on attribution (White, 2009). Attribution establishes a direct causal link between an intervention and its effects based on a counterfactual, i.e. answering the question as to what would have happened without the intervention. For all impact assessments, appropriate indicators for determining impact have to be identified. These can be grouped as either economic (such as different poverty or wealth proxies, efficiency, equity or risk), social (such as health or working conditions) or environmental indicators (such as water and soil use and management, amongst others). Finding measurable proxies for these indicators can be a challenge in particular for environmental indicators. Overall, tracing and understanding the causal chain of development, from intervention to impact, is a prerequisite and integral part of any impact evaluation (Deaton, 2009, 2010; White, 2009). The chapter departs from the OECD definition of impact by also considering short- to medium-term impacts and accounting for the fact that the standards under consideration do not chiefly focus on achieving development effects, for example, as defined in the Millennium Development Goals.

Impact evaluation: establishment of causal chains

An impact evaluation often uses the logical framework for establishing the causal chain from intervention to impact. The logical framework is a well-established tool for project planning and evaluation and has been adopted by many development organizations. Its structure follows a vertical, as well as a horizontal, logic. The vertical logic stretches from inputs, to outputs, to purpose, to contribution, to the overall goal, while the horizontal logic stretches across the evaluation domain from the vertical axis, to verifiable indicators, to the means of verification given important assumptions (McLean, 1988) (Table 3.1). This framework has been further specified to expand from outputs to outcomes, which are defined as everything that happens as a consequence of a project's interventions.

Outcome mapping (Earl *et al.*, 2001) and participatory impact pathway analysis (Douthwaite *et al.*, 2007, 2008) are further developments of the logical framework. In outcome mapping, the project team and its partners clarify the selected project's desired long-term impacts and identify key boundary partners (partners that control change; Earl *et al.*, 2001), as well as progress makers, at the start of the project and during implementation. Key boundary partners, as well as progress makers, are relied on for tracking performance (Walker *et al.*, 2008). In contrast, in participatory impact assessment stakeholder workshops establish hypotheses about cause-and-effect linkages (i.e. impact pathways) that connect project outputs to outcomes and, under some circumstances, to impacts (Walker *et al.*, 2008). Both of these approaches are more participant driven than the logical framework.

Further approaches to establish the causal chain between intervention and impact have been defined in the field of adoption and diffusion of innovation theory (e.g. David, 1969; Rogers, 1995; Waibel and Zilberman, 2007), which focuses on the uptake of an innovation such as a technology or practice through a target group. This process includes the stages of information,

Table 3.1. The logical framework (from NORAD, 1999).[5]

Goal: the higher-level objective towards which the project is expected to contribute (mention target group)	Indicators: measures (direct or indirect) to verify to what extent the goal is fulfilled (means of verification should be specified)	Assumptions: important events, conditions or decisions necessary for sustaining objectives in the long term
Purpose: the effect that is expected to be achieved as the results of the project (mention target groups)	Indicators: measures (direct or indirect) to verify to what extent the purpose is fulfilled (means of verification should be specified)	Assumptions: important events, conditions or decisions outside the control of the project that must prevail for the development objective to be attained
Outputs: the results that the project management should be able to guarantee (mention target groups)	Indicators: measures (direct or indirect) to verify to what extent outputs are produced (means of verification should be specified)	Assumptions: important events, conditions or decisions outside the control of the project management necessary for the achievement of the immediate objective
Activities: the activities that have to be undertaken by the project in order to produce the outputs	Inputs: goods and services necessary to undertake the activities	Assumptions: important events, conditions or decisions outside the control of the project management necessary for the production of the outputs

decision and implementation. Implementation covers full, partial and non-adoption of an innovation. This approach traces the decision-making framework and stages among potential users of an innovation, which in the end determine the impact of a project. For example, when an intervention is not adopted, little impact will be realized among the potential target group, while partial adoption may imply that indicators for measuring impact have to be adjusted to the modifications of the intervention through the target group.

Such causal chains can be used in a wide range of contexts and project stages – for *ex ante* and *ex post* assessment, while quantitative impact assessments that formally address attribution are only applicable for *ex post* analysis.

Impact evaluation and attribution: what would have happened without the intervention?

This section briefly summarizes the current approaches to impact evaluation establishing attribution based on several references, including Baker (2000), Ravallion (2001), Leeuw and Vaessen (2009) and White (2009). It does not strive to give a full insight into the technical detail of approaches to impact assessment design and analysis, but rather highlights key issues, which, in the next section, will be applied to the case of standards, certification and quality assurance systems. All of these have in common that they strive to deal with attribution, counterfactual and selection bias.

Attribution assigns the output and outcome to an intervention and addresses the question, 'What would have happened without the intervention?' This is a complex issue since changes take place constantly and other changes may lead to a similar impact to that caused by the intervention in question; linkages to other sectors and multiplier effects may further increase the complexity of the context. Interventions are not 'stand alone' changes but have to be seen in context, i.e. changes of other external factors and sometimes in a sequence of linked/related interventions. Interventions may be fungible (Elbers *et al.*, 2009). Attribution requires a counterfactual, which strives to address this complexity by establishing a comparison group to which households who receive the intervention are compared. A comparison group should have identical characteristics to the households who received the intervention but should not have been 'contaminated' through, for example, spillovers from the treatment group (IEG, 2006). Often, receipt of the intervention is not random, e.g. if richer households are more likely to receive a technology or more likely to decide to participate in a programme. Such non-random assignment is called (self-) selection bias and may result in biased impact estimates. Selection bias may be due to both measurable and non-measurable (observable and non-observable) effects, which have to be controlled for by using appropriate statistical methods.

Different methods for quantitative impact assessment are available, and a good overview and introduction are provided in Ravallion (2001). The approach to impact assessment depends on the stage in the intervention cycle at which the assessment is initiated. When an impact assessment is commissioned before an intervention, the intervention itself can be designed to follow a randomized controlled trial, which is similar to clinical trials, and has become more popular in the agricultural development literature in recent years. In this case, the intervention is allocated randomly among the target population.[6] In such experimental settings, the impact is quantified by estimating the difference in levels of impact indicators between recipients and non-recipients. Indicators can be monitored by surveys. Randomization is the best approach for controlling for selection bias in the presence of unobservables, i.e. non-measurable effects (White, 2009). Alternatively and in the case of constraints to appropriate randomization or political feasibility, the impact assessment can follow a double-difference approach. This is survey-based, using a baseline (data collected prior to the intervention) and a follow-up survey. The baseline, as well as the follow-up survey, covers households that will receive the treatment and a comparison group of households that will not receive the project. Such surveys include variables on the indicators to be monitored as well as further control variables. This design is also referred to as the 'before and after' and the 'with and without' designs (Ravallion, 2005). In the case of situations where baselines do not exist and impact assessment is commissioned in hindsight or after the start of the interventions, quasi-experimental designs with surveys across with and without intervention households are used for impact assessment.

Analysis is often based on a triangulation of different statistical approaches, such as matching, or two-step econometric models, such as treatment effect and instrumental variable[7] analysis. The latter two have the added benefit over matching in that they also allow for the control of further external factors apart from the intervention, which allows the establishment of the relative weight of other factors of influence on the impact indicator. (Self-) Selection bias is controlled in the latter two methods by first establishing an estimate for selection bias and then controlling for it in a second step when impact is estimated. Such regression-based techniques are flexible and can deal with heterogeneity in treatment, multiple interventions as well as interaction between interventions (Leeuw and Vaessen, 2009). Even multiple interventions through multiple, sector-wide programmes can be assessed using such econometric approaches (Elbers *et al.*, 2009), thereby exploiting the heterogeneity in interventions and the intensity of an intervention.

In conclusion, different impact assessment approaches that address attribution have different advantages; therefore, a combination of approaches has been recommended. The overall approach has to ensure internal (attribution), construct (use of appropriate variables and indicators), external (ability to generalize findings under different settings) and statistical conclusion validity (the degree of confidence in the link between intervention and impact) (Leeuw and Vaessen, 2009).

Moving from a logical framework to a quantitative impact assessment statistically implies the use of structural models, which model the causal link along each output–outcome–impact pathway. However, depending on the complexity of the intervention, this implies potentially large numbers of instruments to control for selection bias (or a large number of randomized interventions), so that Elbers *et al.* (2009) recommend using logical frameworks (and impact pathways) for organizing data, and actual impact assessment based on the reduced form, which skips such intermediate steps along the impact pathway and concentrates only on the link between intervention and indicators for output (and outcome). However, while this satisfies the demand for establishing quantitative evidence on impact addressing accountability needs, it does not necessarily contribute to understanding mechanisms of development, which would also trace questions of combined interventions and sequencing.

Quantitative impact assessments have the benefits of providing information on the cost-effectiveness of the intervention (White, 2009), contributing to deriving lessons on what worked and what did not work, as well as contributing to transparency and accountability for the spending of public funds. However, as Deaton (2009, 2010) outlines, establishing causality as well as understanding mechanisms of development is more informative in the long term than the evaluation of single projects, i.e. understanding why projects worked rather than whether projects worked (Pawson and Tilley, 1997; Deaton, 2009). Regarding understanding of causalities and impact pathways, a combination of quantitative and qualitative approaches can give the best insights, as demonstrated by Place *et al.* (2007). They showed that, especially for complex and multidimensional concepts such as poverty, participatory impact assessment can help to elicit indicators closely fitting the local understanding of poverty.

Selected standards applied in developing countries in the primary sector: Agriculture and forestry

Mandatory public standards with effect on Kenyan horticulture have been mostly part of EU legislation, because the EU is the major export market. EU legislation can be differentiated in regulations and directives, which are both legislative acts of the European Union but differ in the degree of prescriptiveness and their strategy for enforcement. Regulations automatically turn into enforceable law in member states and prescribe the result, as well as the means to get there. Directives have to be translated into national law and prescribe the outcome, but not the means of getting there (Table 3.2).

In the Kenyan horticulture context, EU regulations as well as GlobalGAP have parallels to process standards, while the EU directives such as that on MRLs and traceability more resemble performance standards.

GlobalGAP

GlobalGAP is a private sector body initially founded by large European retailers; its legal form is a GmbH (which is similar to a limited liability company) registered in Germany and uses a modular structure to cover crops, livestock and aquaculture. The crop module includes not only horticultural crops such as fruit, vegetables and flowers, but also perennial crops such as green coffee and tea. GlobalGAP has several internationally accredited certification bodies that audit and certify producers. It has the provision for benchmarking for the equivalent national and other standards.

GlobalGAP, with its smallholder representation[8], vast consultation forum and membership, is a mixture of a voluntary consensus standard and a de facto mandatory standard due to its high and increasing adoption rate. The number of certified producers globally increased from 18,000 to 94,000 in 2008 (GlobalGAP, 2009a). In Africa, as of July 2009, South Africa held the largest number of certified producers (1951), followed by Kenya (207), Ghana (149), Côte d'Ivoire (157) and Senegal (97) (GlobalGAP, 2009b).

The standard is structured into five major blocks, each with a set of complementary elements, including (i) the General Regulations, which include the system rules; (ii) the Control Points and Compliance Criteria,

Table 3.2. Public and private standards relevant to the Kenyan horticulture sector.[a] Source: Based on Cosmas Kyengo (Syngenta, formerly FPEAK, Nairobi), personal communication 2008 and Ruth Nyagah (AfriCert, Nairobi), personal communication 2008.

Year[b]	Public and private standards.
2000	Directive 2000/29/EC: Control of the introduction of pests and diseases harmful to plants and plant products
2001	Directive 2000/42/EC: Maximum Residue Levels (MRL)
2000	First work on EurepGAP
2001	Packhouse and large-scale farm British Retail Consortium (BRC) and Tesco Nature's Choice (TNC) certified
2004/2005	First formal EurepGAP certification for small-scale farmers (large-scale farms EurepGAP, BRC, TNC already certified)
2005	FAO Guideline 15: Wooden packaging and material in international trade (2000/29/EC)
2005	Regulation EC/178/2002 Food Law & Regulation Hygiene of Foodstuffs EC/852/2004 (process-control/traceability)
2006	Regulation EC 1148/2001 & Regulation (EC) 431/2006: Certificate of Conformity (Kenya Plant Health Inspectorate Service (KEPHIS) laboratory accredited to international standard)
2007	KenyaGAP[c] benchmarked to EurepGAP
2008	Regulation (EC) No. 396/2005: Harmonization of MRL
2008	KenyaGAP Domestic[c]
2009/2010	Auditing and certification of first farmers for KenyaGAP Domestic

[a] List not exhaustive – amendments include: Directive 2000/42/EC of 22 June 2000 amending the Annexes to Council Directives 86/362/EEC, 86/363/EEC and 90/642/EEC on the fixing of maximum levels for pesticide residues. EurepGAP founded in 1997, EurepGAP, Version 2.1-Oct04 relevant in 2005. EurepGAP renamed GlobalGAP in 2007 to promote the harmonization of good agricultural practices schemes beyond Europe. 2004: Amendment 2002/98 to 2000/29/EC because of interceptions. 2006: 882/2004 food and feeds control, complementing 178 food law.

[b] Year of implementation in Kenya.

[c] KenyaGAP is fully equivalent to GlobalGAP, while KenyaGAP Domestic maintains all compliance criteria related to food safety.

which define the specific global good agricultural practice requirements; (iii) the checklist, providing the inspection documents; (iv) the Approved National Interpretation Guidelines, specifying the national good agricultural practices; and (v) the harmonization tools for other standards referred to as the Benchmarking Cross-Reference Checklist (http://www.globalgap.org/cms/front_content.php?idcat=3). Implementation of management practices and adherence to control points is ensured through a checklist and third party auditing by accredited certification bodies.

GlobalGAP is an industry/private sector standard, while the Rainforest Alliance (RA) and the Forest Stewardship Council (FSC) are NGO-based standards (see Box 3.1 for their characteristics). GlobalGAP and the Sustainable Agriculture Network (SAN) are not labelled on the product, thus acting at the business-to-business level, while RA and FSC are labelled on the product and can be used for product differentiation. Standards, apart from Fairtrade,[9] do not primarily aim at economic or social development outcomes, but rather are driven by consumer concerns, retailer goals and/or the importing country's public health or environmental concerns. For example, GlobalGAP defines as its goal 'to reassure consumers about how food is produced on the farm by minimizing detrimental environmental impacts of farming operations, reducing the use of chemical inputs and ensuring a responsible approach to worker health and safety as well as animal welfare'. The standard's tangible output is that it 'serves as a practical manual for Good Agricultural Practice (GAP) anywhere in the world. The basis is an equal partnership of agricultural producers and retailers who wish to establish efficient certification

Box 3.1. Characteristics of further standards relevant to smallholder producers in developing countries. Source: Forest Stewardship Council (undated); Rainforest Alliance (undated).

Other standards: Rainforest Alliance (RA) and Forest Stewardship Council (FSC)

The Rainforest Alliance works under the Sustainable Agriculture Network (SAN) and is a coalition of several conservation groups. One accredited certification body (Sustainable Farm Certification International) certifies and authorizes farms that comply with SAN standards. Audit services are authorized by Sustainable Farm Certification, International and provided by inspection bodies, which are typically local NGOs who provide their services to farmers in their respective countries. Inspection bodies also offer their knowledge and experience continually to help refine and develop the Sustainable Agriculture Standard. RA is a member of the Global Association for Social and Environmental Standards (ISEAL). RA also supports smallholders through an option for group certification. In agriculture, Rainforest Alliance mostly works with perennial crops such as fruit, cocoa and coffee, and in forestry RA works with the smart wood programme, which certifies to the Forest Stewardship Council Standard.

The Forest Stewardship Council was established to promote the responsible management of the world's forests and is also driven by conservation groups. FSC does so in providing standard setting, trademark assurance and accreditation services and market access for companies and organizations interested in responsible forestry. Products carrying the FSC label are independently certified to assure consumers that they come from forests that are managed to meet the social, economic and ecological needs of present and future generations.

standards and procedures' (GlobalGAP, undated).

Overall, for crop management, GlobalGAP aims to follow the principles of Integrated Crop Management (ICM), which are defined as:

> a farming system that meets the requirements of long-term sustainability. It is a whole-farm strategy that involves managing crops profitably, with respect for the environment, in ways that suit local soil, climatic and economic conditions. It safeguards the farm's natural assets in the long-term. ICM is not a rigidly defined form of crop production but is a dynamic system that adapts and makes sensible use of the latest research, technology, advice and experience.
>
> (EurepGAP, 2004a, Annex 10, p. 4)

At the farm level, the standard follows the principles of Integrated Farm Management, which is defined as

> an approach to farming which aims to balance production with economic and environmental considerations by means of a combination of measures including crop rotation, cultivations, appropriate crop varieties and careful use of inputs.
>
> (EurepGAP, 2004a, Annex 10, p. 4)

Lastly, the principle of pest control also follows an integrated principle, which is defined thus:

> Integrated Pest Control is the rational application of a combination of biological, biotechnical, chemical, cultural or plant breeding measures whereby the use of chemical plant protection products is limited to the strict minimum necessary to maintain the pest population at levels below those causing economically unacceptable damage or loss. Integrated Pest Management (IPM) refers to the careful consideration of all available pest control techniques and subsequent integration of appropriate measures that discourage the development of pest populations and keep plant protection products and other interventions to levels that are economically justified and reduce or minimize risks to human health and the environment. IPM emphasizes the growth of a healthy crop with the least possible disruption to agro-ecosystems and encourages natural and or non-chemical pest control mechanisms.
>
> (EurepGAP, 2004a, Annex 10, p. 4)

GlobalGAP operationalizes its principles through control points, which are organized as (i) major musts, i.e. compulsory requirements, which must all be adhered to;

(ii) minor musts, of which 95% must be adhered to; and (iii) recommendations. The crop protection section contains 30% of all GlobalGAP control points, followed by the produce-handling section. Jointly, these sections have the highest number of major musts, which all refer to the handling of agrochemicals (EurepGAP, 2004b). Appropriate handling of agrochemicals is further organized in ten sub-components. The comprehensiveness of GlobalGAP and its control points is illustrated in Fig. 3.1.

The impact pathway of GlobalGAP in Kenyan horticulture

GlobalGAP, as well as RA and FSC, primarily targets sustainable natural resource use which, in the case of GlobalGAP, is chiefly driven by for-profit private sector bodies. These private retailer goals of good agricultural practice are a risk reduction tool linked to European liability legislation. Such a driving force can lead to different selection criteria for target farmers than those for a development intervention through a public institution. A randomized selection of farmers for inclusion in certification schemes is unlikely due to company-strategic reasons. Consequently, impact analysis must test for and deal with potential selection bias.

Fritz and Schiefer (2008) suggest economic evaluation for micro (firm), meso (supply chain) and macro (society) levels. GlobalGAP key standard components (Fig. 3.1) can be classified according to their primary point of direct intervention; some of them affect the whole supply chain, such as

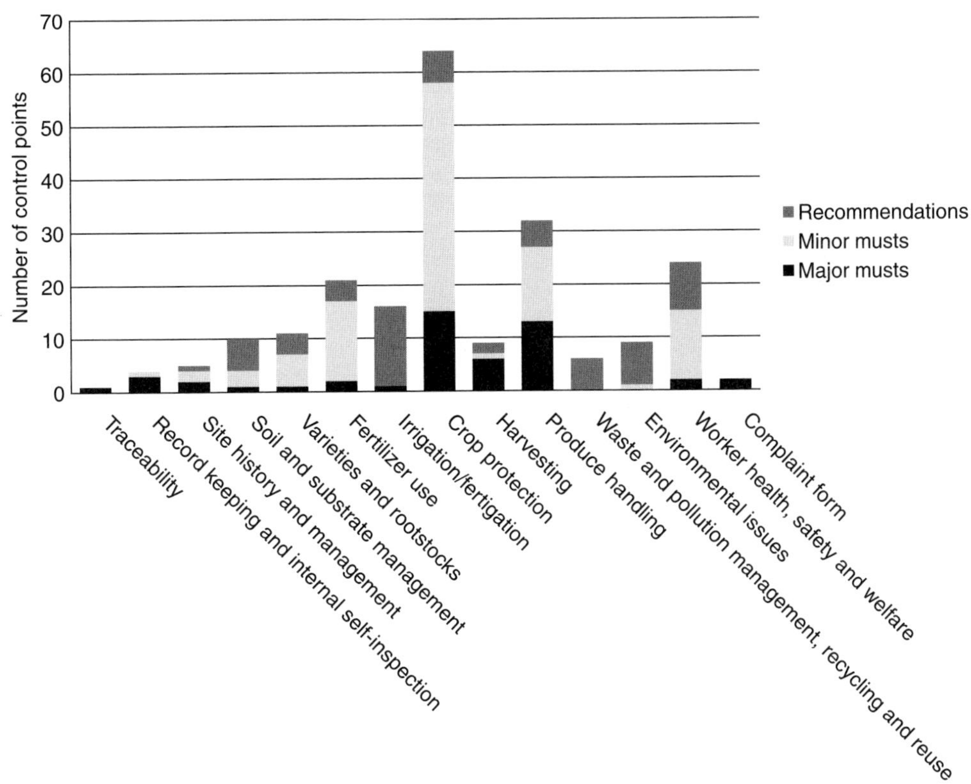

Fig. 3.1. Number of control points for major and minor musts and recommendations, by section (based on EurepGAP Version 2.1, October 2004).

traceability, while others intervene only at the farmer level in the production process (Table 3.3). Others, such as fertilizer use and crop protection, may have impact from input supplier to farmer to group level in the case where different products are required and farmer groups switch to purchasing in bulk.

Although studies on the impact of GlobalGAP in Kenya (see Chapters 4, 5 and 6 of this book as well as the synthesis in the next section) have formally addressed attribution, the attribution of impact of GlobalGAP can only be done within the context of development and history of changes preceding GlobalGAP (Table 3.2). For example, GlobalGAP benefited from the introduction of previous EU regulations and directives, which reduced the costs of compliance, because the system was already closer to the benchmarks set through the standard, compared with a country with no export production and less exposure to European market requirements. For example, EU Directive 2000/42/EC on maximum residue levels prepared farmers, as well as exporters, for tighter residue standards and the subsequent closer monitoring.[10] On the other hand, all regulations following GlobalGAP will have benefited from coming later, since this particular standard raised the bar for good agricultural practices and their implementation in Kenya.

Natural resource management and environmental impact

Due to the main environmental goal, conceptual and methodological issues highlighted by Shiferaw *et al.* (2005) for impact assessment of natural resource management projects are relevant. As GlobalGAP relies on integrated pest and crop management practices, impact assessment of the standard can further learn from impact assessments of IPM projects. IPM is a complex concept, for which so far no conceptual framework for impact assessment has been agreed (van den Berg and Jiggins, 2007).

Frequently, natural resource management (NRM) projects are multidimensional and address the management of economic externalities and/or public goods, such as genetic resource conservation, environmental services and other loss-avoidance objectives (CGIAR, 2006) that are not readily captured in markets; thus, indicators for impact assessment have to go beyond simple economic efficiency measures (Swinton *et al.*, 2005). NRM interventions are often knowledge and information based and do not come in an easily observable unit (Freeman *et al.*, 2005). Often for NRM, impact assessment challenges are beyond attribution, including the measurement of

Table 3.3. GlobalGAP control points and anticipated first point of intervention.

Control point	Point of intervention
Traceability	Supply chain
Record keeping and internal self-inspection	Farmer and group level
Site history and site management	Farmer level
Soil and substrate management	Farmer level
Varieties and rootstocks	Farmer level
Fertilizer use	Input supplier/farmer/group level
Irrigation/fertigation	Farmer level
Crop protection	Input supplier/farmer/group level
Harvesting	Farmer level
Produce handling	Farmer and group level
Waste and pollution management, recycling and reuse	Farmer and group level
Environmental issues	Farmer level
Worker health, safety and welfare	Worker and farmer level
Complaint form	Worker and farmer level

multidimensional environmental effects, spatial and temporal scales, multidimensional outcomes and valuation (Freeman *et al.*, 2005). On the first and the last points, finding appropriate indicators, methods for valuation (since NRM projects often target non-market goods and services) and appropriate aggregation can be major challenges (Freeman *et al.*, 2005). Frequently, impacts of natural resource management projects are to produce public goods or to deal with public 'bads'. In this case, incentives for the adoption of appropriate NRM techniques further complicate the analysis. Even before valuation, measuring changes in the flow and stock of natural resources and services can be challenging as indicators may vary, as well as time frames, before impact can be observed (Swinton *et al.*, 2005). Understanding of the link between agricultural activities, natural resource management and their environmental impact is limited (Altieri, 2002; Swinton *et al.*, 2005), further complicating attribution and impact assessment. Natural resource management projects often necessitate a change of farmers' (or other users') practices as a first step, which then impacts natural indicators targeted through the project (Pender, 2005).

Lessons learned from impact assessment studies of natural resource management projects commissioned through the CGIAR are that the rates of return of these specific NRM projects are sufficient to justify the investment, but the rates are not at the upper end of the spectrum of rates of return associated with agricultural research such as crop genetic improvement. Further lessons include (CGIAR, 2006, p. 30):

- Rates of return would probably be higher if environmental benefits were included (also confirmed in van den Berg and Jiggins (2007) for IPM impact assessments).
- More effort is needed to quantify and value environmental benefits (also confirmed in van den Berg and Jiggins (2007) for IPM impact assessments).
- Costs of adoption are often difficult to quantify (farmer-time, learning).
- Lack of baseline data leads to assumptions, especially regarding the counterfactual.
- NRM projects include micro- (technology) and macro- (policy) projects. Micro-projects in NRM are often small and local, although in some cases there are comparative studies across regions and countries that can produce more general knowledge.
- Natural resource management projects often include involvement in extension.

Contrary to IPM implemented through farmer field schools, GlobalGAP schemes vary greatly in their incentive and enforcement structure. While the former relies much on adult learning and appreciation of the positive environmental changes, among other benefits (van den Berg and Jiggins, 2007), the latter employs internal control and monitoring systems, annual third party audits and penalties, such as exclusion from the certification schemes for achieving integrated pest management as defined under the standard.

Empirical Evidence on the Impact of Standards

At the micro level, efficiency and equity impacts of standards (and vertical integration in general) have been listed. Their magnitude remains an empirical issue differing from case to case depending on competition and contract enforcement (Swinnen and Vandeplas, 2009). So far, few impact assessments of standards addressing attribution have been conducted. To date, all of these have been conducted at farmer and worker level looking at selected economic and social indicators, including efficiency; the results are synthesized below. These studies followed a 'with' and 'without' design, comparing standard holders with a control group, and document the difference in indicators between the two groups. Most of these studies triangulated results with different statistical methods. Studies on equity effects have been done by focusing more on analysis of causal chains rather than attribution, which is a consequence of a lack of panel data or in general observations over a longer time period on households with and without the standard. Attribution is difficult to capture at meso and macro levels due to difficulties in establishing a comparison group and maintaining it; therefore, these studies focus more on analysing causal chains of impact.

Overall, vertical integration and production standards have a positive impact on direct efficiency effects, such as enhanced productivity, better access to inputs and timely payments, as shown by case studies analysed in Swinnen and Maertens (2007) as well as findings by Asfaw in Kenyan vegetable production (Asfaw *et al.*, 2009a,b) and Minten *et al.* (2009) in vegetable production in Madagascar. Further micro-level effects include indirect effects on efficiency, such as cross-company, household and farm spillover effects, as well as a decreased marketing risk through guaranteed sales (see case studies reviewed in Swinnen and Maertens, 2007). In addition, Minten *et al.* (2007) show spillover efficiency effects from high-value crop export production to staple food crop production for the domestic market. Efficiency gains may be due to overcoming market imperfections through access to high-quality inputs under standard schemes, which are often supplied through the export company and access to management advice (Minten *et al.*, 2009). The positive impact of certification may be due to price effects, such as those shown by Bolwig *et al.* (2009) for organic coffee, which are a function of the intra-sectoral competition and enforcement structures (Swinnen and Vandeplas, 2009). With respect to social indicators, at the micro-farmer household level, GlobalGAP in Kenyan horticulture has been shown to lead to a shift to less hazardous pesticides (Asfaw *et al.*, 2009b) and reduced health impairments related to pesticide use (Asfaw *et al.*, 2010).

With respect to equity, access to and participation in markets, the evidence is not clear cut. It is recognized overall that market access and participation result in positive (economic) development and welfare gains (see, for example, Barrett, 2008, for an overview). Concerns have been raised that standards may constitute a technical barrier to trade (see, for example, Wilson, 2008, for a general overview) and may exclude smallholders from export markets due to the high cost of compliance to standards and increased vertical integration (Dolan and Humphrey, 2000; Reardon *et al.*, 2003). In addition, less well-off farmers may not be able to comply and make the necessary investments (Okello, 2005, unpublished doctoral thesis; Okello and Swinton, 2006). In Kenya, dropouts from certification schemes have been reported (Graffham *et al.*, 2007), with less clear evidence on overall export market participation (Blackmore and MacGregor, this book, Chapter 12; Mithöfer *et al.*, 2008). In addition to dropouts, negative equity effects can arise due to the fact that standards and vertical coordination in general can exacerbate the existing weaknesses of a sector (Jaffee and Henson, 2004; Diaz Rios and Jaffee, 2008). However, the net effects of exclusion from high-value export markets are an empirical issue, because producers may shift to fringe markets with lower levels of standards (Fulponi, 2007). Opportunities for countering exclusion are seen in collective action to meet standard requirements (Okello and Swinton, 2006; Narrod *et al.*, 2009); however, this may have cost implications at the smallholder level, as well as further secondary effects, for example in terms of power distribution at the community level and/or transaction costs at the community level.

Further impact at the micro level includes potential labour market effects, such as creating employment for unskilled labour, which results in positive equity effects (Maertens and Swinnen, 2009). Standards can result in upgraded workplace conditions, which can have positive welfare effects on farm workers (Ehlert *et al.*, this book, Chapter 6). The positive labour market effects of increasing formalization of the horticulture sector have also been documented in the formal domestic market in Kenya (Neven *et al.*, 2009).

At the meso level, standards may affect input and service supply, giving rise to new business opportunities to small service providers, such as private extension systems specializing in advice to achieve standard compliance (Nyambo *et al.*, 2009).

Further secondary effects in service provision include new or additional business opportunities for certification bodies, including the creation of a private Kenyan certification company in 2003 through international research and development partner initiative (Munyi and Nyagah, 2007), which subsequently gained full international accreditation to various global standards. The impact for all actors along the supply chain may include shifts of transaction costs from the retail end, where produce sourcing is more transparent due to clearly defined quality criteria (Jaffee and Henson, 2004), to the production level, where standards are implemented and have to be monitored (Mausch *et al.*, 2009).

At the macro level, private standards can substitute for missing public standards (Henson and Reardon, 2005; Fulponi, 2007). Vertical coordination is a private institutional response to the void in public institutions (Swinnen and Maertens, 2007). If governments take the initiative in setting a minimum standard, firms occasionally raise the standard higher to be on the safe side and pre-empt further regulation (Fulponi, 2007). Further, at the macro level, the standards set by the World Trade Organization and international standard-setting bodies have resulted in greater discipline, transparency and definition of a more common vocabulary (Henson, 2006).

The impact of standards on trade is difficult to trace methodologically, due to the scarcity of data and challenges in attribution. Such models often predict huge effects, whereas case studies predict rather modest effects, which is due to the multitude of responses to standards as well as repercussions (see Diaz Rios and Jaffee (2008) for Africa's groundnut exports). Such trade effects are, therefore, not further reviewed here. A recent review of trade studies has been carried out by Korinek *et al.* (2008).

A number of effects create impact at all levels through linkages, e.g. the impact on the distribution of rents depends on specific input use needs, structures of enforcement institutions and enforcement costs (Swinnen and Vandeplas, 2009). This aspect links the impact realized at the micro level to changes at the macro level. The overall impact on the economy and poverty alleviation depends on multiplier effects and linkages among sectors, as well as on the size of the base for the sector under consideration (see, for example, Diao and Dorosh (2007) on the contribution of traditional versus non-traditional export crops for income growth).

Most of the studies synthesized in this section used economic indicators for impact, and thus evidence on the environmental benefit of standards is lacking. Further to the lessons from impact assessment of the NRM project listed above, appropriate identification and characterization of the counterfactual are crucial in order appropriately to capture the benefits from the introduction of environmentally oriented standards. Such environmental benefits would need to be balanced against the cost of compliance and maintenance. In this respect, appropriate valuation of environmental (non-market) benefits may balance out potential negative distributional effects, depending on the situation without the standard. For example, certification of adherence to organic standards may actually lead to considerable improvement and efficiency gains in the production system, as well as to natural resource management in cases where extractive 'organic-by-default' systems prevail (Bennett and Franzel, 2009). In a situation when few inputs are used and the environmental risk is small a systematic approach, such as GlobalGAP, might be too cumbersome and inflexible, with compliance and certification costs potentially exceeding the environmental benefit achieved (Mattson, 2009).

Summary and Conclusions

Due to the multidimensional nature of many standards, impact assessment is not straightforward. Impact evaluation strives to establish causal chains from an intervention to its outcome dealing with attribution. Impact evaluation of standards requires the use of a multitude of indicators to capture the potential benefit, as well as the cost side. Studies on the impact of standards formally addressing attribution have so far been concentrated mostly at the micro level for selected economic and social indicators. A number of these studies showed positive efficiency effects. Since, in most settings, data prior to standards introduction are lacking, formal evidence including attribution of the impact on equity is less clear. All of these micro-level studies are short- to medium-term studies; the long-term impact of standards has not been assessed. Most assessments focused on rewards and less on their distribution over time and the risks involved. The seemingly contradictory discussion on impact pathways (whether or not standards constitute a barrier to trade or a catalyst for modernization) focuses on two different impact pathways. The first focuses on equity via the cost of compliance and access to markets, while the second focuses more on efficiency gains and the potential (environmental) benefits. However, neither of these explicitly factor in a range of other potential effects at the meso and macro levels.

At the meso level, fewer impact studies have been conducted and none of these have formally addressed attribution. The upgrading of supply chains in the sense of achieving a better balance between risk and rewards (Ponte and Ewert, 2009) can be structured around four categories – product, process, functional and inter-sectoral upgrading (Humphrey and Schmitz, 2002). Such upgrading needs a nuanced analysis and a broader definition for 'reaching a better deal' (Ponte and Ewert, 2009). For impact assessments of production standards this implies a more detailed analysis of a sector response, including impact pathways at product, production, process, functional and inter-sectoral levels, with a definition of indicators, including indicators for risk analysis at each level.

Following arguments in the impact evaluation discourse, general equilibrium effects (Thorbecke, 2006), as well as fungibility of interventions (Elbers *et al.*, 2009), need to be taken into account. The latter raises the question of the cost-effectiveness of an intervention and poses the question as to whether the intended impact of a sustainability standard could have been achieved through a different approach more cost-effectively. It does not answer the question of the effectiveness of employing standards towards achieving economic development as compared with other approaches. At the production level, standards can be interpreted as 'binary' intervention, i.e. producers adopt the standard or not. However, due to traceability requirements, standards have an impact beyond the producer level and can, therefore, be interpreted as sector-wide interventions, including multiple 'sub-'interventions, such as service provision, input provision, training of staff for the monitoring of production, and the monitoring of transport and trade at different levels of the supply chain. Additivity across units within the sector is questionable because general equilibrium effects have not been analysed. Furthermore, each intervention is affected by other policy interventions and their intensity (Elbers *et al.*, 2009). For example, in Kenya, the introduction of private production standards from about 2000 developed parallel to emerging medium-sized horticultural producers supplying to the formal domestic market,[11] which might have resulted in synergies that have not been documented and assessed. On a global scale, Kenya was one of the first countries to implement GlobalGAP under smallholder conditions; thus, lessons from this early adopter case – for donors as well as certification bodies and their benefit to later adopters – also need to be taken into account.

Overall, the impact of standards is an empirical issue depending on local conditions. International standards that aim to achieve globally applicable systems with comparable results, as well as consumer confidence, face the challenge of taking into account local specificity in international standard-setting and in globally operating verification systems. Standards, therefore, need to maintain some flexibility for interpretation and implementation, while assuring credibility (Poisot, 2003).

Further research

Further research needs can be classified in empirical and methodological advancement. At the micro level, in addition to the empirical short-term impact evaluation including attribution, medium- to long-term assessments, if possible based on panel data, would be valuable, especially including a focus on potential environmental benefits, as well as poverty effects via the tracing of households. Tracing of households would allow analysis of dynamics in market participation and would shed light on the long-term development impact of export versus alternative market participation. It would also shed light on the resilience and flexibility of households, and thus on their adaptive capacity to changes and risk management. At the farmer group level, the cost-effectiveness of quality management and the internal monitoring systems (see also Humphrey, 2008) should be investigated, as well as social changes triggered by the institutional change with standard introduction, thereby adding to the research of Paalhaar and Jansen (this book, Chapter 7). Due to the knowledge-intensive nature of the integrated pest management component, cost-effectiveness of the training structure as provided under GlobalGAP structures needs to be assessed against alternative IPM training approaches, such as through farmer field schools (Mithöfer, 2009). Methodological advancement would be valuable for environmental impact assessments.

Impact assessment at the meso and macro levels is lacking. Riisgaard's (2009) analysis stresses the point that understanding needs to go beyond the producers' level to include other stakeholders along the value chain, a point also supported by Barrett (2008), who highlights the need for more evidence on the function of service providers and middlemen. Therefore, at the meso level, the impact of standards on the reshaping of value chains, transaction costs, furtherance of vertical integration and the concentration of value chain actors at each level is relevant. Furthermore, the impact on service and input providers would be informative in assessing secondary effects. Following Riisgaard (2009), the governance structures of value chains, as well as power distribution, and their association with labour conditions along the value chain, need to be elaborated. Furthering analytical approaches for addressing attribution at this level would be valuable.

At the macro level, the potential of standards to reshape and modernize the institutional and policy landscape needs to be assessed. In support of Riisgaard's (2009) findings, research on synergies between standards and the existing institutions would be valuable, in particular the question, 'How can standards best be made use of for promoting environmental and development goals?' It needs be taken into account that standards are not a neutral market intervention (Riisgaard, 2009), but rather that all interventions take place in a system of interrelated policies and policy changes. In specific country contexts, alternative incentive structures for achieving sustainability goals, as well as the cost-effectiveness of public and private interventions, need to be factored in. The impact of standards on trade, including reviews of the analytical approach, needs to be expanded (Henson, 2006).

Overall, conceptual work needs to contribute to disentangling economic, social and environmental goals and associated appropriate interventions. Such interventions need to be assessed individually and in combination whereby the latter needs to account for issues of appropriate sequencing.

Endnotes

[1]In this chapter, 'standards' refer to systems that are compliant to a standard, as well as systems that are certified.

[2]As shown in the following sections, standards are an institutional change with sector- and supply chain-wide implications. Developing a general economic theory of standards taking such a sector- and supply change-wide view is beyond the scope of this book. Any such approach would benefit from a narrower focus on selected actors and indicators, with new institutional economics providing a good starting point.

[3]http://www.worldbank.org/ieg/nonie/

[4]http://impact.cgiar.org/. The impact assessments in the CGIAR system have covered methodological challenges and case studies for impact assessment of crop genetic improvement, natural resource management research, policy research and *ex post* impact assessment (http://www.sciencecouncil.cgiar.org).
[5]Many more examples and applications of the logical framework exist, so the one cited serves as one example.
[6]See Deaton (2009) on the challenges and Imbens (2009) on the strengths of randomized interventions for impact assessment.
[7]A recent critical assessment of the limitations to the use of randomized trials (Bamberger and White, 2007; Deaton, 2009) as well as the instrumental variable approach when dealing with heterogeneity, exogeneity and external effects can be found in Deaton (2009), and a response to Deaton (2009) by Imbens (2009) elaborating on the strengths of randomized interventions for impact assessment.
[8]http://www.africa-observer.info/
[9]http://www.fairtrade.org.uk/
[10]This development led to some smallholder dropout from export production (Löhr (Horticulture Programme, International Centre of Insect Physiology and Ecology), personal communication 2006).
[11]The emerging middle class of horticultural producers supplying the domestic formal market is documented in Neven *et al.* (2009).

References

Altieri, M. (2002) Agroecology: The science of natural resource management for poor farmers in marginal environments. *Agriculture, Ecosystems and Environment* 93, 1–24.

Antle, J.M. (2000) The cost and quality in the meat industry: Implications for HACCP regulation. In: Unnevehr, L.J. (ed.) *The Economics of HACCP – Costs and Benefits*. Eagan Press, St Paul, Minnesota, pp. 81–96.

Aragrande, M., Segre, A., Gentile, E., Malorgio, G., Giraud Heraud, E., Robles Robles, R. *et al.* (2005) *Food Supply Chains Dynamics and Quality Certification*. EU/DG Joint Research Centre, Brussels.

Asfaw, S., Mithöfer, D. and Waibel, H. (2009a) Investment in EU private food-safety standards compliance: Does it pay off for small-scale producers in sub-Saharan Africa? *Quarterly Journal of International Agriculture* 48(4), 337–362.

Asfaw, S., Mithöfer, D. and Waibel, H. (2009b) Food-safety standards, pesticide use and farm level productivity: The case of high-value crops in Kenya. *Journal of Agricultural Economics* 60(3), 645–667.

Asfaw, S., Mithöfer, D. and Waibel, H. (2010) Health impact of EU private-sector standards. Evidence from Kenya's export vegetable growers. *Agricultural Economics* 41(3), 251–263.

Augier, P., Gasiorek, M. and Lai Tong, C. (2005) The impact of rules of origin on trade flows. *Economic Policy* 20, 567–623.

Baker, J.L. (2000) *Evaluating the Impact of Development Projects on Poverty: A Handbook for Practitioners*. The World Bank, Washington, DC.

Bamberger, M. and White, H. (2007) Using strong evaluation designs in developing countries: Experiences and challenges. *Journal of Multidisciplinary Evaluation* 4(8), 58–73.

Barrett, C.B. (2008) Smallholder market participation: Concepts and evidence from eastern and southern Africa. *Food Policy* 33, 299–317.

Bennett, M. and Franzel, S. (2009) *Can Organic and Resource-conserving Agriculture Improve Livelihoods? A Meta-analysis and Conceptual Framework for Site-specific Evaluation*. World Agroforestry Centre (ICRAF), Nairobi, Kenya.

Bolwig, S., Gibbon, P. and Jones, S. (2009) The economics of smallholder organic contract farming in tropical Africa. *World Development* 37(6), 1094–1104.

CGIAR (Consultative Group on International Agricultural Research) (2006) *Natural Resources Management Research Impacts: Evidence from the CGIAR*. Science Council Secretariat, Standing Panel on Impact Assessment (SPIA), Consultative Group on International Agricultural Research, Rome (www.sciencecouncil.cgiar.org).

Cho, B.-H. and Hooker, N.H. (2009) Comparing food safety standards. *Food Control* 20, 40–47.

David, P. (1969) A contribution to the theory of diffusion. Stanford Center for Research in Economic Growth. Memorandum No. 71, Stanford University, Stanford, California.

Deaton, A. (2009) *Instruments of Development: Randomization in the Tropics, and the Search for the Elusive Keys to Economic Development*. Research Program in Development Studies, Center for Health and Wellbeing, Princeton University, Princeton, New Jersey.

Deaton, A. (2010) *Understanding the Mechanisms of Economic Development*. Woodrow Wilson School of Public and International Affairs, Department of Economics, Princeton University, Princeton, New Jersey.

Diao, X. and Dorosh, P. (2007) Demand constraints on agricultural growth in East and southern Africa: A general equilibrium analysis. *Development Policy Review* 25(3), 275–292.

Diaz Rios, L.B. and Jaffee, S. (2008) Barrier, catalyst, or distraction. Standards, competitiveness, and Africa's groundnut exports to Europe. Agriculture and Rural Development Discussion Paper 39. The World Bank, Washington, DC.

Dolan, C. and Humphrey, J. (2000) Governance and trade in fresh vegetables: Impact of UK supermarkets on the African horticulture industry. *Journal of Development Studies* 37(2), 147–176.

Douthwaite, B., Alvarez, B.S., Cook, S., Davies, R., George, P., Howell, J. *et al.* (2007) Participatory impact pathways analysis: A practical application of program theory in research-for-development. *Canadian Journal of Program Evaluation* 22(2), 127–159.

Douthwaite, B., Alvarez, S., Thiele, G. and Mackay, R. (2008) Participatory impact pathways analysis: A practical method for project planning and evaluation. ILAC Brief 17. Institutional Learning and Change (ILAC) Initiative (www.cgiar-ilac.org), hosted by Bioversity International, Rome.

Earl, S., Carden, F. and Smutylo, T. (2001) *Outcome Mapping: Building Learning and Reflection into Development Programs*. International Development Research Centre, Ottawa, Canada.

Elbers, C., Gunning, J.W. and de Hoop, K. (2009) Assessing sector-wide programs with statistical impact-evaluation: A methodological proposal. *World Development* 37(2), 513–520.

EurepGAP (2004a) General regulations: fruit and vegetables. Version 2.1-Oct04, EUREPGAP c/o FoodPLUS GmbH, Cologne, Germany.

EurepGAP (2004b) Control points and compliance criteria: fruit and vegetables. Version 2.1-Oct04, EUREPGAP c/o FoodPLUS GmbH, Cologne, Germany.

Forest Stewardship Council (undated) www.fsc.org/

Freeman H.A., Shiferaw, B. and Swinton, S.M. (2005) Assessing the impacts of natural resource management interventions in agriculture: Concepts, issues, challenges. In: Shiferaw, B., Freeman H.A. and Swinton, S.M. (eds) *Natural Resource Management in Agriculture: Methods for Assessing Economic and Environmental Impact*. CAB International, Wallingford, UK, pp. 3–16.

Fritz, M. and Schiefer, G. (2008) A multi-level cost benefit approach for regulatory decision support in food safety and quality assurance scenarios. In: Fritz, M., Schiefer, G. and Rickert, U. (eds) *2nd International Forum on System Dynamics and Innovation in Food Networks* (endorsed by the European Association of Agricultural Economists EAAE as the 110th Seminar), Innsbruck-IGLS, Austria, 18–22 February.

Fulponi, L. (2007) The globalization of private standards and the agrifood system. In: Swinnen, J.F.M. (ed.) *Global Supply Chains, Standards and the Poor*. CAB International, Wallingford, UK, pp. 5–18.

GlobalGAP (2009a) Facts and figures. Cologne, GlobalGAP (www.globalgap.org/cms/upload/Resources/Publications/Flyer/090814_GG_Flyer_ENG_FL_Web.pdf, accessed 9 December 2009).

GlobalGAP (2009b) Annual Report 2009 and Special Tour 2009 Edition. Cologne, GlobalGAP (www.globalgap.org/cms/upload/Resources/Publications/Newsletter/090915-GLOBALGAP-Annual-Report-2009_WEB-Doublepages.pdf, accessed 9 December 2009).

GlobalGAP (undated) http://www.globalgap.org/cms/front_content.php?idcat=9

Graffham, A., Karehu, E. and Macgregor, J. (2007) *Impact of EurepGAP on Small-scale Vegetable Growers in Kenya*. Fresh Insights 6, International Institute for Environment and Development, London.

Henson, S. (2006) The role of public and private standards in regulating international food markets. Paper prepared for the *IATRC Summer symposium 'Food Regulation and Trade: Institutional Framework, Concepts of Analysis and Empirical Evidence'*, Bonn, Germany, 28–30 May.

Henson, S. and Humphrey, J. (2008) Understanding the complexities of private standards in global agrifood chains. Paper presented at the *International Workshop on Globalization, Global Governance and Private Standards*, Leuven, Belgium.

Henson, S.J. and Reardon, T. (2005) Private agrifood standards: Implications for food policy and the agrifood system. *Food Policy* 30, 241–253.

Humphrey, J. (2008) Private standards, small farmers and donor policy: EurepGAP in Kenya. Working Paper 308, Institute of Development Studies, University of Sussex, Brighton, UK.

Humphrey, J. and Schmitz, H. (2002) How does insertion in global value chains affect upgrading in industrial clusters? *Regional Studies* (36)9, 1017–1028.

IEG (2006) *Impact Evaluation: The Experience of the Independent Evaluation Group of the World Bank*. Independent Evaluation Group, The World Bank, Washington, DC.

Imbens, G. (2009) Better LATE than nothing: Some comments on Deaton (2009) and Heckman and Urzua (2009). Department of Economics, Harvard University, Cambridge, Massachusetts.

Jaffee, S. and Henson, S. (2004) Standards and agrofood exports from developing countries: Rebalancing the debate. Policy Research Working Paper No. 3348, The World Bank, Washington, DC.

Jensen, M.F. (2004) Developing new exports from developing countries: New opportunities and new constraints. Doctoral thesis, The Royal Veterinary and Agricultural University, Copenhagen, Denmark.

Korinek, J., Melatos, M. and Rau, M.-L. (2008) A review of methods for quantifying the trade effects of standards in the agrifood sector. OECD Trade Policy Working Papers 79, Organization for Economic Cooperation and Development, Paris.

Leeuw, F. and Vaessen, J. (2009) *Impact Evaluations and Development. NONIE Guidance of Impact Evaluation.* Independent Evaluation Group, The World Bank, Washington, DC.

Maertens, M. and Swinnen, J.F.M. (2008) Standards as barriers and catalysts for trade, growth and poverty reduction. *Journal of International Agricultural Trade and Development* 4(1), 47–61.

Maertens, M. and Swinnen, J.F.M. (2009) Trade, standards and poverty: Evidence from Senegal. *World Development* 37(1), 161–178.

Mattson, E. (2009) *Comparative Study on the GlobalGAP Fruit and Vegetables Standard and the EU Organic Agriculture Regulation.* Grolink AB, commissioned by the UNCTAD secretariat, UNCTAD, Geneva, Switzerland.

Mausch, K., Mithöfer, D., Asfaw, S. and Waibel, H. (2009) Vegetable production in Kenya under EurepGAP standard: Is large 'more beautiful' than small? *Journal of Food Distribution Research* 40(3), 115–129.

McLean, D. (1988) The logical framework in research planning and evaluation. International Service for National Agricultural Research (ISNAR), The Hague, Netherlands (http://pdf.usaid.gov/pdf_docs/PNABA228.pdf, accessed 18 August 2010).

Minten, B., Randrianarison, L. and Swinnen, J.F.M. (2007) Spillovers from high-value agriculture for exports on land use in developing countries: Evidence from Madagascar. *Agricultural Economics* 37, 265–275.

Minten, B., Randrianarison, L. and Swinnen, J.F.M. (2009) Global retail chains and poor farmers: Evidence from Madagascar. *World Development* 37(11), 1728–1741.

Mithöfer, D. (2009) Linking smallholders to high value crop markets: How does the group approach work? In: Borot de Battisti, A., MacGregor, J. and Graffham, A. (eds) *Standard Bearers – Horticultural Exports and Private Standards in Africa.* International Institute of Environment and Development, London, pp. 118–121.

Mithöfer, D., Nang'ole, E. and Asfaw, S. (2008) Smallholder access to the export market: the case of vegetables in Kenya. *Outlook on Agriculture* 37(3), 203–211.

Munyi, P. and Nyagah, R. (2007) From science to market: Transferring standards certification know-how from ICIPE to Africert Ltd. In: Krattiger, A., Mahoney, R., Nelson, L., Thompson, J., Bennett, A., Satyanarayana, K. *et al.* (eds) *Intellectual Property Management in Health and Agricultural Innovation: A Handbook of Best Practices.* MIHR, Oxford, UK and PIPRA, Davis, California (available online at www.ipHandbook.org).

Narrod, C., Roy, D., Okello, J., Avendaño, B., Rich, K. and Thorat, A. (2009) Public–private partnerships and collective action in high value fruit and vegetable supply chains. *Food Policy* 34, 8–15.

Neven, D., Odera, M.M., Reardon, T. and Wang, H. (2009) Kenyan supermarkets, emerging middle class horticultural farmers and employment impacts on the rural poor. *World Development* 37(11), 1802–1811.

NORAD (1999) *The Logical Framework Approach (LFA). Handbook for Objectives-oriented Planning.* Norwegian Agency for Development Cooperation, Oslo, Norway.

North, D.C. (2005) Institutions and the performance of economies over time. In: Menard, C. and Shirley, M.M. *Handbook of New Institutional Economics.* Springer, Dordrecht, Netherlands, pp. 21–31.

Nyambo, B., Sief, A., Varela, A.M., Löhr, B., Cooper, J. and Dobson, H. (2009) Private extension-service provision for smallholder horticultural producers in Kenya: An approach. *Development in Practice* 19(1), 94–102.

OECD (2002) *Glossary of Key Terms in Evaluation and Results Based Management.* Development Co-operation Directorate (DCD-DAC), Organization for Economic Cooperation and Development, Paris.

Okello, J.J. and Swinton, S.M. (2006) Do international food safety standards marginalize poor farmers? Evidence from Kenyan green bean family farms. *Journal of Food Distribution Research* 37(1), 187.

Overdevest, C. and Rickenbach, M.G. (2006) Forest certification and institutional governance: An empirical study of forest stewardship council certificate holders in the United States. *Forest Policy and Economics* 9(1), 93–102.

Pawson, R. and Tilley, N. (1997) *Realistic Evaluation.* Sage Publications, London.

Pender, J. (2005) Economic methods for measuring natural resource management impacts: Theoretical issues and illustrations from Uganda. In: Shiferaw, B., Freeman, H.A. and Swinton, S.M. (eds) *Natural Resource*

Management in Agriculture. Methods for Assessing Economic and Environmental Impact. CAB International, Wallingford, UK, pp. 127–155.

Place, F., Adato, M. and Hebinck, P. (2007) Understanding rural poverty and investment in agriculture: an assessment of integrated quantitative and qualitative research in Western Kenya. *World Development* 35(2), 312–325.

Poisot, A.S. (2003) Summary analysis of codes, guidelines, and standards related to good agricultural practices. Background paper for the *FAO Expert Consultation on a Good Agricultural Practice Approach*. Agriculture Department, Food and Agriculture Organization of the United Nations, Rome.

Ponte, S. and Ewert, J. (2009) Which way is 'up' in upgrading? Trajectories of change in the value chain for South African wine. *World Development* 37(10), 1637–1650.

Rainforest Alliance (undated) www.rainforest-alliance.org/agriculture.cfm?id=san

Ravallion, M. (2001) The mystery of the vanishing benefits: An introduction to impact evaluation. *The World Bank Economic Review* 15(1), 115–140.

Ravallion, M. (2005) Evaluating anti-poverty programs. World Bank Policy Research Working Paper 3625. World Bank, Washington, DC.

Reardon, T., Timmer, C.P., Barrett, C.B. and Berdegué, J. (2003) The rise of supermarkets in Africa, Asia, and Latin America. *American Journal of Agricultural Economics* 85(5) (proceedings issue), 1140–1146.

Reardon, T., Barrett, C.B., Berdegué, J.A. and Swinnen, J.M. (2009) Agrifood industry transformation and small farmers in developing countries. *World Development* 37(11), 1717–1727.

Riisgaard, L. (2009) Global value chains, labor organization and private social standards: Lessons from East African cut flower industries. *World Development* 37(2), 326–340.

Rogers, E.M. (1995) *Diffusion of Innovations*. The Free Press, New York.

Shiferaw, B., Freeman, H.A. and Swinton, S.M. (eds) (2005) *Natural Resource Management in Agriculture: Methods for Assessing Economic and Environmental Impact.* CAB International, Wallingford, UK.

Swinnen, J.F.M. and Maertens, M. (2007) Globalization, privatization and vertical coordination in value chains in developing and transition countries. *Agricultural Economics* 37, 89–102.

Swinnen, J.F.M. and Vandeplas, A. (2009) Rich consumers and poor producers: Quality and rent distribution in global value chains. IFPRI Discussion Paper 00932, International Food Policy Research Institute, Washington, DC.

Swinton, S.M., Shiferaw, B. and Freeman, H.A. (2005) Towards comprehensive approaches in assessing NRM impact: What we know and what we need to know. In: Shiferaw, B., Freeman, H.A. and Swinton, S.M. (eds) *Natural Resource Management in Agriculture: Methods for Assessing Economic and Environmental Impact.* CAB International, Wallingford, UK, pp. 361–375.

Thorbecke, E. (2006) The evolution of the development doctrine, 1950–2005. Research Paper No. 2006/155, World Institute for Development Economics Research, United Nations University, Helsinki.

van den Berg, H. and Jiggins, J. (2007) Investing in farmers – the impacts of farmer field schools in relation to integrated pest management. *World Development* 35(4), 663–686.

Waibel, H. and Zilberman, D. (eds) (2007) *International Research on Natural Resource Management.* CAB International, Wallingford, UK.

Walker, T., Maredia, M., Kelley, T., La Rovere, R., Templeton, D., Thiele, G. *et al.* (2008) Strategic guidance for ex post impact assessment of agricultural research. Science Council Secretariat, Standing Panel on Impact Assessment (SPIA), Consultative Group on International Agricultural Research, Rome (www.sciencecouncil.cgiar.org).

Weatherspoon, D.D. and Reardon, T. (2003) The rise of supermarkets in Africa: Implication for agrifood systems and the rural poor. *Development Policy Review* 21(3), 333–356.

White, H. (2009) *Some Reflections on Current Debates in Impact Evaluation.* International Initiative for Impact Evaluation, Global Development Network, New Delhi, India.

Wilson, J.S. (2008) Standards and developing country exports: A review of selected studies and suggestions for future research. *Journal of Agricultural Trade and International Development* 4(1), 35–45.

4 The Impact of Food Safety Standards on Rural Household Welfare

Solomon Asfaw

Introduction

Today, about 1.1 billion people live in extreme poverty on less than US$1 per day, while another 1.6 billion live on US$1–2 per day. Three out of four poor people in developing countries (83 million people) were living in rural areas in 2002 (WDR, 2008). Most depend on agriculture for their livelihoods, directly or indirectly. Therefore, a more dynamic and inclusive agriculture is required to accomplish the Millennium Development Goals (MDGs) and to relieve the effects of poverty worldwide. The first MDG, to eradicate extreme poverty and hunger, in particular depends on raising the productivity of agriculture. However, in today's more integrated world economy, success in productivity-based agricultural growth crucially depends on the expansion of market opportunities. Improving the competitiveness of developing countries' agricultural products in international, regional and domestic markets is the key to expanding market opportunities.

In recent years, governments and development agencies have sought to promote the diversification of agrifood exports in order to accelerate economic growth, expand employment opportunities and reduce rural poverty. Particular attention has been given to facilitating the exports of higher-value foods, including fruit and vegetables. For these commodities, international trade has exhibited considerable growth in recent decades. According to Labaste (2005), high-value products provide an opportunity for farmers in developing countries to compete for a share of this lucrative export market. Trade in horticultural products is often considered an example of successful exports in some developing countries, with some of them managing to gain access to the horticultural value chains.

Based on the most recent trade statistics (UNCTAD, 2008), it is estimated that total sub-Saharan African exports of fresh fruit and vegetables (FFV) amounted to some US$2 billion in 2006. South Africa, with FFV exports worth almost US$1.2 billion, accounted for almost two-thirds of the region's FFV exports in value terms. The Côte d'Ivoire (US$195.2 million in 2006) and Kenya (US$178.2 million in 2004, estimated at US$215 million for 2006 by extrapolating the 2004 figure) are also important exporters, followed by Cameroon, Ethiopia, Ghana, Madagascar, Namibia, Senegal, Tanzania, Zambia and Zimbabwe, each of which exported FFV exceeding US$25 million. In several sub-Saharan African countries, the FFV sector contributes significantly to the total exports of agricultural products. For example, FFV exports accounted for 29% of South Africa's agricultural exports in 2006.

For all sub-Saharan African countries as a group, FFV represented 11.7% of all agricultural exports (UNCTAD, 2008).

Domestic markets for high-value crops are also the fastest-growing agricultural markets in most developing countries, expanding by up to 6–7% per year (WDR, 2008). As incomes rise, supermarkets become more dominant in the domestic retail sales of agricultural products – reaching 60% in some Latin American countries (WDR, 2008). The poverty impact of this growth in the domestic and global market depends on how the rural population participates in high-value markets, either directly as producers, as in Bangladesh, or through the labour market, as in Chile.

Enhancing smallholder participation requires market infrastructure, upgrading farmers' technical capacity, risk management instruments and collective action through producer organizations (Labaste, 2005). Addressing the stringent sanitary and phytosanitary standards in global markets is an even bigger challenge. In the past few years the challenge of international competitiveness in higher-value food trade has become increasingly linked to the development of the capacity to manage food safety. In the European Union (EU), official food safety requirements are becoming more stringent, while new standards are being applied to address previously unknown or unregulated hazards. Parallel to the changes in official standards, public oversight has been accelerated by the private sector to address food safety risks and otherwise address the environmental and social concerns and preferences of consumers and civil society organizations. Supermarket chains in Europe have developed prescriptive, production-oriented standards, such as the EU Retailers Protocols on Global Good Agricultural Practices (GlobalGAP), which are intended for growers of fresh fruit and vegetables and require certification by an independent, internationally accredited certification body.

This proliferation and enhanced stringency of food safety standards represents potential barriers to farmers seeking to expand their trade in higher-value foods. Yet they may also represent a catalyst for the upgrading of production operations, and for improved collaboration between the public and private sectors. According to Henson and Jaffee (2006), a major implication of adopting the standards as catalysts for development is the need to view compliance as a strategic issue, so that the opportunities and challenges are managed to a competitive advantage. In this context, it is essential to understand the current status and likely trajectory of agrifood standards, the feasible range of commercial, administrative and technical options available to African farmers, firms and governments, and the underlying economics of such responses.

The organization of this chapter is as follows: the following section (Emerging Food Safety Standards) discusses the regulatory EU public standards in comparison with private voluntary food safety standards; the third section (Conceptual Framework) presents the general theoretical framework for assessing the welfare impact of GlobalGAP certification; the fourth section (Empirical Evidence from Kenya) presents some empirical findings from Kenyan survey data; and, in the fifth section (Summary and Conclusions), conclusions and policy implications of the study are pointed out.

Emerging Food Safety Standards

Food safety is a matter of primary importance, because foods that are unsafe can easily affect consumers' health and destroy their trust in the supplier, with significant negative effects on its future sales (Fulponi, 2006). What makes it even more important is the fact that food safety problems can affect a large number of consumers simultaneously throughout several regions, since food is often distributed nationwide. The food safety scandals of the 1990s have led governments of the EU and retailers to enact strict food safety standards covering four broad areas: pesticide residue limits, worker safety, packer hygiene and traceability. In order to ensure that the food safety standards are met, various quality assurance systems have been developed. These can be classified into systems of Good Practices,[1] Hazard Analysis and Critical Control Points (HACCP)[2] systems and international standards set by the International Organization for Standardization (ISO)[3] (Henson and Reardon, 2005). The compliance of food suppliers with these systems is ensured through either first-party auditing (suppliers audit themselves), second-party auditing (the retailer performs the audit) or third-party auditing, which is supposed to be independent from the other parties.

Although a matter of public importance, globalization in the food sector with supply chains crossing many national borders, growing product differentiation and new technologies have made it increasingly difficult for national governments to regulate matters of food safety and quality in detail (Hatanaka *et al.*, 2005). In response to the problems of governmental regulation bodies to keep pace with new developments in the food sector, private food safety and quality standards emerged during the 1990s. One common formula in the discussion about public and private food standards is that public standards are more concerned with physical product characteristics and, to a lesser extent, with processes, while private standards are concerned with physical and process attributes, to ensure both product characteristics and the production process.

Regulatory public standards

In the case of the EU, the tightening of public product standards has to be seen in the context of various 'food scares' (Table 4.1) that have shattered public trust in food production, as well as in the context of harmonization and rationalization of standards across countries and product areas (Humphrey, 2006). One of the most important changes in public product standards was established with EU directive 42/2000/EC on Maximum Residual Levels, which came into force in July 2001 as part of the EU pesticide regulation harmonization programme. This programme requires safe levels of pesticide use to be established by scientific testing, usually based on data submitted by agrochemical companies. Apart from some other important changes (Humphrey, 2006), the new EU regulations require inspection of all imports of plant products, which creates new costs for developing countries' producers, particularly for producers sending only small batches to the EU.

Furthermore, a significant shift has occurred in the way food production is conceptualized. It has moved away from a pure product-control to a process-control approach, as clearly expressed in EU regulation EC 178/2002, which came into effect in January 2005. Food safety is viewed as an outcome of the value chain as a whole, stretching from primary production to the final consumer. This approach largely builds on the

Table 4.1. Examples of major food scares in industrialized countries (from Jaffee *et al.*, 2005).

Year(s)	Event	Country(ies)
1987/1988	Beef hormone scare	Italy/EU
1988	Poultry salmonella outbreak/scandal	UK
1989	Growth regulator scare for apples	USA
1996/1997	Microbiological contamination, berries	USA, Canada
1995/1997	Avian flu spreads to humans	Hong Kong, Taiwan
1999	Dioxins in animal feed	Belgium
2000	Large-scale food poisoning, dairy	Japan
2001	Contaminated olive oil	Spain
2001	Foot-and-mouth disease	UK/EU
2007	Foot-and-mouth disease	UK

UK Food Safety Act of 1990, which shifted the responsibility for safe food to retailers and required them to show 'due diligence', which means they have to show that they have employed all reasonable means in the stages of production, transport, storage and preparation of food to prevent health risks. This principle of due diligence has spread to the supranational level with the new EU regulation, which stipulates that food operators should 'primarily' have 'responsibility for ensuring food safety' and, therefore, must have a system in place to identify and respond to safety problems and are required to ensure the traceability of products. The latter refers to the requirement that food operators be able to trace products to the respective exporter in supplying countries, but usually not beyond this point. These process-oriented shifts in food safety largely follow the HACCP system. This approach is widely applied as a sector-specific standard in food and other sectors in order to contain risks at critical points within the production process (Humphrey, 2006). These important shifts in public regulation have had a particular influence on the development of private standards.

Private food safety standards

The line between public and private actors is frequently blurred, and private standard-setting processes usually involve different actors such as corporations, industry associations, NGOs, development organizations and the public sector operating on different geographical scales (Humphrey, 2006). While producers may have driven food chains in the past, often through spot-market arrangements, retailers have gained control over supply chains with the ongoing concentration on the marketplace over the last two decades. For instance, at the end of the 1990s, only five supermarket chains accounted for 75% of all grocery sales in the UK (Jaffee *et al.*, 2005). Private retailers have become much more involved in imposing requirements as to how food is produced throughout the commodity supply chain, even to the degree of monitoring and controlling production in developing countries (Dolan *et al.*, 1999).

Today, the motivation for private food safety standards is much more strategic. Retailers have moved into domains such as product development, branding/private label development, distribution and supplier selection as corporate strategies in order to increase rents in a highly competitive market. Fierce competition among retailers in Europe has spurred on not only innovation, but also the need to express social responsibility and accountability as a strategy of corporate benchmarking in times of increased societal reflexivity. The standards help firms to gain access to new markets and to coordinate international production due to standardization of product requirements across suppliers, thereby helping to reduce their transaction costs (Hatanaka *et al.*, 2005; Henson and Reardon, 2005). Furthermore, the standards help retailers to protect their reputation and, therefore, to maintain customer loyalty, which can be easily jeopardized by a food safety problem (Fulponi, 2006). Achieving these goals required a restructuring of value chains, with the most significant impacts on the sourcing of fresh produce. Initially they were aimed at addressing the problem of microbial contaminants in food, but later evolved to cover three broad areas: (i) pesticide residue standards, including pesticide usage, handling and storage as well as disposal of pesticide containers and unused pesticides; (ii) hygiene standards, including sanitation of grading and storage facilities, as well as general personal hygiene; and (iii) traceability requirements, including documentation of production activities, especially pesticide usage, planting and spraying dates, and labelling of graded produce (Jaffee *et al.*, 2005).

The private food safety standards that have an international scope and are applied in the fresh produce vegetable and fruit sector are very similar, because many of these are built around the same principles, such as those found in HACCP, ISO, GAP and GMP. Tesco's Nature's Choice (UK), Marks and Spencer's Field to Fork (UK) and Carrefour's Qualité Supérieure (France) are prominent examples of company-based standards that are applied exclusively to value chains governed by these retailers and are known as value chain-specific standards. Despite the significant market share of some large retailers, the most profound impact arises from the development of private collective standards being set and enforced by global standard networks, the most prominent of which have been the British Retail Consortium (BRC), the Global Food Safety Initiative (GFSI), the Ethical Trading Initiative (ETI), the International Food Standard (IFS), ISO 9000 and 14000, the Forest Stewardship Council (FSC) and GlobalGAP. These are usually applied sector-specifically. Fulponi (2006) refers to these standards as 'private voluntary standards' (PVS), which arise on three levels. First, they are a response to public regulatory pressure such as the BRC, a standard that governs issues of Good Manufacturing Practices (Processing) and was largely a response to the UK Food Safety Act. Secondly, they may develop from civil society's pressure or from a development context, as in the case of the ETI. Thirdly, they may be regarded as an effort to harmonize standards themselves, as is the case with the the GFSI or GlobalGAP. To enhance credibility and externalize monitoring costs, most of the standards such as GlobalGAP, ISO 9000 and 14000, BRC and ETI are enforced through third-party certification, such as an independent, accredited certification body, which inspects and audits the respective producers on an annual basis.

GlobalGAP standards for fruits and vegetables

The GlobalGAP guidelines reflect a harmonization of the existing safety, quality and environmental guidelines of the major European retailers, and are a response to increasing consumer interest in food safety and environmental issues (GlobalGAP, 2004). The detailed production protocols were first developed for fruit and vegetables and now also cover flowers and grains. GlobalGAP has a growing membership of retailers, including leading food retailers such as Sainsbury's, Tesco, Safeway, Coop Italia, Belgian Wholesale Markets, Waitrose and Kesko (GlobalGAP, 2004). The Euro-Retailer Produce working group established its own standard for good agricultural practices, GlobalGAP, in 1997 based on retailer needs to reassure consumers that food was being produced in a safe and sustainable manner within the context of a globalized food economy. Originally developed with reference to fruit and vegetables, it has expanded to include integrated assurance schemes for farms and aquaculture, as well as protocols for flowers and ornamentals. A protocol for green coffee will soon be completed; its organization has evolved from being uniquely retailer dominated to one of partnership with producers, and it consults regularly with consumer groups, NGOs and governments in the development of its protocols.

GlobalGAP is a quality and safety management system, providing tools for verifying best practices in a systematic and consistent way through the use of product protocols and compliance criteria. The GlobalGAP schemes are based on compliance with four main criteria: food safety, environmental protection, occupational health and safety and animal welfare. The food safety criteria are based on the application of HACCP principles, while criteria for the environment are designed to minimize the negative effects of agricultural production. Although a minimal level of occupational health and safety criteria is part of GlobalGAP, these are not to be considered with regard to in-depth audits of social conditions. All GlobalGAP schemes require compliance with national and international legislation.

Compliance with GlobalGAP is assessed on the grounds of control points, which are classified into three levels of importance, ranging from 'major musts' to 'minor musts' to recommendations. Major musts have to be fulfilled with a compliance rate of 100% (GlobalGAP, 2004). All control points that are viable for food safety and some points related to occupational safety belong to this category. A compliance of 95% is sufficient for the sum of minor musts. Those major and minor musts that are not fulfilled, but were deemed as not being applicable for certain farms, do not enter into the calculation of compliance. In contrast, recommendations have no minimum percentage that has to be complied with. Recommendations have, therefore, mainly the function of raising awareness for the topics in question. The control points for environmental protection belong to this category.

In the case of non-compliance with one of the major musts that was not declared to the certification body, the certification is suspended for three months. A second violation of the same control point leads to a withdrawal of the GlobalGAP certificate. Declared violation of major musts compliance leads to a rejection of the produce that was affected by the non-compliance. In the case of compliance with less than the required 95% of the minor musts, corrective measures have to be taken by the producer within one month. These corrections are subject to a subsequent inspection by the certification body (GlobalGAP, 2004). Repeated violation of more than 5% of the minor musts leads to a suspension of the certificate for up to six months. Within these six months, the issues have to be resolved, otherwise the GlobalGAP certificate is withdrawn (GlobalGAP, 2004).[4]

Although approximately 66% of certified producers are located within the EU, GlobalGAP is gaining in importance all over the world (GlobalGAP, 2004), especially in North and South America where GlobalGAP is already broadly implemented with certified growers in almost every country, whereas, in Africa, certified growers are mainly located on the eastern coast and, in Asia, GlobalGAP is mainly implemented in South-east Asian countries.

GlobalGAP offers four options to producers who seek to obtain certification under the standard. Under option 1, an individual farmer applies for certification. The farmer must carry out an internal self-inspection and undergo an external inspection by a certification body, which is a certification enterprise accredited by GlobalGAP. Under option 2, a group of farmers applies for a group certificate. Under options 3 and 4, individual farmers or farmer groups that have already implemented another standard can apply for a 'GlobalGAP benchmarked scheme certificate', i.e. GlobalGAP recognizes the existing standards scheme as being equivalent to the GlobalGAP standard (GlobalGAP, 2004). The four key elements for group certification are described in the GlobalGAP General Regulations (GlobalGAP, 2004).

First, the group must have an operating Internal Management and Control System (ICS). This system includes quality control through written protocols and a procedure manual, which is supposed to guarantee that internal inspections are undertaken in a competent way. It has a traceability system, which allows GlobalGAP-certified products to be separated from non-certified products and enables them to be traced back to the farm where the products were obtained. All registered members of the farmer group must be operating under the same management and be subject to a central management review. Farmers belonging to the farmer group must register for GlobalGAP certification for at least one whole year, and all farmer groups must have internal audit procedures that establish a minimum of one annual inspection for each registered farmer.

Secondly, each registered farmer should complete a farmer internal self-inspection, and this includes a completed internal self-inspection based on the GlobalGAP checklist, which must be available on each registered farm for review by either the internal or external inspector. Internal self-inspection must be carried out at least once per year. Thirdly, qualified staff must complete an internal inspection of all registered farms, i.e. a minimum of one internal inspection per year of each registered farm must be carried out by qualified staff within the farmer group, or subcontracted to an external verification body different from the certification body responsible for the external verification. The annual internal inspection must be based on the GlobalGAP checklist. Fourthly, external verification should be performed by a GlobalGAP-approved certification body. The external inspection is annual and is carried out by taking a random sample that is, as a minimum, the square root of the total number of GlobalGAP-registered farmers within the farmer group. An overview of selected GlobalGAP compliance criteria and investments required by small-scale producers is presented in Table 4.2.

Conceptual Framework

The production of food products in developing countries for export markets to industrialized countries is constrained primarily by the lack of information. Producers are often unaware of the consumer preferences and regulatory standards of the importing country. Policies to enhance higher environmental and health standards of food production rarely exist, due to weak institutions and poor capacities of regulatory enforcement. Likewise, consumers in importing countries are often misinformed about the actual production conditions in the developing countries. Their perception can be heavily influenced, for example, by reports of misuse of outdated pesticides, child labour or otherwise poor labour conditions, etc. Nobel Laureate Spence (1974) suggests that a mechanism is needed that will solve the problem of asymmetric information. Certification is a means of providing information for the buyer so that the product can conform to clearly defined standards. At the same time, certification can be a vehicle to induce producers to adopt safer and more environmentally benign crop management practices. In this chapter, an attempt is made to highlight both the general conceptual framework for understanding the adoption process from the producer perspective and, at the same time, the potential impact of GlobalGAP certification on both producers and consumer welfare.

Innovation and adoption of standards

Adoption and diffusion of the innovations theory (David, 1969; Rogers, 1995; Sunding and Zilberman, 2001; Zilberman and Waibel, 2007) have been widely used to identify factors that influence an individual's decision whether to adopt or reject an innovation. 'An innovation is an idea, practice or object that is perceived as new by individuals or other units of adoption. The perceived newness of the idea for the individual determines his or her reaction to it' (Rogers, 1995). Rogers identifies five characteristics of an innovation that affect an individual's adoption decision; these are (i) relative advantage, which is the degree to which an innovation is perceived as being better than the idea it supersedes; (ii) compatibility, or the degree to which an innovation is perceived as being consistent with the existing values and beliefs, past experiences and the needs of potential adopters; (iii) complexity, which is the degree to which an innovation is perceived as relatively difficult to understand and use; (iv) trialling potential, or the degree to which an

Table 4.2. Selected GlobalGAP compliance criteria and investments required (from GlobalGAP, 2004).

Compliance criteria	Investments required
Record keeping and internal self-inspection	Office construction, office furniture, writing materials, technical staff, etc.
Crop protection	Chemical store, use of integrated pest management (IPM), shift to approved chemicals, etc.
Worker safety, health and welfare	Construction of toilet and bathroom, first aid equipment, protective clothing, disposal pit, potable water, etc.
Soil management and quality of water	Soil and water testing, etc.
Product handling	Grading shed, charcoal cooler, etc.

innovation may be used experimentally on a limited basis; and (v) observability, which is the degree to which the results of an innovation are visible to others. The relative advantage and observability of an innovation describe the immediate and long-term economic benefits from using it, whereas compatibility, complexity and trialability indicate the ease with which a potential adopter can learn about and use an innovation (Rogers, 1995). Although this concept is relatively old, it is relevant for most current adoption studies. However, it is important to note that it may not altogether be applicable to the case of adoption of standards.

The adoption and certification of standards such as GlobalGAP cannot be seen as a single event that takes place on a farm; rather, they must be described as a process over time with different stages from the first knowledge of the standard until its implementation. The process of compliance can be described as a process consisting of at least three stages: (i) information; (ii) decision; and (iii) implementation. At the information stage, the producer obtains information on and knowledge of the standard. He or she becomes aware of the existence of the standard and gains knowledge about how the standard works. The information stage is essential to progress to the subsequent steps of the compliance process, as certain knowledge of the standard is necessary to form an attitude towards the standard and to make a decision. It is vital to emphasize the critical importance of this stage in developing countries. In such countries, the largest number of producers face great difficulties in accessing information due to limitations, including the lack of formal education and poor infrastructure. These limitations create obstacles to information access. At the second stage, the decision stage, the producer makes a decision on the implementation of the standard. Once the decision to implement the standard is made, the producer enters the third stage, the implementation stage, which consists of the actual adoption of the standard and the introduction of the standard's requirements on the farm.

The decision to adopt safety standards is an investment decision, which may involve sizeable fixed costs, such as a grading shed, pesticide store, office, etc., while the benefits will be realized over time. The choice of whether to adopt emerging standards will, therefore, be based on a careful assessment of a large number of technical, economic and social factors. The technical feature of the standards may have a direct consequence on the decision-making process. The potential capability of the standards, in terms of enhancing yield, reducing the cost of production and giving rise to higher profit, is substantially important. The problem, however, is that, when a technology is first introduced, uncertainty with respect to its functioning under local settings is often high and it is difficult to foretell its economic outcome with certainty. However, over time, as farmers adopt and become familiar with the new technology, the uncertainty and the associated cost will fall. Some farmers may fail to adopt the standards totally if they think that the standards simply do not function well under their circumstances, or if the size or type of their farm operation is not suited to the technology in question.

The economic theory of adoption is generally based on the assumption that the potential adopter makes a choice based on the maximization of expected utility subject to different constraints. Faced with the high cost of compliance and the complexity of the standard, farmers generally examine the perceived benefit vis-à-vis the expected cost before making any kind of decision to adopt the standard.

Welfare effects of GlobalGAP certification

The analysis of the economic consequences of GlobalGAP certification in welfare terms can be illustrated by considering some simplified assumptions. Considering the case where there is no certified vegetable market to begin with, partial equilibrium implications are apparent on the right-hand side of Figure 4.1 that shows the non-certified quantity demanded, which represents the marginal willingness to pay (MWP) and quantity supplied, which in turn represents the marginal private cost (MPC) of fresh vegetables, where a* and q_o* are the pre-certification equilibrium price and quantity of vegetables, respectively. However, every unit of production is associated with negative health and environmental externalities, which represent the marginal external cost (MEC). A certification scheme such as GlobalGAP is introduced, and a portion of the vegetable producers change their practices to meet the new requirements. Two markets are now relevant, one for certified vegetables and the other for uncertified ones, assuming that there exist producers and consumers who maintain the new market. The certified and uncertified markets are denoted by the subscripts c and u, respectively.

If indeed a market for certified vegetables appears and consolidates, where agents participate voluntarily, it is because both suppliers and consumers of this market obtain some benefits. However, although it may be likely that they will be better off than with only the non-certified products market situation, this is not necessarily the case: it will depend on the specific market conditions. Assuming they were better off, both producers and consumers of this market would incur welfare gains. The externals benefiting from the certified market would gain, since now the negative externalities are smaller per unit of vegetables provided. The net result would be an improvement in welfare, depending on the effect of certification on the consumers and producers of uncertified vegetables.

The existence of a certified market affects the outcome in the uncertified market. There is likely to be a decrease in the demand for uncertified vegetables, represented in Fig, 4.1 with a shift of the marginal willingness-to-pay curve to the left, from MWP to MWP_u. When the marginal private cost in the non-certified market remains unchanged, the equilibrium quantity and the marginal price are represented by b* and q_u*. It is assumed that the marginal external cost (MEC_u) in the non-certified market segment remains the same as the original one (MEC). In terms of welfare, consumers' surplus refers to A, producers' surplus to B and externalities to C.

The certified vegetable market is represented on the left-hand side of Fig. 4.1. The marginal private cost for a certified market

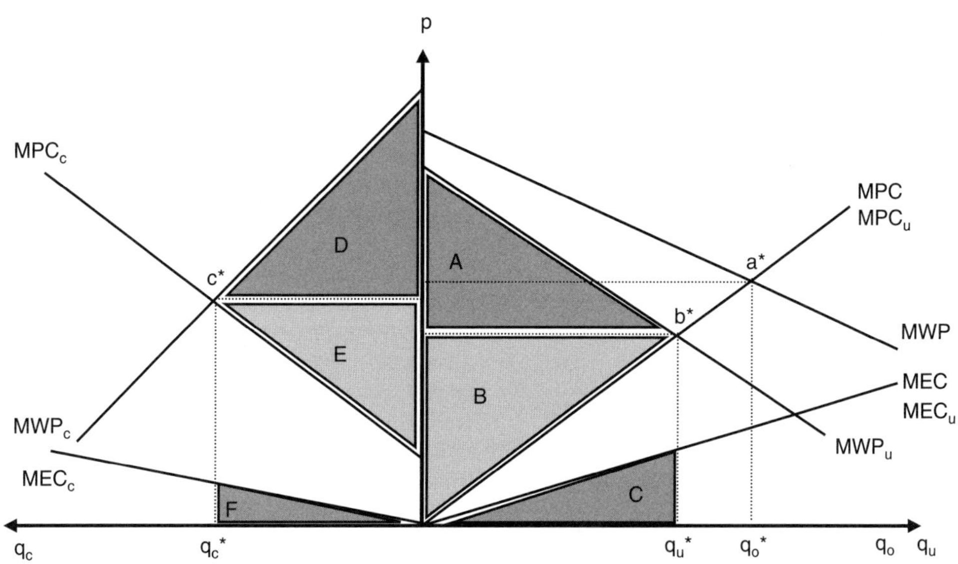

Fig. 4.1. The theoretical welfare effects of GlobalGAP certification (see text for explanation; from Asfaw, 2008).

(MPC_c) intersects the vertical axis above MPC_u, showing the additional costs producers would have to incur when following certification standards. Similarly the marginal willingness to pay for a certified market (MWP_c) intersects the vertical axis at a higher value than MWP_u, which captures the idea that willingness to pay for the first unit is higher when the product is certified. To simplify the presentation, the marginal private cost functions are represented with the same slope in certified and uncertified markets, even if in general they could be different. It is plausible to assume that negative externalities of certified production are lower than under conventional production schemes, as represented by MEC_c. In this situation, the price in the uncertified market (b*) is lower than its counterpart in the certified market (c*). The welfare surplus for consumers, producers and externals in the certified market is represented by areas D, E and F, respectively.

The segmentation of the market, caused by the introduction of a certification policy, generated a surplus equivalent to area (A + B + D + E – C – F) to be compared with the surplus before market segmentation. The comparison can also be broken down by type of agents. Thus, the sum of the producers' surpluses B and E can be compared with the original producer surplus. Likewise, consumers' surpluses A and D can be weighted against the producer surplus before the market segmentation, and C and F compared with the original external costs.

This general framework for assessing the welfare impact of GlobalGAP certification illustrates the societal gains that, in principle, can be expected from certification. In reality, of course, the welfare analysis of the economic consequences of certification can be more complex, as pointed out by Sedjo and Swallow (2002), for instance, if the assumptions presented above are relaxed. However, it is important to note that this general framework helps to identify the major research questions emerging from such interventions. For this chapter, it is not possible to provide an estimate of the overall welfare impact of standards on producers and consumers at the national level, but an attempt is made to estimate one component of producers' welfare.

Empirical Evidence from Kenya

Performance of the Kenyan horticultural sector

Horticultural exports have grown dramatically in many sub-Saharan African countries while many other agricultural export commodities have faced stagnation and declining world market prices. The Kenyan horticultural export industry has been most successful and is now by far the largest exporter of vegetables to the EU (GoK, 2006). The exports of vegetables have increased quickly over the last decade, surpassing coffee – historically Kenya's most prosperous export crop – as the nation's second major source of foreign exchange in the agricultural sector, next to tea.

Between 1980 and 2000, the value of horticultural exports increased from US$50 million to nearly US$250 million (Jaffee, 2003). After the year 2000, this development continued and the value of horticultural produce, in terms of gross market production value, increased to over US$550 million in 2005 (GoK, 2006). The value share of horticultural products in the percentage of total export value grew between 2001 and 2005 from 31% to 41%, underlining the growing importance of horticulture for the Kenyan economy (GoK, 2006). As shown in Fig. 4.2, the volume of fresh produce exports has not dropped since 2003. The horticultural industry, both export and domestic, currently contributes about 19% of Kenya's annual Gross Domestic Product (GDP) (GoK, 2006).

The major export vegetable crops are green beans, peas and Asian vegetables (such as karella, chillies, aubergines and okra), with beans and peas mostly being exported to the EU. The main flowers exported are roses, carnations, statice and a variety of summer flowers (voor den Dog, 2003).

The vast majority of this vegetable produce (89%) is destined for Europe, with the UK market absorbing the major share. Kenya also exports Asian vegetables to Middle East

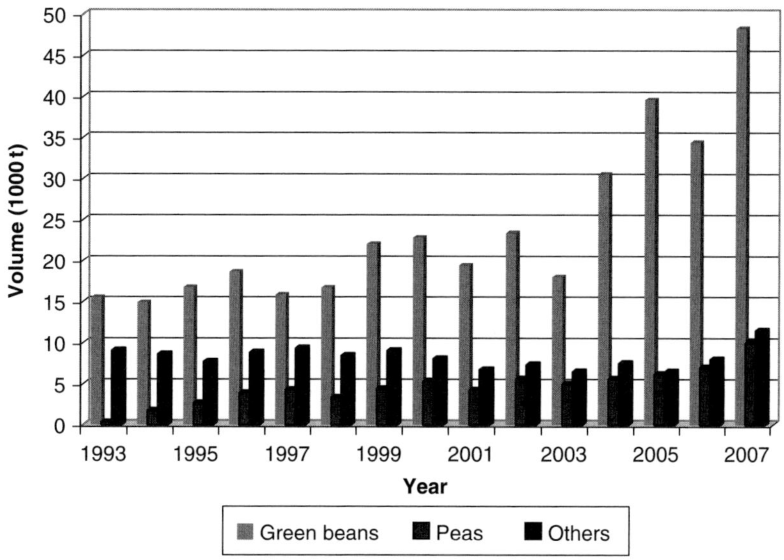

Fig. 4.2. Trends in Kenyan vegetable exports, 1993–2007 (from HCDA, 2004).

markets (Harris *et al.*, 2001). Apart from the UK, most of the Kenyan exports go to the Netherlands, France and Germany, as demonstrated in Table 4.3. For fruit the picture is more complex, with passion fruit mainly going to the UK, avocado mainly going to the Netherlands and France, while mango goes almost exclusively to the Middle East.

The concentration of products in a small number of markets implies that the development in these markets becomes crucial. In Europe in general and in the UK in particular, large supermarket chains play a dominant role in the retail of fresh produce. In the UK, the supermarkets sell around 70% of the fresh produce total while the remaining 30% is channelled through wholesale markets (Barrett *et al.*, 1999). In general the supermarkets have a similar share of trade in northern Europe, while they are much less dominant in southern Europe where much fresh produce is still sold in traditional markets and through independent greengroceries. France is somewhere in the middle. Everywhere, though, supermarkets are gaining market shares.

The growing dependence of Kenyan exporters on the UK market has had important consequences for the production and processing of horticultural products. For example, in the 1960s horticultural products were grown by smallholder farmers, sold to a small number of exporters and channelled through wholesale markets to UK retailers. According to Gray and Kleih (1997), in that period wholesalers controlled 90% of the UK fresh horticultural trade. When the UK supermarkets entered the fresh vegetable trade, they too purchased products from the wholesale market. However, this system could not achieve the quality and quantity that supermarkets required. As a result, by the 1990s, this loose network between the UK importers and an array of exporters became a coordinated value chain dominated by a few UK supermarkets (Dolan and Humphrey, 2000). The strong link to the UK can be seen as an advantage for the export industry, because the UK's supermarkets have maintained a product differentiation strategy that allows higher margins (Jaffee, 2003).

Participation of small-scale farmers in export markets

Although solid quantitative evidence concerning the actual number of small-scale

Table 4.3. Markets for Kenyan fresh horticultural exports by crop type and country (percentage[a] of export volume; from HCDA, 2004).

Crop type	UK	Netherlands	France	Germany	Rest of Europe	Middle East	South Africa
Total, fresh horticulture	35	33	10	7	3	7	5
Cut flowers	16	65	1	9	3	4	1
Green beans	56	0	29	3	3	0	10
Snow and snap peas	73	9	5	7	3	0	3
Asian vegetables	66	1	4	12	1	0	16
Avocado	1	46	32	9	2	7	3
Mango	4	1	1	1	1	92	1
Passion fruit	49	18	16	6	9	1	0

[a] Percentages may not add up to 100 as they have been rounded to the nearest digit.

farmers involved in fresh vegetable production for export in Kenya is controversial, there is a general consensus that smallholder farmer participation in export markets has been significant. Estimates from the early 1990s suggest that smallholders supplied over half of the export vegetable production (Kimenye, 1993; Jaffee, 1995). More recently, the Horticultural Crops Development Authority (HCDA) estimated that 70% of exported vegetables are produced by smallholders (Harris *et al.*, 2001). According to interviews with four leading exporters, Dolan and Humphrey (2000) conclude that just 18% of vegetables for export come from smallholders. They found that smallholders are unable to comply with the food safety and quality requirements imposed by supermarkets and other buyers. They argue that these requirements are leading exporters to grow their own produce or purchase from large-scale commercial farms. On the other hand, exporters may wish to under-report the share of production that comes from smallholders to satisfy European buyers who are suspicious of smallholder quality control (Harris *et al.*, 2001). Jaffee (2003) interviewed several exporters and estimated that smallholders account for 27% of exported fresh vegetables. A recent census estimated the number at about 12,000 smallholders producing for the vegetable export market in nine districts of Kenya by the end of 2005 (Mithöfer *et al.*, 2008). The data from the 2000 Rural Household Survey suggest that almost all farmers, large and small, rich and poor, participate in some form of horticultural production.

Costs and benefits of compliance with GlobalGAP standards

The implementation of GlobalGAP necessitates changes of production practices and investment in infrastructure. This can impose substantial costs on smallholder export farmers and may be a constraint to adoption. Table 4.4 presents the estimated costs of compliance with GlobalGAP standards incurred by individual farmers and donors and/or exporters contracting the farmers. Estimates for expenses at the household level for the adopters are obtained from the household survey,[5] whereas the total certification cost, including the share covered by exporters as well as donors, is acquired from AfriCert.[6] The estimates show that the costs of compliance with GlobalGAP standards for small-scale export vegetable producers operating under the option 2 certification scheme is about KSh36,600[7] per individual member of the group and about KSh8390 per group member by the exporters and/or donors. The investment cost borne by individual farmers accounts for approximately 30% of their total annual crop income.

The bulk of costs incurred by individual farmers (about 90%) are for investment

Table 4.4. Costs of compliance with GlobalGAP standards (from Asfaw *et al.*, 2009).

Costs incurred by individual farmer			Costs incurred by exporter per farmer	
Requirement	Cost (KSh)	Time span (years)	Requirement	Cost (KSh/year)
Application as a legal business (group)	100	Annually	Internal audits	530
Application for water permit	1,500	Annually	Pre-audits (paid once)	1,250
Toilet	9,000	5–10	Certification audits	1,650
Bathroom	3,500	5–10	Training	2,333
Grading shed	1,500	5–10	MRLs testing	875
Fertilizer store	3,000	5–10	Water analysis	28
Chemical store	4,500	5–10	Soil analysis	28
Waste disposal pit	750	3	Organic fertilizer test	33
Pesticide disposal	2,500	3	Quality controller's salary	417
Charcoal cooler	1,500	5–10	Technical assistant's salary	1,250
Office	667	5–10		
First aid kit	900	3		
Protective clothing	4,000	3		
Knapsack sprayer	2,575	3		
Harvesting buckets	195	3		
Record keeping	420	Annually		
Total	36,597			8,394

in infrastructure and equipment that farmers must have as a precondition for implementing standards. These represent the non-recurring costs and are primarily intended for record keeping, and are in support of internal self-inspection (e.g. office construction and furniture), crop protection (e.g. chemical store, pesticide disposal pit), worker safety, health and welfare (e.g. waste disposal pit, toilet and bathroom) and product handling (e.g. grading shed and charcoal cooler). Some of these structures, such as the grading shed, office and charcoal cooler, are established at the farmer group level and, hence, the costs incurred per individual are relatively low compared with the costs incurred for other structures. Such investments are more of a constraint for small-scale farmers as compared with large-scale farmers, who generally face fewer financial restrictions to purchasing equipment and building facilities. This is reflected in the respective pay-off periods for such investments. Mausch *et al.* (2009) calculated static break-even points for the returns of investments related to certification, which for a typical large-scale producer is reached 1 year after attaining certification status, while a typical smallholder needs more than 2 years to break even.

The cost incurred by exporters and/or donors is primarily for auditing, training, testing for maximum residue levels of toxic substances (MRLs), soil and water analysis, and salaries for technical assistants and staff in charge of quality control. These constitute the recurring costs and largely accrue annually. Mostly exporters who receive support from donors to subsidize the certification for smallholders pay for these costs.

Figure 4.3 highlights a number of wider benefits from compliance with GlobalGAP as perceived by the survey respondents. Smallholder growers who adopted the protocol appreciated being part of a group going through the GlobalGAP compliance process. They also perceived that adoption would assure them of markets and higher prices, as well as timely payment by the exporters. Many perceived that the implementation of GlobalGAP at the farm level increased the quality of production and reduced the amount of reject by the buyer. Under GlobalGAP, agrochemicals are stored and handled by trained individuals and many growers felt that their health is better protected. Likewise, the installation of disposal pits for the waste generated on the farm, clean toilets, baths and hand-washing facilities were perceived by the respondents as reasons for better hygienic conditions. In addition, GlobalGAP adopters expressed pride in the neatness of their farms compared with the situation before compliance. Finally, another perceived benefit to the farmers is improved bargaining power with their buyers, which enables them to switch more easily from one buyer to another. The question remains as to whether these benefits are large enough to offset the investments associated with GlobalGAP compliance, which will be addressed in the following section.

Profitability of investment in GlobalGAP standards compliance

The incremental financial benefit from investment in GlobalGAP certification was estimated using different econometric models.[8] Adoption of GlobalGAP standards is strongly and positively associated with household net income in all the econometric models. *Ceteris paribus*, adoption of GlobalGAP protocol results in an increase in net income from export vegetables of KSh8727[9] per cropping season. Note that the net income computation does not include the investment cost; it is calculated as gross income less the variable cost including family labour. Thus, it is important to address the question of whether these benefits are sufficiently large to cover non-recurring and recurring costs of obtaining and maintaining the certification standard and rendering the investment profitable. This is analysed by considering two scenarios, taking into account the planting schedule of smallholders in Kenya. Scenario one assumes that smallholders plant three export crops per year, which is the most frequent case in Kenya, and scenario two considers the worst-case situation of two cropping seasons only. Assuming a

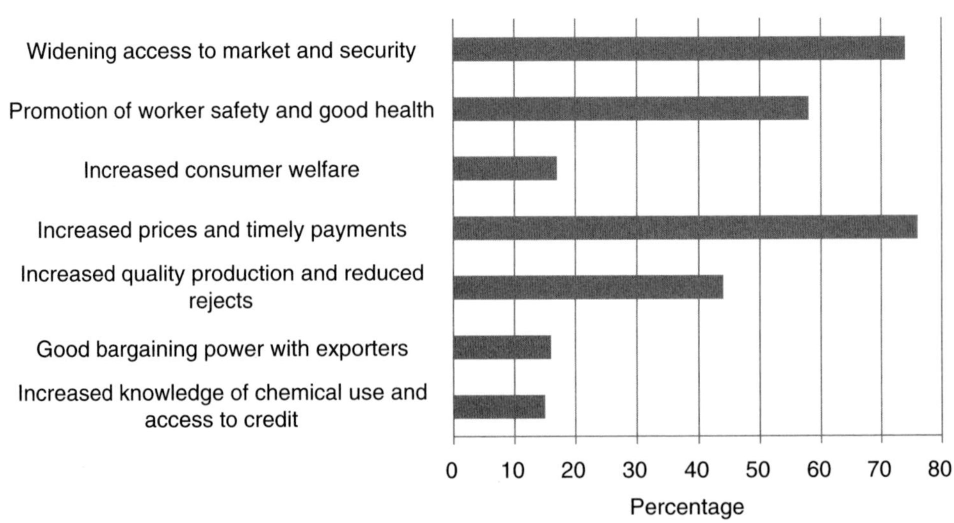

Fig. 4.3. Adopters' perceptions (expressed as a percentage) of the benefits of GlobalGAP protocol (n = 149).

constant impact of GlobalGAP on net income in all cropping seasons of KSh8727, the annual net income attributable to GlobalGAP adoption is approximately KSh22,443 under the three-cropping season scenario and KSh14,962 under the two-cropping season scenario. Using the cost data presented in Table 4.4, the financial internal rate of return (FIRR), net present value (NPV), benefit–cost ratio (BCR) and pay-off period are computed and presented in Table 4.5.[10]

First, it is assumed that farmers pay all the costs including auditing, training and testing. Considering three cropping seasons per year and a constant net income over the lifespan of the investment, the estimated FIRR is 33% for the conservative 5-year lifespan and 42% for the upper, 10-year lifespan of the investment. However, when two cropping seasons per year are considered the IRR declines to –1% for the 5-year lifespan and 15% for the 10-year lifespan of the investment.

In the second scenario, it is assumed that external agencies (donors or exporters) cover the annual audit fees, training and the tests as has been the case for small-scale farmers in Kenya (see Table 4.4). Under all scenarios the FIRR ranges from 30% for a two-crop season to up to 66% for the three-crop season scenario (see Table 4.5 for details). The pay-off period analysis demonstrates that smallholders can recover their investment cost in 2–3 years under the three-crop season scenario. However, if a two-crop season without any donor/exporter support scenario is considered, it takes up to 7 years to recoup the initial investment cost.

Due to lack of information, this analysis did not incorporate the risk inherent in the investment, nor did it compare the findings against alternative investment options that are available to smallholders. However, when comparing the FIRR with the medium-term lending rate by banks in Kenya, which is about 12%, investment in EU private

Table 4.5. Profitability of investment in GlobalGAP compliance (from Asfaw *et al.*, 2009).

		Three-cropping seasons		Two-cropping seasons	
Lifespan of the investment (years)	Decision criteria[a]	Farmer pays all the costs	Donor or exporter pays some of the costs[b]	Farmer pays all the costs	Donor or exporter pays some of the costs
5	FIRR	35	60	–1	30
	NPV	28,721	59,181	–7,062	23,398
	BCR	1.37	2.23	0.91	1.49
7	FIRR	41	65	10	37
	NPV	47,343	90,327	3,624	43,294
	BCR	1.56	2.78	1.04	1.85
10	FIRR	44	66	15	40
	NPV	74,528	125,873	13,232	64,577
	BCR	1.68	3.17	1.12	2.11
	Pay-off period (years)	3	2	7	3

[a] Discount rate used for the computation of NPV and BCR is 7%.

[b] Exporter/donor pays about KSh8394 per group member for auditing, training, MRLs testing, soil and water analysis and salaries for technical assistants and staff in charge of quality control (see Table 4.4).

FIRR, financial internal rate of return; NPV, net present value; BCR, benefit–cost ratio.

food safety standards compliance pays off for small-scale producers in that country even in the absence of external support. Yet the question remains whether or not many small-scale farmers in Kenya can finance the initial cost of about KSh36,000 in year zero to start up the implementation of the protocol and, at the same time, have the donor/exporter continue their financial and technical support.

Summary and Conclusions

Over the past five decades, production and trade of agricultural commodities have played a major economic role in many developing countries, especially in the small and least-developed ones. This progress is partially attributed to the lowering of the barriers to trade through a removal of quotas, a reduction in tariffs and preferential trade agreements for developing-country imports in Western economies. As a result, many sub-Saharan African countries have been diversifying their export portfolios away from primary commodities into non-traditional exports with more auspicious market trends. In particular, horticultural production has been indicated as a sector that can provide real opportunities for enhancing farm incomes and reducing poverty and, since the export of vegetables involves a much higher rate of labour inclusion than traditional export crops, it also suits the donors' support of labour-intensive trade.

However, it is argued that the gains from trade liberalization are offset by increasing food standards that are mainly imposed by high-income countries and that increasingly dominate the world's food trading system. Exporters are required to conform to increasingly stringent regulations and, in many cases, private sector standards. There has been an increasing awareness of health risks related to food consumption in many developed countries and, as a result, it has become a clear objective of governments and the private sector in these countries to keep these risks at a low level. Developing countries, Kenya in particular, are generally more concerned with food security (i.e. there being sufficient nutrients available for the population) than food safety for consumers. However, from the developing-country perspective, food safety issues have important implications for export opportunities to countries with a low risk tolerance. To reduce the risk intrinsic in the global trading of food and other agricultural products, these products are confronted with stringent technical requirements at the country and business levels.

Thus, there are concerns that the enhanced stringency of food safety standards that are imposed by high-income countries can negatively affect the competitiveness of producers in developing countries. First, small-scale producers may be squeezed out of the vegetable value chains because exporters prefer to work with larger-scale farmers who can be coordinated more easily. Secondly, private standards may contribute to a shift away from procurement from independent producers through contract farming towards increased agroindustrial production in estates. Yet, in some cases, contrary to this hypothesis, others argue that such standards can play an important and positive role, providing the catalyst and incentives for the modernization of export supply and regulatory systems, and the adoption of safer and more sustainable production practices. By certifying the quality of their products through standards schemes, developing countries' farmers can add value to their products, differentiate them and climb the value chain.

Results of the case study from Kenya show that GlobalGAP adopters enjoy a substantial income benefit. The FIRR, computed for different cost and benefit scenarios, demonstrates that investment in EU private food safety standards pays off for small-scale producers in Kenya. The aforementioned analysis has dealt only with the direct financial benefit from the adoption of standards, but nevertheless, as noted earlier, the adoption of standards has indirect benefits such as potential spillover effects on domestic production, stronger bargaining positions with exporters, domestic food safety, farmers' health and the environment (Asfaw *et al.*, 2010). Thus, the economic internal rate of return, which could not be calculated here, would be higher if positive externalities are accounted for.

Generally the empirical results support the notion that small-scale farmers who adopt GlobalGAP standards enjoy significant financial and non-financial benefits, in turn supporting the argument that standards can also serve as a catalyst to change and improve the production systems of farmers in developing countries. Thus, institutional arrangements that enhance small-scale farmers' physical, social and human capital are vital in influencing farm household decisions towards the adoption of emerging standards. Both public and private sector support for small actors in the supply chain is important in the adoption of a strategic perspective in addressing the challenges presented by high-value agricultural and food markets in the context of evolving food safety standards. Governments could promote awareness of the benefits of good agricultural practices and promote their wider use, improve the necessary infrastructure, develop an enabling legal/regulatory framework to facilitate compliance with standards control points and compliance criteria, provide and strengthen extension services, and support private sector activities. It is important, too, that governments provide support to strengthen well-functioning groups of smallholders and self-help groups, as well as use various tools to reduce the compliance costs of emerging private standards.

The opportunities for smallholders to remain actively involved in the lucrative export market also depend on the strategies chosen by export companies. It is important that companies adopt strategic planning that minimizes the negative impact of enhanced standards in marginalizing the poorest segment of the rural producers. Donors and other private sector actors have a key role to play in enhancing small-scale producers' capacities to comply with private sector standards. In the light of these challenges, consideration needs to be given to the policies that shift small-scale producers away from the most demanding global markets. It is important for smallholders to diversify their product categories, invest in better postharvest qualities and partake in domestic and south–south trade, the market that might be growing the fastest in the next two decades. From the standard-setter point of view, it is crucial that the emerging private standards are/will be smallholder friendly, which is acceptable to both buyers and producers and could be implemented without significant donor support.

Further research

A follow-up study is recommended in order to assess the impact of compliance with GlobalGAP on fresh produce traded in developing countries' domestic markets. During the survey already completed, it was observed that the adoption of standards not only has direct impact but also results in positive externalities in terms of enhanced food safety in both the domestic market and non-certified export market, which are mostly attributable to the more judicious use of pest control strategies. Such spillover effects may affect the production as well as the policy level, with the two levels being interlinked. Thus, there are some research questions that need to be answered in future studies, at both the policy and production level. It is not clear how the international debate has influenced the awareness of food safety issues and the standards of Kenyan policy makers and whether the increased awareness has been translated into the adoption of new strategies/policies. It is important to investigate to what extent the certified export producers apply their new knowledge to domestic production, and its impact in the production system.

Another question for further study is whether or not the non-certified smallholder export producers have adopted production technologies from certified farmers and, if so, what is their impact on production. Future research should investigate whether there is a significant shift from smallholder contract-farming towards estate farm worker, caused by increasing food safety standards. Despite the recognition of emerging private standards as an entry barrier for asset-poor smallholder farmers, it is not yet clear what farmers who are dropping out of the export markets are doing. Thus, further research should investigate what alternative livelihood strategies are available for drop-out farmers in Kenya and other countries.

Acknowledgements

This chapter is based on PhD research of the author with the University of Hannover, Germany and the International Centre of Insect Physiology and Ecology, Nairobi. The author gratefully acknowledges financial support from the German Development Cooperation (BMZ/GTZ) and technical support from the Kenyan Ministry of Agriculture and export companies. I thank the farmers involved in the survey and the enumerators who collected the data. I would further sincerely like to thank the editor and the anonymous referees for their helpful comments on earlier versions of this chapter. Any remaining shortcomings are my own.

Endnotes

[1]Good practices are guidelines in ensuring minimum standards for food processing and storage. They include Good Manufacturing Practices (GMP), Good Agricultural Practices (GAP) and Good Hygienic Practices (GHP) (Luning *et al.*, 2002).
[2]HACCP was developed in 1959 and deals with the prevention of food safety failures in food production. It covers the whole process of food production, from growing to preparation for consumption (Luning *et al.*, 2002).
[3]For example, the ISO/9000 series specifies quality system requirements and the ISO/22000 deals with food safety management systems and the requirements for any organization in the food chain (Færgemand and Jespersen, 2004).
[4]For a listing of the specific criteria according to product scheme, see http://www.globalgap.org
[5]Data were collected by means of farm household surveys in five export vegetable-producing districts of Kenya from September 2005 to August 2006. Overall, 21 sub-locations were randomly selected from the five districts and a total of 439 export vegetable producer households were chosen randomly for the interviews. Of these, 149 respondents were GlobalGAP adopters and 290 non-adopter export farmers.
[6]AfriCert is one of the few certification companies operating in Kenya to carry out certification services for mainly agricultural production and processing systems.
[7]The exchange rate at the time of the survey was US$1 = KSh72.
[8]We used a treatment effect model, regression based on propensity scores and matching techniques. For detail specifications of the analytical models, please refer to the annex and Asfaw *et al.* (2009).
[9]The incremental net income represents the average estimated income across all econometric models.
[10]The lifespan of most of the investments is assumed to range from 5 to 10 years, and hence the benefit–cost analysis is conducted considering the 5-, 7- and 10-year lifespans of the investment.

References

Asfaw, S. (2008) Global agrifood supply chain, EU food safety standards and African small-scale producers: the case of high-value horticultural export from Kenya. Pesticide Policy Project Publication Issue No. 13, Leibniz University of Hannover, Germany.

Asfaw, S., Mithöfer, D. and Waibel, H. (2009) Investment in EU private food safety standards compliance: Does it pay off for small-scale producers in sub-Saharan Africa? *Quarterly Journal of International Agriculture* 48(4), 337–362.

Asfaw, S., Mithöfer, D. and Waibel, H. (2010) Agrifood supply chain, private-sector standards, and farmers' health: Evidence from Kenya. *Agricultural Economics* 41(3–4), 251–263.

Barrett, H.R., Ilbery, B.W., Browne, A.W. and Binns, T. (1999) Globalization and the changing networks of food supply: The importance of fresh horticultural produce from Kenya into the UK. *Transactions of the Institute of British Geographers* 24(2), 159–174.

David, P.A. (1969) A contribution to the theory of diffusion. Memorandum 71, Stanford Center for Research in Economic Growth, Stanford, California.

Dolan, C. and Humphrey, J. (2000) Governance and trade in fresh vegetables: Impact of UK supermarkets on the African horticulture industry. *Journal of Development Studies* 37(2), 147–176.

Dolan, C., Humphrey, J. and Harris-Pascal, C. (1999) Horticulture commodity chains: The impact of the UK market on the African fresh vegetable industry. Working Paper 96, Institute for Development Studies, Brighton, UK.

Færgemand, J. and Jespersen, F. (2004) ISO 22000 to ensure integrity of food supply chain. *ISO Management Systems* September/October, 21–24.

Fulponi, L. (2006) Private voluntary standards in the food system: The perspective of major food retailers in OECD countries. *Food Policy* 31(1), 1–13.

GlobalGAP (2004) GlobalGAP protocol for fresh fruit and vegetables, September version. GlobalGAP Secretariat, Cologne, Germany (available at http://www.globalgap.org).

Government of Kenya (GoK) (2006) *Central Bureau of Statistics. 2006 Statistical Abstract*. The Government Printer, Nairobi, Kenya.

Gray, A. and Kleih, U. (1997) *The Potential for Selected Indian Horticultural Products on the European Market*. Marketing Series 11, Natural Resources Institute, Chatham, UK.

Harris, C., Hegarty, P.V., Kherallah, M.X., Mukindia, C.A., Ngige, J.A., Sterns, P.A. *et al.* (2001) The impacts of standards on the food sector of Kenya. Report organized by the Institute of Food and Agricultural Standards at Michigan State University, East Lansing, Michigan.

Hatanaka, M., Bain, C. and Busch, L. (2005) Third-party certification in the global agrifood system. *Food Policy* 30(3), 354–369.

HCDA (2004) *Export Statistics for Horticultural Crops. Annual Report 2004*. Horticultural Crops Development Authority, Nairobi, Kenya.

Henson, S. and Jaffee, S. (2006) Food safety standards and trade: Enhancing competitiveness and avoiding exclusion of developing countries. *The European Journal of Development Research* 18(4), 593–621.

Henson, S. and Reardon, T. (2005) Private agrifood standards: Implications for food policy and the agrifood system. *Food Policy* 30(3), 241–253.

Humphrey, J. (2006) Policy implications of trends in agribusiness value chains. *The European Journal of Development Research* 18(4), 572–592.

Jaffee, S. (1995) The many faces of success: The development of Kenyan horticultural exports. In: Jaffee, S. and Morton, J. (eds) *Marketing Africa's High-value Foods*. The World Bank, Washington, DC, pp. 319–374.

Jaffee, S. (2003) From challenge to opportunity: The transformation of the Kenyan fresh vegetable trade in the context of emerging food safety and other standards. Agricultural and Rural Development Working Paper 2, The World Bank, Washington, DC.

Jaffee, S., van der Meer, K., Henson, S., de Haan, C., Sewadeh, M., Ignacio, L. *et al.* (2005) *Food Safety and Agricultural Health Standards: Challenges and Opportunities for Developing Countries' Exports*. The World Bank, Washington, DC.

Kimenye, L.N. (1993) The economics of smallholder French bean production and marketing in Kenya. PhD dissertation, Michigan State University, East Lansing, Michigan.

Labaste, P. (ed.) (2005) The European horticulture market: Opportunities for sub-Saharan African exporters. Working Paper 63, The World Bank, Washington, DC.

Luning, P., Marcelis, W. and van der Spiegel, M. (2002) Quality assurance systems and food safety. In: Luning, P., Devlieghere, F. and Verhé, R. (eds) *Safety in the Agrifood Chain*. Wageningen Academic Publishers, Wageningen, The Netherlands, pp. 249–297.

Mausch, K., Mithöfer, D., Asfaw, S. and Waibel, H. (2009) Vegetable production in Kenya under EurepGAP standard: Is large 'more beautiful' than small? *Journal of Food Distribution Research* 40(3), 115–129.

Mithöfer, D., Nang'ole, E. and Asfaw, S. (2008) Smallholder access to the export market: The case of vegetables in Kenya. *Outlook on Agriculture* 37(3), 203–211.

Rogers, E.M. (1995) *Diffusion of Innovations*, 4th edn. The Free Press, New York.

Rosenbaum, P.R. and Rubin, D.B. (1983) The central role of the propensity score in observational studies for causal effects. *Biometrika* 70(1), 41–55.

Sedjo, R.A. and Swallow, S.K. (2002) Voluntary eco-labelling and the price premium. *Land Economics* 78(2), 272–284.

Smith, J.A. and Todd, P.E. (2005) Does matching overcome LaLonde's critique of experimental estimators? *Journal of Econometrics* 125, 305–353.

Spence, M. (1974) Competitive and optimal responses to signals: An analysis of efficiency and distribution. *Journal of Economic Theory* 9(3), 296–332.

Sunding, D. and Zilberman, D. (2001) The agricultural innovation process: Research and technology adoption in a changing agricultural industry. In: Bruce, L.G. and Rausser, G. (eds) *Handbook of Agricultural and Resource Economics*. Elsevier Science, Amsterdam, pp. 207–261.

UNCTAD (2008) Private-sector standards and national schemes for good agricultural practices: Implications for exports of fresh fruit and vegetables from sub-Saharan Africa. In: *Experiences of Ghana, Kenya and Uganda*. United Nations Conference on Trade and Development (UNCTAD), Geneva, Switzerland.

voor den Dog, T. (2003) Export chain of French beans from Kenya. MSc thesis, Wageningen University, Wageningen, The Netherlands.

WDR (2008) *World Development Report 2008: Agriculture for Development*. The World Bank, Washington, DC.

Zilberman, D. and Waibel, H. (2007) Productivity enhancement and natural resource management. In: Waibel, H. and Zilberman, D. (eds) *International Research on Natural Resource Management: Advances in Impact Assessment*. CAB International, Wallingford, UK, pp. 21–56.

Annex: Analytical Model

Two-stage standard treatment effect model

Unlike the conventional selectivity model in which the effects of adoption are calculated using the sub-samples of adopters and non-adopters separately, the standard treatment effect model uses all the observations. In this model, the observed indicator variable, G_i, indicates the presence or absence of treatment, which in this case refers to the adoption of GlobalGAP standards by household *i*. Formally, given the unobserved or latent variable, G_i, and its observed counterpart, G_i (dummy for adoption of GlobalGAP), the treatment–effect equation can be expressed as:

$$G_i^* = \beta X_i + u_i \tag{4.1}$$

$$Y_i = a V_i + \gamma G_i + e_i \tag{4.2}$$

$$G_i = 1 \text{ if } G_i^* > 0, \text{otherwise } G_i = 0 \tag{4.3}$$

where X_i is a non-stochastic vector of observed farm and non-farm characteristics determining adoption; Y_i denotes the vegetable export production net-income; V_i is a vector of exogenous variables thought to affect farm financial performance; and u_i and e_i are random disturbances associated with the adoption of GlobalGAP and the impact model.

Propensity score methods and matching techniques

The other methods used in the literature to correct for selectivity bias are propensity score methods and matching techniques. These methods are applied for the analysis to complement the results of the two-stage estimates. For these techniques to be valid, the fundamental assumption is ignorable treatment assignment (Rosenbaum and Rubin, 1983), and can formally be represented by

$$(Y_1, Y_2) \perp G_i / X \tag{4.4}$$

where Y_1 and Y_2 are net income from export vegetables for adopters and non-adopters, respectively. This assumption states that, conditional on a set of observables *X*, the respective treatment outcome is independent of actual treatment status (adoption of GlobalGAP).

Considering the underlying assumption of ignorability of treatment, the propensity score is used as the control function in the second model to overcome the endogeneity problem of adoption variable. The propensity score is estimated using a probit or logit model and indicates the conditional probability of adoption given observable regressors *X*. The structural equation then is expressed as

$$Y_i = aV_i + \gamma G_i + \mu Pscore + e_i \tag{4.5}$$

where

$$Pscore(X) = \Pr(G_i = 1/X) \tag{4.6}$$

The third model is based on matching techniques, which have to deal with the challenge of defining an observationally similar group of non-adopters to that of the adopters. Smith and Todd (2005) demonstrate that impact estimates calculated using matching methods are highly sensitive to the matching method itself, but robustness can be improved by restricting matches only to those adopters and non-adopters having a common support in the distribution of propensity scores. Therefore, the impact is estimated by applying the common support condition and further checking for robustness by using four different methods for selecting matched non-adopters, namely stratification matching, nearest neighbour matching, radius matching and Kernel matching.

5 The Impact of Compliance with GlobalGAP Standards on Small and Large Kenyan Export Vegetable-producing Farms

Kai Mausch and Dagmar Mithöfer

Introduction

Horticultural production is a major driver of the Kenyan economy, with a large export component as well as domestic production. Kenyan export production is mostly for the European market, and therefore international standards such as GlobalGAP[1] (Global Good Agricultural Practice) present a challenge to the Kenyan horticulture sector. GlobalGAP is a private standard that emerged from an initiative of UK retailers who recognized that the existing quality assurance system was not sufficient to satisfy consumer demand for food safety. It is an international quality system that primarily addresses food safety, as well as environmental and occupational health and safety issues (Galdos, 2004). GlobalGAP offers four options of certification. Under option 1, individual farmers apply for certification, while under option 2 a group of farmers applies for a group certificate; options 3 and 4 apply to individuals and groups of farmers, respectively, who are certified under other standards benchmarked to GlobalGAP. As compliance with the standard requires various investments in both long-term fixed structures, such as a grading shed, charcoal cooler and pesticide storage, and changes in the use of variable inputs, such as a switch to approved pesticides (Okello and Swinton, 2006; Asfaw *et al.*, 2009), smallholders generally opt for group, rather than individual, certification. At the time of this study in 2005/2006 Kenyan smallholders were mostly certified under option 2, whereas medium- to large-scale farms had chosen option 1 certification.

In 2001 and 2002 smallholder farms supplied 27% of the total export produce, while large-scale contracted farms supplied 29% and exporter-owned farms produced 44% (Jaffee, 2003).[2] The major export vegetable product from smallholders is green beans which, in 2006, accounted for 55% of the total vegetable exports from Kenya and for 22% of all horticultural exports (HCDA, 2009).

The effects of international production standards on developing countries' agricultural sectors are the subject of a lively debate. Some studies conclude that standards may have a negative impact on equity measures, since poor smallholder farms may drop out of the export market as they do not have the resources to comply with stringent standards. Thus, standards can contribute to increased poverty and vulnerability in rural communities, while the richer smallholder farms are able to meet these requirements (Dolan and Humphrey, 2000; Jaffee, 2003; Okello, 2005). On the other hand, the studies of Manda (1997),

Humphrey *et al.* (2004) and Maertens (2006) all conclude that an increase in employment on large farms may offset the income loss of smallholder production, although the employment opportunities are mostly temporary or casual (Henson and Jaffee, 2004; Maertens, 2006). The studies cited above agree that standards result in the concentration of production, with fewer and larger farms operating in the sector in order to reduce transaction costs; however, smallholders may continue export market participation due to the limited growth options of large-scale farms due, for example, to land or land ownership constraints (Humphrey, 2008). Even if produce could be sourced from large producers, companies may prefer to source from smallholders, since contracting them may be less risky and they are often in a better position to produce labour-intensive crops. In order to maintain a smallholder supplier base, the collective action of smallholders is promoted and contracts are used to provide services tackling resource constraints particular to smallholders (Reardon *et al.*, 2009).

This chapter aims to contribute to this discussion by presenting empirical findings on the effects of standards on three different types of Kenyan farms producing export vegetables: smallholder, large-scale contracted and large-scale exporter-owned farms. The objectives are threefold: (i) to assess the economic performance of the three different certified farm types; (ii) to compare the financial cost of compliance with GlobalGAP standards as well as the risk involved in the certification investment; and (iii) to analyse the transaction costs based on the farm type. This chapter expands on the findings and conclusions of Narrod *et al.* (2009), who conclude that small-scale farmers can maintain participation in profitable high-value crop markets through collective action; it also provides some indicator figures on the cost of running a monitoring standard-compliant smallholder production system and, thereby, addresses the research gap highlighted by Jaffee (2003).

The chapter is organized as follows: the next section describes the methodology of

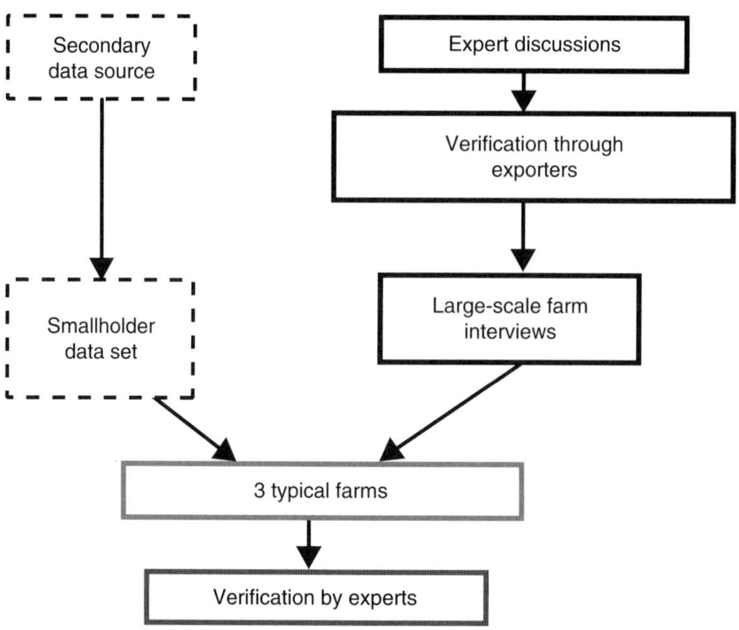

Fig. 5.1. The model validation process.

the survey design and data collection procedures; the third section describes the methodology used for the analysis of the empirical data and presents the results and discussion, while conclusions are drawn in the last section along with recommendations for further research.

Methodology

Typical farm models and data collection

To compare the effects of the GlobalGAP standard on Kenyan vegetable production, the concept of typical farm models (Hemme, 2000) is utilized. This concept follows a multi-step procedure, as shown in Fig. 5.1.

In the first step, an expert consultation is carried out in order to obtain an overview of a sector through the identification and quantification of a set of indicators. In the second step, values for the indicators are established based on multiple data sources including sector statistics, case study reports, additional communication with experts and primary data through farm surveys. Based on these indicators, the typical farm models are derived, which in a final step are verified and validated by experts. The process allows triangulation of information throughout.

The three typical farm models of this study are the large-scale exporter-owned, large-scale contracted and smallholder farms producing vegetables for the export market. Initial, informal meetings with representatives of export companies were conducted with the aim of creating a general understanding of the system of vegetable export production in Kenya and introducing the aim of this survey. Furthermore, future strategies concerning the composition of supplies sourced from the different farm types were discussed. Finally, representatives of the export companies were asked to identify some of their large producers to facilitate their participation in this study. The large-scale contracted and exporter-owned farms were contacted via the exporters. These two farm types in the sample are linked to the seven biggest exporters in Kenya. The sample covers 18 large-scale contracted farms and 8 exporter-owned farms out of the 34 certified large-scale farms (> 2 ha).

Smallholder farms were selected via a multi-stage sampling procedure[3] by selecting districts, sub-locations and smallholder vegetable producers, respectively. First, 5 districts were selected from the major vegetable-producing provinces, based on the intensity of export vegetable production, agroecology, the types of crop produced and accessibility. These districts represent the major export vegetable-producing areas which, according to the current update on the number of smallholders producing for the vegetable export market by Mithöfer *et al.* (2008), cover approximately half of the smallholder vegetable export producers. Secondly, 21 sub-locations were selected from these 5 districts based proportionately on the size of the export vegetable producer population and, thirdly, a total of 439 households producing export vegetables were selected randomly for the interview (Asfaw *et al.*, 2009). Of these, 46 are GlobalGAP-certified producers and were included in the present analysis, while at the time of the survey in Kenya a total of 201[4] smallholder farms (< 2 ha) were GlobalGAP certified.

Data for the large-scale farms were collected during a single visit to the 18 contracted and 8 exporter-owned large-scale farms from December 2005 to February 2006, whereas data from the smallholders were collected in multiple visits during two planting seasons between September 2005 and August 2006. Figure 5.2 shows the locations of the sampled farms.

The technical parameters, as well as the input prices paid and product prices received by the typical farms, were given by the mean or median of the survey data. Indicators for which of the survey data were approximately normally distributed are described by survey means, while indicators based on data with a different distribution are described by the median. Indicators included were land size, allocation of land to different crops and varieties, number of

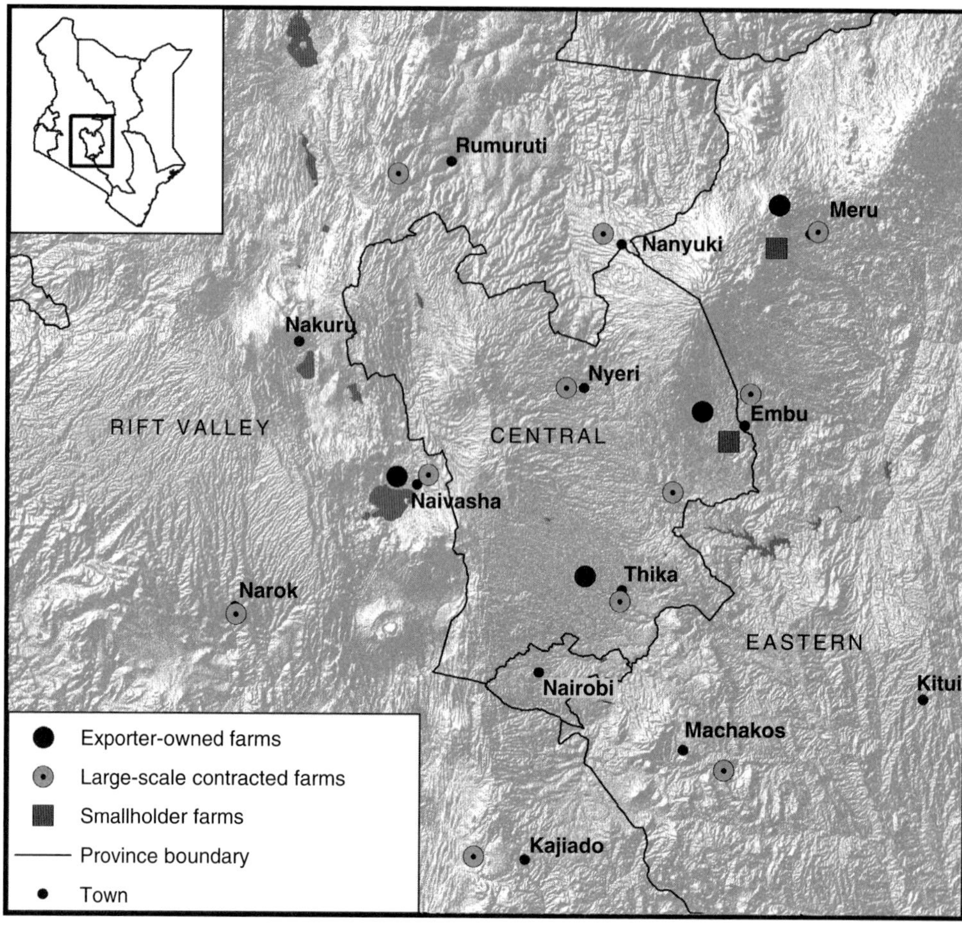

Fig. 5.2. Map of study and sampling sites.

employees and wage rates, value of machinery and buildings, as well as the marketing channel. Extreme values of the primary data set were discussed with experts – in this case mostly the agronomists employed by the sampled farms or exporter companies during a validation workshop. Because of some extreme values, two indicators, namely cultivated land and the number of workers employed for the large-scale contracted and exporter-owned farms, were based on the median. Finally, the typical farm models were validated overall with an expert workshop.

Data analysis and performance indicators

For this study the net revenue, the crop portfolio and cost structure are used as indicators to assess the farms' management decisions and performance. Break-even point analysis is included to assess the relative risk for each farm type when investing in standard compliance. The transaction cost considerations shed some light on the exporters' perspective on the costs associated with supplier choice.[5] The financial performance of the three farm types is analysed only for bean production, as this is the crop produced on all three types of farm and is the major Kenyan export vegetable crop.

All three types of typical farms pursue market-oriented rather than subsistence horticultural production and, therefore, aim at net revenue maximization. The annual net revenue for a vegetable farm is defined as the sum over net revenue of all crops grown by the farm as a function of each crop's respective area in hectares. Net revenue of each crop is calculated as revenue less variable cost, including opportunity cost of family labour.[6] Annual net revenue does not consider fixed costs due to the unavailability of information on fixed costs during primary data collection. One could expect higher fixed costs for the large-scale contracted and exporter-owned farms than for the smallholder farms. Nevertheless, this is unlikely to alter the results as scale effects decrease fixed costs on a per-output unit basis.

Based on the annual net revenue and the cost of GlobalGAP compliance investments, break-even points are computed. In this chapter, we follow the concept applied by Fleischer *et al.* (2002) in an analysis of extension systems in Egypt. 'Break-even points' refer to the months required to accrue benefits matching investment, as well as maintenance costs to that date. In addition to the variable costs of production, the entrepreneurial wage is deducted to account for the opportunity costs of the farm owners' labour. It is calculated at the rate of the best alternative wage employment for each farmer. The best alternative wage employment for a smallholder farmer would be a supervisor position on a large farm. Consequently, the typical supervisor wage on the contracted large-scale farm is used, as this is the highest possible wage. This is the upper bound of the opportunity costs but, as the smallholder farmer is an independent self-employed person, the cost of giving up his independence has to be considered as well. The same consideration holds for the large-scale contracted farmer but, as the farm owner of the large-scale farm is likely to be highly educated, the optimal alternative wage employment would be a farm managerial position on another farm. Therefore, the typical wage for this position on the exporter-owned farm is used. The same holds for the exporter-owned farm; although the owner is a firm and not a person, the calculation is based on the same principle.

For the two types of large-scale farm that plant the main crops, beans and peas, on a weekly basis, this approximation is likely to represent the real distribution of the net revenue flows over the year. Smallholder farms have three planting seasons per year; in each season the total area allocated to vegetable export production is divided into smaller plots that are planted consecutively. Recurring costs, i.e. those that occur annually, and non-recurring GlobalGAP investments are incorporated in month zero. Overall, this analysis compares the investment costs with the revenue flows for each of the farm types and, thus, illustrates financial implications, including potential liquidity constraints across the farm types. Furthermore, this analysis is taken as a proxy for relative risk associated with investment in GlobalGAP compliance, since the longer the time taken to break even the greater the likelihood of an event that leads to non-recovery of investment costs.

Transaction costs are important considerations in the analysis of imperfect markets. Two categories of transaction costs can be distinguished: market-based costs such as time for information gathering, bargaining, contracting or monitoring and enforcement of agreements and, secondly, managerial and administration costs that are a consequence of operating an organizational entity.

The major impact of GlobalGAP on transaction costs is expected to be associated with market-based activities. Exporters contracting smallholder farms face additional costs of information gathering, in searching for either certified farms or farmers willing to undergo training and certification. Furthermore, the standard increases transaction costs due to the increased costs of monitoring production to ensure compliance with the standards. The latter point, especially, can be expected to be higher when comparing production of 1 ha of beans under smallholder conditions with that of large-scale farmers, since each farmer is a separate decision maker because this includes crop management decisions such as pest management. One hectare on a large-scale farm is under one decision maker, while under smallholder conditions 1 ha of export vegetables is distributed across 3.2 farms, which all need to be monitored. Furthermore, smallholder production is distributed across a larger area overall; smallholder farms are on average farther from Nairobi, and hence movement of smallholder products to the market is associated with higher transport costs.

At the other end of the supply chain a GlobalGAP certificate can help to reduce transaction costs from the retailers' perspective, because uncertainty about production practices is reduced and the market partners, including consumers in Europe, are in a better position to make an informed decision as to where to purchase horticultural produce. Although GlobalGAP is designed to be a business-to-business certificate, European supermarkets do use the standard as a quality indicator for their customers.[7]

Results and Discussion

Typical farm description

The exporter-owned farms are concentrated in four different areas of Kenya that have the highest agroclimatic potential as well as good access to infrastructure, while many of the large and small contracted farms are located in places that are less favourable for vegetable production, with less irrigation infrastructure.

Table 5.1 shows descriptive indicators for the three typical farm types. Total land size varies tremendously among the three types, starting from 0.65 ha under production

Table 5.1. Key indicators for characterization of typical farms (from Mausch *et al.*, 2009).

Indicator	E-O farms[a]	L-S C farms[a]	SH farms[b]
Workers (*N*)	340	77	1
Permanent workers (%)	49	31	0
Value of machinery (thousand KSh)	7500	4000	25
Export vegetable area (ha)	101	14.2	0.31
Vegetable crops (*N*)	6	5	3
Export vegetable area (%)	100	54	48
Bean production indicators			
Seed (kg/ha/year)	183	223	49
(KSh/kg)	460	860	560
Fertilizer (thousand KSH/ha/year)	134	96	77
Chemicals (thousand KSh/ha/year)	54	45	96
Irrigation (thousand KSh/ha/year)	77	62	0
Labour (proportion of supervisors, %)	26	18	0
(KSh/day/worker)	117	110	100
(workers/ha)	3.4	5.4	N/A
(thousand KSh/ha/year)	307	319	77
Output (kg/ha/year)	30	25	15
(KSh/kg)	48	44	38

[a] Authors' survey data; [b] smallholder survey and data are further described in Asfaw *et al.* (2009); E-O farms, large-scale exporter-owned farms; L-S C farms, large-scale contracted farms; SH farms, smallholder farms. US$1 equalled about KSh74 during the time of the survey.

for smallholders to 101 ha under production for exporter-owned farms; the large-scale contracted farm is at an intermediate level, producing on 27 ha. The number of crops grown, the number of employees and the value of the farms' machinery vary accordingly, as does the share of land under export vegetable production.

Large-scale exporter-owned farms have more diversified crop portfolios in comparison with the large-scale contracted farms, as shown by the number of crops produced and the area allocated to each crop. Large-scale contracted farms produce few crops on relatively large areas, whereas exporter-owned farms produce a higher number of crops and different varieties on smaller areas in order to guarantee a constant supply of all crops to the exporter owner. Smallholders have the least number of vegetable crops in their portfolio, which includes mostly French beans for the export market in addition to two vegetable crops such as kale and tomatoes for the domestic market. Small-scale farms have the largest proportion of their farm allocated to production for the domestic market.

Some individual exporter-owned and large-scale contracted farms are very large. The largest farm from the sample is a contracted farm, which overall has 2900 ha under agricultural production, of which around 300 ha are allocated to export vegetables. This farm employed 850 staff.

For the area allocated to vegetable production as well as the number of employees, the survey data were characterized by mean values higher than median values. In the validation workshop, experts confirmed median values to be more representative than survey means, and hence the former were used for the two indicators and are shown in Table 5.1.

Table 5.1 further shows selected indicators characterizing production of the lead crop (beans) for the three typical farms. The typical smallholder farm uses the lowest number of inputs across all types of inputs, while the two types of large-scale farms have similar production costs for most inputs.

Large-scale contracted farms have the highest cost for seed due to the large amount of seed used and the high price of this seed. The smallholder accrues lower seed costs and uses much less seed per hectare and year, since they plant fewer rotations than do the two types of large-scale farms. The exporter-owned farm ranges at the bottom end of the seed costs and in the upper middle for the amount of seed per hectare and year. Overall, smallholder farms receive seed from the exporters and at lower prices as compared with the large-scale farms, which is later recovered by lower output prices. Provision of seed through the exporter ensures the use of high-quality seed and, consequently, higher production levels. Provision of seed to smallholders serves as a means not only of reducing production risk by the use of high-quality inputs, but also of ensuring that known amounts of seed and, therefore, plants are under production. Previous research showed that smallholders can successfully use collective action to purchase seed in bulk, as well as being successfully linked to resource-providing contracts, two factors favourable for maintaining successful smallholder export market participation (Reardon *et al.*, 2009).

The comparison of labour costs in bean production by farm type is more complex, since the organizational structure of the smallholder farm is different from that of the two large-scale farm types. The smallholder farm mostly depends on family labour; additional casual labour is mainly employed for harvesting and, in some cases, for pesticide application. For the purpose of this study, family labour in the case of smallholder farms is valued at its opportunity costs, represented by the wage rate in the nearest village and is, therefore, accounted for in the calculation of the labour costs.

Another difference, due to the huge variation in farm size, is the administrative overheads. The exporter-owned farm and the large-scale contracted farm employ a high share of workers outside primary field work. Of the total workforce, 26% on the exporter-owned farm are in administration, supervision or management, while for the large-scale contracted farm this share is around 18%. In this study, these costs are attributed to crop production according to the share of land allocated to each crop.

The wage rate for casual workers differs among the farm types. The smallholder farm pays the lowest rate; the large-scale contracted farm pays a higher wage and the exporter-owned farm pays the highest wage. The large-scale farms of both types employ permanent workers who receive higher wages and who are not found on the smallholder farm. Overall, the labour costs in bean production are the highest on the large-scale contracted farm, due to its more labour-intensive production as compared with the exporter-owned farm, and the additional administrative overhead as compared with the smallholder farm. Second is the exporter-owned farm, due to its high wage rates and its administration costs, and the lowest is the smallholder farm based on low wages and low number of administration tasks at the farm.

Fertilizer costs per hectare increase from smallholder to large-scale contracted to exporter-owned farm. Chemical costs are highest for the smallholder farm, being more than twice as much as the large-scale contracted farm spends. The exporter-owned farm uses slightly more than the large-scale contracted farm. Although seed is provided by the exporter, smallholders usually source their own agrochemicals following the instructions of field supervisors and adhering to the list of approved pesticides. Due to the smaller scale of production, smallholders require agrochemicals of smaller package size with relatively higher prices. Irrigation expenditures include the costs of using mechanical pumps or machinery. These are relevant only on the large-scale farms as the typical smallholder irrigates manually, which is included in labour costs. No fees are charged for the use of irrigation water. Irrigation expenditures for the two large-scale types are almost equal.

The output of beans per hectare and year increases with the size of the typical farm type. The smallholder farm realizes only 50% of the annual output per hectare of the exporter-owned farm, and the large-scale contracted farm ranges between these two. Taking the previous discussion into account, this might be attributed to the difference in input use and intensification.

Table 5.2 shows the history of certification, as well as investments and costs linked to certification for the three farm types. The exporter-owned large-scale farms were the first ones to be GlobalGAP certified. During the same period, exporter-owned farms obtained certification for further standards such as the Ethical Trade Initiative or Tesco Nature's Choice, and hence operate a system of multiple standards. The major investment necessary for attaining GlobalGAP certification differs between the farm types due to the different levels of adherence to production standards before compliance with GlobalGAP. Large farms owned by exporters had management structures and infrastructure in place that made upgrading to the GlobalGAP standard relatively simple and cheap in comparison with contracted and small-scale farms. Small-scale farmer certification was the subject of much support by donors and non-governmental organizations (Humphrey, 2008); in our survey only that share borne by the small-scale farmers themselves was factored in. Hence, the figures shown in Table 5.2 include that portion of investment in infrastructure and buildings such as construction of a grading shed, charcoal cooler and pesticide store that were incurred by small-scale farmers producing 1 ha of export vegetables. Recurring costs are not yet factored in, since small-scale farms have just been certified and, at the time of the survey, it was not clear what share small-scale farmers would have to pay in the long term and which share would be covered by the exporter.

Financial performance of production

Based on data in Table 5.1, the annual net revenues for all crops of the exporter-owned farm, of the large-scale contracted farm and of the smallholder farm are KSh27.6

Table 5.2. Characteristics of compliance with production standards within the three farm types (from Mausch, 2007).

	Exporter-owned	Large-scale contracted	Small-scale
Year of GlobalGAP certification	2003	2005	2005
Other certificates	Tesco (Nature's Choice) NEMA	None	None
Major investments	Buildings and other infrastructure Registration and audit fees Protection kits	Buildings and other infrastructure Registration and audit fees Protection kits	Buildings and other infrastructure N/A Protection kits
Costs (KSh/ha)	8,614	76,051	117,212
Major recurring costs	Analysis Additional workers Maintenance	Analysis Additional workers, mainly for record keeping Maintenance	Not yet factored in, since at the time of the survey smallholders had just been certified
Total costs (KSh/ha)	2,397	9,904	

NEMA, National Environment Management Authority.

million, KSh5.5 million and KSh135,000, respectively. For the exporter-owned farms, this figure represents the net revenue from export vegetable production, as they do not produce domestic crops. For the smallholder farms and the large-scale contracted farms, this figure includes revenues from domestic crops. On the large-scale contracted farms, the crops for the domestic market generate 80% of the net revenue per hectare and year of the export crops. The same proportion was assumed to be applicable to the smallholder farm, whose total net revenue from export vegetable production is KSh78,000 per year.

Table 5.3 compares the three farm types in terms of financial performance and efficiency using a few simple indicators such as the revenue:cost ratio. Farms are compared based on bean production data. Comparison of the revenue:cost ratio across the three typical farms shows that the smallholder farm performs best and is the most efficient, although it has the lowest output per hectare and year and receives the lowest price for its produce. Its low level of input use and costs more than compensates for low output and price received for beans. The comparison further shows that the large-scale contracted farm has the lowest revenue:cost ratio, the highest output per hectare and year and receives higher prices for its produce compared with the smallholder farm, but high production costs turn this to an overall less favourable performance on the efficiency measure.

Differences in performance can partly be attributed to economies of scale, as the price received increases and the price paid for some inputs decreases with farm size due to bargaining power and vertical integration. The level of intensification and optimization of input allocation is another factor that drives net revenue. The smallholder farm might be able to increase the net revenue per hectare and year by intensifying its production, while the large-scale contracted farm might be able to gain from optimizing its input use.

Table 5.3. Financial performance and efficiency of the three typical export farm types based on bean production (from Mausch *et al.*, 2009).

	Output (t/ha and /year)	Production costs (KSh/kg)	Revenue (thousand KSh/ ha and /month)	Total cost (thousand KSh/ ha and /month)	Revenue: cost ratio
Smallholder farm	15	16	47	20	2.35
Large-scale contracted farm	25	29	93	60	1.55
Exporter-owned farm	30	2	121	58	2.09

Financial benefits from certification and break-even analysis

Since the area covered for export vegetables varies from 0.025 to 0.46 ha for the smallholder farms, from 2 to 281 ha for the large-scale farms and from 30 to 303 ha for the exporter-owned farms, the break-even period can vary tremendously among farms of one type. If the net revenue from beans only is factored in for the calculation of break-even points, the smallholder farm breaks even after 25 months, while the large-scale contracted and large-scale exporter-owned farms break even after 13 months and 1 month, respectively. When considering net revenues from the whole farm planting programme, i.e. even factoring in net revenue from domestic crops, the break-even point switches to 6 months for both the smallholder and the contracted large-scale farm due to the difference in net revenue shares from domestic crops. The break-even period does not change for the large-scale exporter-owned farm, since this produces exclusively for the export market.

Both types of contracted farms own land that is not cultivated; therefore, they are able to increase the cultivated area within a reasonable time frame without purchasing or renting additional land. Thus, a sensitivity analysis is conducted that recalculates the break-even points depending on both the area allocated to export vegetable production and the area currently not under cultivation for each of the farm types. Figure 5.3 shows break-even points for the large-scale contracted farm, comparing the net revenue from export production only, as well as the factoring in of whole-farm production. Figure 5.4 compares the elasticity of break-even points with changes in area allocated to export production across farm types.

Based on current production and the cost structure given above, the sensitivity analysis shows that the minimum profitable area under bean export production for a typical contracted large-scale farm is 3.6 ha, while it is 0.24 ha for the smallholder[8] and 1.6 ha for the exporter-owned large-scale farm. Accounting for the opportunity cost of the farm manager, any area below this threshold results in an overall loss; farm operations move away from the break-even point instead of reaching it, as indicated by the left-hand side of Fig. 5.3.

In this static analysis, the smallholder farm's break-even point is much more sensitive to changes in the area allocated to export vegetable production than that of the two large-scale farms. With an area decrease of 5%, the time to break even increases by 12%; with a decrease of 20%, it takes 5.5 times longer until the break-even point is reached. In the latter scenario the smallholder farm would require more than 11 years before it broke even. Compounding this, factors other than land allocated to export production affect the net revenue from production. For example, adverse climatic conditions can easily result in a similar effect to produce the worst-case scenario above. Thus, for the smallholders the investment in certification

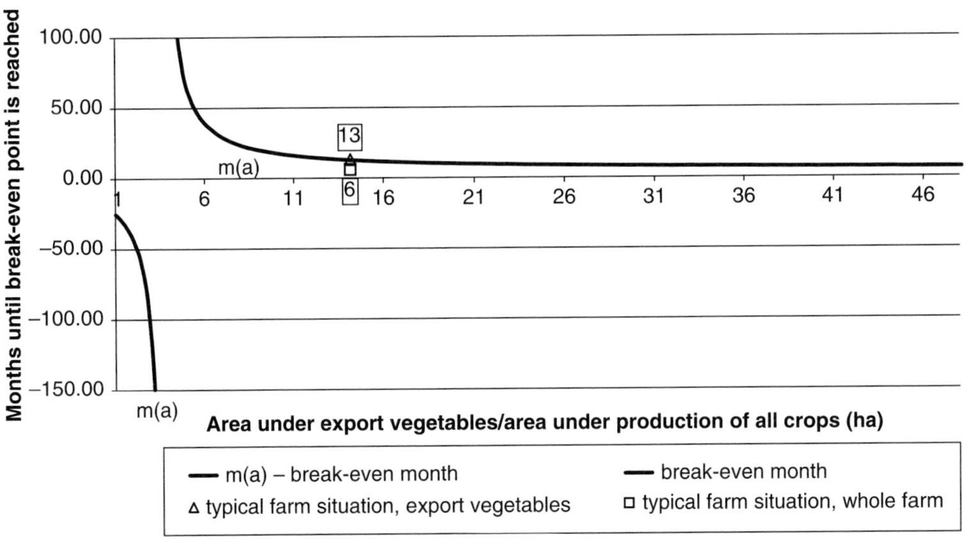

Fig. 5.3. Impact of area allocated to export production on break-even time in the case of the typical large-scale farm.

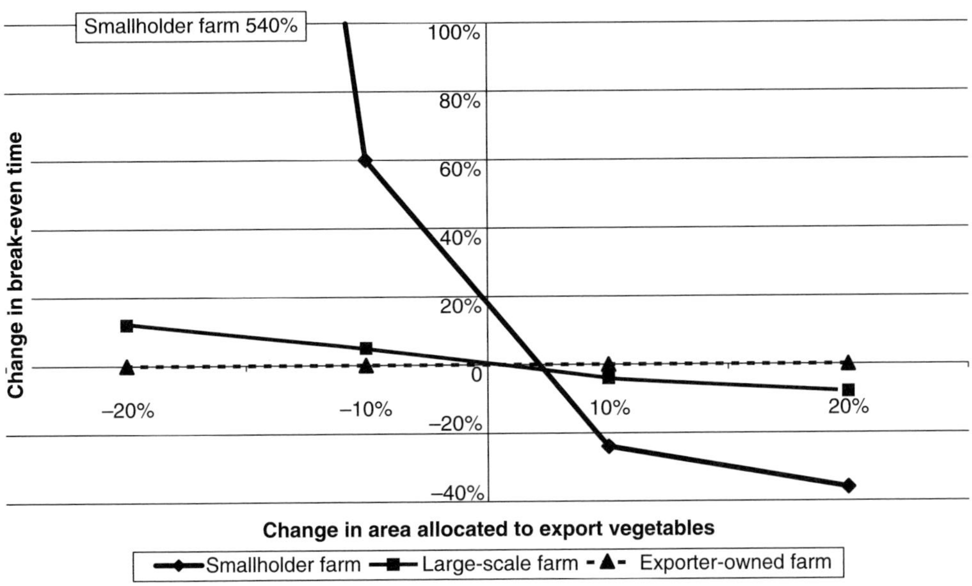

Fig. 5.4. Sensitivity analysis of break-even points across farm types.

involves the largest risk. On the other hand, an increase in the export vegetable area would be accompanied by a disproportionately large shortening of the time frame to reach the break-even point. Here, an increase of just 5% (20%) of the acreage allocated to export vegetables would result in a 12% (36%) decrease in the break-even period.

For the large-scale contracted farm, the lower limit for profitable export vegetable production is approximately 3.6 ha. By including the whole-farm planting programme, the break-even month is reached in 50% of the time of the base case of current production practice. Comparing this with the smallholder farm illustrates that the income of smallholder farms is affected more by changes in export production area and that currently smallholders depend to a larger extent on domestic crop production. The contracted farms are closer to their optimal farm size than the smallholder farms, as shown by the less elastic response and smaller effect of changes in acreage allocated to export production on the break-even period. Even a 20% decrease in export vegetable area leads to only a 12% increase in the break-even period. Hence, the large-scale contracted farms' investment is less vulnerable to financial setbacks that may easily be caused by adverse climatic conditions, resulting in some crop failure and other possible changes. The same holds true for an increase in land allocated to export production, which leads to a disproportionately shorter period to reach break-even point.

Finally, the figures for the exporter-owned farms' break-even period show that the profit threshold of the exporter-owned farm is at 1.6 ha. The exporter-owned minimum acreage is likely to be underestimated more than that of the large-scale contracted farm, as the exporter-owned farms work on a much higher level of sophistication and thus have a higher share of fixed costs, which would increase the minimum acreage for profitable production. Nevertheless, this would not affect the sensitivity of the break-even point significantly. At the current level of production, the break-even point is almost inelastic to changes in the acreage, and is unlikely to alter, even when factoring in the fixed costs. Here, a 5% decrease in the area is accompanied by a change in the time frame of less than 0.5%, and even a reduction of 50% leads to only a 1.7% increase in the period until break-even. Hence, exporter-owned farms do not incur much financial risk in not recovering their investment.

This calculation shows that the smallholder farm faces the highest risk, but on the other hand it has the greatest opportunity to improve its income situation by fully utilizing its production possibilities, e.g. by allocating more land to horticultural export production. However, the decision on the acreage allocated to the export crop, as well as the export crop choice, is not necessarily the smallholder or contracted large-scale farmer's own choice, but has to be agreed upon with the exporter. This is a complex decision, as additional considerations of production risk and transaction costs have to be taken into account. Nevertheless, as discussed above, domestic crops do generate a significant share of whole-farm revenues and, therefore, even in a situation where the exporter would not allow for a larger export vegetable area, the smallholder farmer is still able to expand vegetable production for the domestic market.

Table 5.4 shows the survey results of the benefits from certification, which go beyond the previously analysed direct financial benefits. The upper part of the table lists realized benefits as perceived by the respondents, while the lower part lists further expected benefits. The comparison between the two large-scale farm types shows that the contracted farms realized benefits on a greater range of aspects than the exporter-owned farms, highlighting the fact that the latter were operating at levels closer to compliance than the former. On this aspect the smallholder survey used different categories for eliciting benefits beyond the directly financial. However, even these show that a major share of smallholders agree that certification led to the increased value of the output, higher physical production, fewer rejects, quality improvements and more timely payments.

Transaction cost analysis

By definition, the total administrative and managerial transaction costs increase with farm size and the degree of diversification, and thus costs are compared per kilogram of beans produced. The most obvious difference in the management structure of the different farm types is the number of levels in the decision-making hierarchy. Table 5.5 gives an overview of selected transaction

Table 5.4. Certification benefits as perceived by farmers and farm managers.

	Exporter-owned farm	Contracted large-scale farm	Smallholder farm
Realized benefits			
Increased value of output	0	0	+
Higher physical production	0	+	+
Quality improvement	+/0	+	+
Cost reduction	–	+	N/A
Field loss reduction	0	+	N/A
Rejection rate reduction	0	+/0	N/A
Gains from mechanization	0	0	N/A
Expected benefits (ranked according to frequency of responses)			
	Management improvements	Price premium	N/A
	Lasting market access	Lasting market access	N/A
	Marketing advantages	Marketing advantages with the possibility of direct export	N/A
	Better-trained workforce	0	N/A

Sources: Mausch (2007) for large-scale exporter-owned and contracted farms; Asfaw data (Chapter 4, this book; Table 4.4) for smallholder farms. 0, no change; +, expected change realized; –, expected change not realized; N/A, not applicable for type of benefit with no information available.

cost indicators across the farm types. The exporter-owned farm has a five-tier hierarchy with 13% of its employees supervisors, while the large-scale contracted farm operates with four steps in the hierarchy and 11.5% supervising staff. As the smallholder farm is a family business, with an average of one casual worker, the hierarchy consists of two levels with the farm owner directly supervising the worker.

However, in contrast to the large-scale farms, the smallholder farm is organized within a grower group and is not individually contracted by an exporter, which leads to other off-farm managerial transaction costs. The grower group has to establish a quality management system, including an internal auditing and monitoring scheme and a centralized management. Decision making for the grower group is based on democratic principles, and consequently becomes much more complex and time consuming compared with that of a single farm owner/manager. Overall, the two types of large-scale farm bear higher direct financial costs for the staff members managing the farm, while the smallholder farm bears higher transaction costs in terms of time for group discussions and meetings. Apart from the smallholder farmers' time spent for the grower group meetings, the costs of on-farm supervision are not fully attributable to GlobalGAP certification. Although certification increases these costs, these would also occur on non-certified farms.

The differences in monitoring time are included, as this is one of the most important factors in ensuring GlobalGAP compliance. This cost item is analysed from the export companies' point of view, as they have to ensure that the produce sent to Europe fulfils the requirements and is not rejected. The exporter typically spends 3 h per week on each large-scale contracted farm to monitor production. The monitoring of smallholder groups is more complex, since the production

Table 5.5. Transaction cost indicators across farm types (from Mausch *et al.*, 2009).

	Exporter-owned farm	Large-scale contracted farm	Smallholder farm
Hierarchy tiers (*N*)	5	4	2
Non-field labour (% of workforce)	13.0	11.5	0
Monitoring time/ha (h/week and /ha)	0	0.2	8.0
Cost of produce, including monitoring costs (KSh/kg)	N/A	44	39

area is spread over all group members, who are 30 per group on average. In this case, a technical staff member of the exporter is permanently based in the group's location to ensure close monitoring throughout the production period. This leads to an average time of 2.8 h spent on each farm of one grower group. On the large-scale contracted farm with 13.8 ha under export vegetables, each hectare is monitored about 12 min per week, while on a smallholder farm with 0.3 ha dedicated to beans each hectare is monitored almost 8 h per week. Thus, areas cultivated by a smallholder group require a much higher monitoring effort by the exporter than the same production area cultivated by a single large-scale farm. Additionally, the output produced by smallholder farms is lower than that of large farms, so the monitoring time per unit of produce is not in favour of small-scale producers.

By factoring in the cost of monitoring in the price that is paid for beans, a comparison between sourcing from small versus large-scale farms from the exporter companies' perspective can be drawn. Staff time for supervisors is valued at their opportunity costs: technical field assistants employed by exporter companies for coordination of smallholder production have similar responsibilities and qualifications to the supervisors working on the exporter-owned farm. Thus, the wage of the technical field assistants is assumed to be at least equal to that of the supervisor position of KSh313 per day, and factored into the price paid for the produce. This leads to the price difference for sourcing produce from each farm type and results in a price of KSh39 for beans produced by smallholder farms and KSh44 for the same product produced by large-scale contracted farms. It transpires that, even by adjusting the price paid for smallholder farms' production for the increased monitoring efforts, it still costs 11% less to acquire beans from a smallholder group rather than from a large-scale contracted farm, assuming that the monitoring system assures that the produce is of equal quality between the two farm types.

All in all, the types of transaction costs and their scale vary among the farm types. A smallholder has to bear high non-financial transaction costs by becoming and being a member of a group, while the large-scale contracted farm has high financial administration and supervision costs, although these are not completely related to GlobalGAP certification. Monitoring costs of smallholders are rather high, but, taking lower prices for smallholder produce into account, exporter companies still source more cheaply from smallholders than from large-scale contracted farms. This is reflected in the statements by some export companies that they do not intend to drop the smallholder groups because the combination of supply from all three types of farms best spreads the risk of production failure caused by unfavourable climatic conditions. Finally, sourcing from smallholders may result in marketing advantages, demonstrating corporate social responsibility and propoor company policies.

Summary and Conclusions

The analyses presented in this chapter compare the financial performance of typical Kenyan GlobalGAP-certified farms: large-scale exporter-owned farms, large-scale contracted farms and smallholder farms. The chapter analyses key indicators, characterizing the typical farms such as the area allocated to vegetable export production, farms' vegetable portfolio, input use and output of bean production. Key indicators in assessing financial performance include the revenue:cost ratio of bean production. The investments to achieve GlobalGAP compliance are assessed in terms of break-even time and their implicit risk. Finally, as well as addressing the selected farm-level indicators on the transaction costs for the different organizational structures, we addressed the exporters' perspective as to where to source their produce.

Results show that apart from farm size, smallholder farms have a larger share of land allocated to vegetable production for the domestic rather than the export market, large-scale contracted farms have more land allocated to export vegetables than to the domestic market and, finally, exporter-owned farms produce vegetables for the export market only. The efficiency of bean production as measured by the revenue:cost ratio does not increase with farm size. Overall, smallholders operate on a low input–low output level and have the highest revenue:cost ratio.

The exporter-owned farm has the lowest investment costs per hectare to attain GlobalGAP compliance and reaches the break-even point within 1 month of certification. The contracted farm does not recover its investments in GlobalGAP until month 13. Despite high donor support (Humphrey, 2008), the smallholder farm has the highest investment cost per hectare, and breaks even after 25 months. The latter's success and time required to reach the break-even point is much more sensitive to changes in the area allocated to export crops, its performance and crop portfolio than is the case for the large farms. Overall, the risk encountered by the large-scale contracted farm is lower than that for the smallholder farm, and the exporter-owned farm faces almost no financial risk for the recovery of their investments in certification given the time to break-even point. Differences in the investment cost between the farm types are attributed to the level of compliance to the standards, even before formal certification. Large-scale exporter-owned farms were certified earlier than the other two farm types and were operating much closer to the standard. This result is reflected in the realized benefits in addition to direct financial ones, which show that large-scale contracted, as well as small-scale, farm managers perceive a much higher count on such.

Export companies can source their produce from different types of farm but have to guarantee compliance of production with GlobalGAP regulations. They, therefore, monitor contracted large-scale, as well as smallholder, farms closely. Monitoring time per unit produced is by far the highest in the case of smallholder groups. This demonstrates the complexity for exporters when relying on produce from groups of small-scale farmers, which is consistent with the findings of Okello and Swinton (2006). It further demonstrates that, even though collective action can facilitate successful market participation (Narrod *et al.*, 2009), collective action does come at a cost for bundling several farmers as decision-making units, as compared with one large farm. Lower output prices to small-scale farmers as compared with large-scale contracted farmers reflect the higher monitoring costs on the buyer's side, as well as the higher bargaining power of the buyer. Overall, higher monitoring expenses are more than compensated for by the lower price that smallholders receive for their produce. Considering the low price that smallholders receive, as well as their low bargaining power in contract negotiations, it is likely that even the costs of establishing reliable groups (at least those borne by the exporter) are factored in by the exporter company. Additional benefits to the exporter are potential image advantages, as well as the scope for spreading production risk across a wider area than would be possible with only large-scale farms. Finally, the expansion of large-scale farms in Kenya is limited due to the limited availability of land; thus, overall exporters have little flexibility in choosing their suppliers (Humphrey, 2008).

The ranking of the performance of the three farm types varies depending on the indicator used. GlobalGAP has increased the monitoring costs of production, varying it among the farm types. Overall, based on the results of this and other studies, one cannot conclude that the standards have had negative effects on Kenyan smallholder participation in vegetable export markets. The smallholder farms included in this analysis are early adopters of the standard. These producers benefited from donor support, and overall adoption of the standard had not reached a steady state by the time of the survey (Asfaw *et al.*, 2009).

Further research

Based on the findings of this study, three areas of further research can be identified. First, since smallholders in this study were early adopters of the standard, medium- to long-term impact assessment and monitoring should be conducted to monitor equity effects in the long term. Such assessment could more reliably contribute to understanding of smallholder entry to and exit from the export market and the implications in terms of welfare gains and losses thereof. Research should include the labour market effects of such fluctuations. Secondly, the establishment of reliable grower groups for GlobalGAP certification under option 2 seems a major challenge for export companies. An in-depth study on this topic might lead to a better understanding of the complexities involved. Collective action, as often recommended, may have to take a different organizational form depending on the external rules of the market to which a farmer group is linked. Further analysis should pursue this in more detail, thus accounting for heterogeneity in farmer groups and external factors, including market conditions. Thirdly, a follow-up study is recommended in order to assess the impact of compliance with GlobalGAP on the probability and profitability of adopting further standards such as TNC (Tesco Nature's Choice), ETI (Ethical Trading Initiative) or Fairtrade. Such a study would facilitate the analysis of the exporter-owned farms' investments which, based on this study, suggests that the marginal costs of compliance with several standards decrease. Additionally, a study at the exporter level could provide further insight into the distribution of total certification, as well as the transaction costs over the different actors in the supply chain, and other constraints faced by exporters.

Acknowledgements

This chapter synthesizes the findings of Mausch (2007) and expands the article by Mausch *et al.* (2009), in which more technical details can be found. The authors gratefully acknowledge financial support from the German Development Cooperation (BMZ/GTZ) and technical support from the Kenyan Ministry of Agriculture and export companies. We thank the farmers involved in the survey and the enumerators who collected the data.

Endnotes

[1]The research for this chapter was conducted when EurepGAP, Version 2.1 (October 2004) was relevant. Since then EurepGAP has been renamed GlobalGAP, better to reflect its global presence and mandate (http://www.globalgap.org/cms/front_content.php?idcat=9&idart=182); thus, in this chapter, we use the term GlobalGAP but refer to EurepGAP Version 2.1 (October 2004).

[2]Jaffee defined smallholder farms as farms with < 5 ha of landholding, but for this study smallholder farms are those with < 2 ha under vegetable production. For further details, see Asfaw *et al.* (2009).

[3]The sampling for the smallholder farms is described in detail in Asfaw *et al.* (2009).

[4]The number of certified farms according to e-mail contact with Foodplus (June 2006). The smallholder farms are organized in ten farmer groups, while the other two farm types are certified as individual farms.

[5]For details on the methodology applied, see Mausch (2007).
[6]Approximated by wage rate in the nearest village.
[7]See online company presentation, e.g. Metro Group (2009); http://www.metrogroup.de/servlet/PB/menu/1108350_l1/index.html) or EDEKA Group (2009); http://www.edeka.de/EDEKA/Content/Unternehmen/Verantwortung/Sortiment/index.jsp).
[8]Note that the calculation for the case of the smallholder farm does account for opportunity costs of family labour, as calculated in Table 5.1.

References

Asfaw, S., Mithöfer, D. and Waibel, H. (2009) Investment in EU private food-safety standards compliance: Does it pay off for small-scale producers in sub-Saharan Africa? *Quarterly Journal of International Agriculture* 48(4), 337–362.

Dolan, C. and Humphrey, J. (2000) Governance and trade in fresh vegetables: The impact of UK supermarkets on the African horticulture industry. *Journal of Development Studies* 37(2), 147–176.

EDEKA Group (2009) Company profile (http://www.edeka.de/EDEKA/Content/Unternehmen/Verantwortung/Sortiment/index.jsp).

Fleischer, G., Waibel, H. and Walter-Echols, G. (2002) Transforming top-down agricultural extension to a participatory system: A study of costs and prospective benefits in Egypt. *Public Administration and Development* 22, 309–322.

Galdos, A.E.B.L. (2004) Implementation of EUREPGAP standards in the agro-export sector of Peru: A case study. Masters thesis, University of Göttingen, Germany.

HCDA (2009) Export statistics from Kenya (http://www.hcda.or.ke/export_statistics.aspx).

Hemme, T. (2000) Ein Konzept zur international vergleichenden Analyse von Politik- und Technikfolgen in der Landwirtschaft. PhD Dissertation, University of Göttingen, Germany.

Henson, S. and Jaffee, S. (2004) Standards and agro-food exports from developing countries: Rebalancing the debate. The World Bank Policy Research Working Paper 3348, World Bank, Washington, DC.

Humphrey, J. (2008) *Private Standards, Small Farmers and Donor Policy: EurepGAP in Kenya*. Institute of Development Studies, University of Sussex, Brighton, UK.

Humphrey, J., McCulloch, N. and Ota, M. (2004) The impact of European market changes on employment in the Kenyan horticulture sector. *Journal of International Development* 16, 63–80.

Jaffee, S. (2003) From challenge to opportunity: The transformation of the Kenyan fresh vegetable trade in the context of emerging food safety and other standards. Agriculture and Rural Development Discussion Paper 1, World Bank, Washington, DC.

Maertens, M. (2006) Trade, food standards and poverty: The case of high-value vegetable exports from Senegal. In: *Proceedings of the International Association of Agricultural Economists Conference*, August, 2006, Gold Coast, Australia.

Manda, D.K. (1997) Labour supply, returns to education and the effect of firm size on wages: The case of Kenya. PhD dissertation, the University of Gothenburg, Gothenburg, Sweden.

Mausch, K. (2007) Do EurepGAP standards favour large-scale vegetable producers in Kenya? Masters thesis, Faculty of Economics and Business Administration. University of Hannover, Germany.

Mausch, K., Mithöfer, D., Asfaw, S. and Waibel, H. (2009) Vegetable production in Kenya under EurepGAP standard: Is large 'more beautiful' than small? *Journal of Food Distribution Research* 40(3), 115–129.

Metro Group (2009) Company profile (http://www.metrogroup.de/servlet/PB/menu/1108350_l1/index.html).

Mithöfer, D., Nang'ole, E. and Asfaw, S. (2008) Smallholder access to the export market: The case of vegetables in Kenya. *Outlook on Agriculture* 37(3), 203–211.

Narrod, C., Roy, D., Okello, J., Avendaño, B., Rich, K. and Thorat, A. (2009) Public–private partnerships and collective action in high value fruit and vegetable supply chains. *Food Policy* 34(1), 8–15.

Okello, J.J. (2005) Compliance with international food-safety standards: The case of green bean production in Kenyan family farms. PhD dissertation, Michigan State University, East Lansing, Michigan.

Okello, J.J. and Swinton, S. (2006) Do international food-safety standards marginalize poor farmers? Evidence from Kenyan family green bean farms. *Journal of Food Distribution Research* 37(1), 187.

Reardon, T., Barrett, C.B., Berdegué, J.A. and Swinnen, J.F.M. (2009) Agrifood industry transformation and small farmers in developing countries. *World Development* 37(11), 1717–1727.

6 Food Production Standards and Farm Worker Welfare in Kenya

Christoph Ehlert, Dagmar Mithöfer and Hermann Waibel

Introduction

The worldwide demand for a high-quality food supply has led to the implementation of various food production standards. These food production standards impose production requirements on food produce, regardless of their original location. In the case of developing-country producers, this development has raised the question of whether or not food production standards, such as GlobalGAP[1] (Global Good Agricultural Practices) for vegetable production,[2] marginalize small-scale farmers by constituting a barrier to trade (for example, Humphrey, 2006; Okello and Swinton, 2007), which may lead to small-scale farmers dropping out of export market participation. Such exit from export vegetable production might have implications for employment opportunities at small-scale export-producing farms. However, as Asfaw *et al.* (2010) show, the success of smallholder participation in export production depends on their technical and socio-economic conditions. Furthermore, the poorest actors in particular can gain from employment opportunities on large-scale farms, as shown in Senegal by Maertens and Swinnen (2009), which may outweigh income losses from small-scale producers who are unable to comply with food production standards.

Standards may influence other welfare dimensions beyond income. These influences are evaluated, using the example of Kenya as one of the leading exporters of vegetables and cut flowers to the EU25 (European Union, 2006). This position is reflected by the value of horticultural exports that quintupled between 1980 and 2000 from US$50 million to nearly US$250 million (Jaffee, 2003), contributing to almost 20% of Kenya's GDP. In 2006, Kenya's horticultural exports accounted for US$700 million, making them the second highest income earner after tourism (FPEAK, 2007). As well as this high monetary contribution, the value shares of horticultural exports (including tea and coffee) increased from 31% to 41% between 2001 and 2005 (Republic of Kenya, 2006), indicating the rising importance of horticulture.[3] For these reasons, the sector is a major source of employment. Estimates show that, in 2005, more than 50,000 people were in paid employment in the sector (Dolan, 2005).[4]

To allow a multidimensional welfare analysis, this study utilizes the Sen's functionings approach to welfare measurement, which is implemented using a MIMIC model, as in Kuklys (2005). This model has the advantage of incorporating the determinants of a functioning and the indicators that compose a functioning, into one model. This, in turn, leads to more precise estimates of functionings. Therefore, the model is suitable for assessing changes in the various welfare dimensions that private food production standards may touch, such as income, health and human capital accumulation, thereby providing answers to the main question of this study: Do private food production standards affect the welfare of farm workers?

The next section, Conceptual Framework and Methodology, describes the application of the Sen's functionings approach in the context of production standards in horticulture. The section Survey Design and Data describes the design of the survey and the process of data collection. An overview of the collected data is given in section Empirical Models and Descriptive results. The descriptive results are followed by the results of the econometric model and a discussion. The chapter concludes with suggestions for policy design regarding the implementation of private food production standards.

Conceptual Framework and Methodology

Agricultural production technologies influence the welfare of labourers (family, as well as non-family, farm workers) engaged in the sector. Welfare includes aspects of income, health, working conditions and human capital. One way to assess these different components of welfare is to use Sen's functionings and capabilities approach. Functionings are achievements or successes of individuals, i.e. they represent what an individual 'manages to do or to be in leading a life' based on the endowment of commodities and characteristics at her or his disposal (Sen, 1993). Therefore, functionings have an intermediate position between the possession of a good and the welfare that might be generated with the good (Sen, 1999). Capabilities, on the other hand, reflect the groups of functionings from among which an individual can choose (Sen, 1993), and are dependent on the individual's personal characteristics and their command over commodities; Sen calls this command 'entitlements' (Sen, 1999). For the assessment of what could be called limited welfare (welfare without the notion of freedom to choose between functionings), the value of the capability set can be derived from the set of functionings achieved, a procedure that Sen calls 'elementary evaluation' (Sen, 1993).

In this study, the value of capabilities is derived by measuring the achieved functionings of individuals, i.e. Kenyan horticultural farm workers. Following Kuklys (2005), functionings are modelled by employing a multiple-indicators and multiple-cause (MIMIC) structural equation model. This analytical approach allows the estimation of functionings simultaneously with variables that explain the level of functioning achievement.[5]

MIMIC models are comprised of two parts – a measurement and a structural model – which are estimated jointly. To take the functioning of physical health as an example, the measurement model aggregates different indicators of physical health into one latent variable (which represents, in our case, the functioning of physical health), while at the same time the variation of this latent variable is explained by variables in the structural part of the model, such as gender, age and the use of protective clothing while using pesticides. The extent to which the indicator contributes to the latent functioning is indicated by the factor loadings. The influence of the explanatory variables on the latent functioning is indicated by the structural parameters.

Because of their relevance for workers' welfare, four functionings were chosen for this evaluation of farm worker welfare on GlobalGAP-certified and non-certified farms: (i) the functioning of 'being trained'; (ii) the functioning of 'earning a decent income'; (iii) the functioning of 'being physically healthy'; and (iv) the functioning of 'being mentally healthy'. These four welfare dimensions allow for a measurement of the different mechanisms through which GlobalGAP is likely to impact workers' welfare. First, it can be expected that certified farms need better-educated workers to comply with the standard. Hence the farms are more likely to invest in the training of farm workers to improve their knowledge and skills, which builds human capital.

Secondly, GlobalGAP farms are expected to pay higher wages to attract and keep more highly skilled workers. Thirdly, with respect to health, GlobalGAP is expected to foster physical and mental health due to improved working conditions, e.g. because of the use of less toxic pesticides and the more appropriate use of protective clothing (EurepGAP, 2004).

Survey Design and Data

Small farms with fewer than 15 employees were drawn from the sample of a previous study (Asfaw *et al.*, 2010). The selected farms all produced horticultural crops and some produced tea. Of the several districts surveyed in Asfaw's survey, Meru Central district was selected because it includes certified and non-certified smallholders spatially separated in different sub-locations.[6] This separation is important in order to minimize the spillover effects of GlobalGAP practices from certified to non-certified farms, because non-certified farms may adopt best practices from neighbouring GlobalGAP-certified farms.[7] Next, workers on small farms were randomly drawn using a sampling frame that contained all workers on all the certified and non-certified farms of Meru Central district used in the sample of Asfaw *et al.* The sampling frame was constructed in July 2006 by a Ministry of Agriculture staff member, who was also involved in the original survey of Asfaw *et al.* From the 477 individuals in the sampling frame in dependent employment on small-scale farms, 216 were randomly chosen. Of these 216 individuals, 94 worked on 42 GlobalGAP-certified farms and 122 worked on 61 non-certified small farms. The farm worker survey, the results of which are presented here, took place from mid-September to the end of October 2006.

Empirical models and descriptive results

Table 6.1 gives an overview of the variables included in the welfare estimation models, as well as the descriptive results. The welfare dimensions and their indicators for the four models were chosen based on theoretical considerations and data quality. Initially, it had been planned to differentiate 'human capital' further into 'general' and 'specific' human capital. This was not feasible due to the survey responses; therefore, human capital is analysed through the functioning of being trained, which was estimated using the quantity of time spent on training (days of training) and its intensity. Intensity is captured by the composite of the number of tasks for which training of more than one day was received on- and off-farm. Combining time and intensity gives a rough measure of the human capital accumulated.

For the functioning of earning a decent income, the hourly wage was calculated by dividing the weekly wage by the hours worked per week. Hence the working hours' influence on weekly wages, for example by accumulating large amounts of overtime, was removed. Additionally, total income per week, including fringe benefits and in-kind payments, was included. Fringe benefits include, among others, health care on the farm, free housing, free transportation to work and free or subsidized meals.

As indicators for the functioning of physical health, a frequency-weighted health symptom aggregate of typical farm work-related health problems was used, differentiating between the years 2004, 2005 and 2006. The question addressed only those symptoms that occurred during or directly following farm work. Headache, dermal and gastrointestinal problems as well as eye irritations were taken as occupational health indicators for pesticide-related problems (Antle and Pingali, 1994; Repetto and Sanjay, 1996). In addition to collecting data on these symptoms, further symptoms included were sneezing, backache, dizziness, blurred vision and diarrhoea. The 3-year time factor does not distort the results, because individuals worked, in almost all cases, exclusively on either certified or non-certified farms. The average tenure on the farms where the interviews took place was around 18 months for casual/seasonal workers and around 5 years

Table 6.1. Descriptive results of variables entering the structural models.

Variable	Small non-certified		Small certified		
	Mean	SD	Mean	SD	*t*-value
Female (1/0)	0.40*	0.49	0.56	0.50	−2.33
Age (years)	31.12	9.72	31.70	8.47	−0.46
Schooling (years)	9.26	3.58	9.59	3.82	−0.65
Training (days)	1.83*	5.49	4.84	12.19	−2.42
Intense training (no. of tasks with > 1 day of training)	0.63	1.54	0.88	1.47	−1.23
Hourly wage (KSh)	13.58	8.74	13.96	9.09	−0.31
Daily wage (KSh)	89.84**	35.64	100.18	47.15	−1.83
Total income (wage + fringe and in-kind benefits; KSh/week)	752.0	407.0	796.0	349.0	−0.83
Health problems (2004, intensity weighted)	0.22	0.35	0.25	0.35	−0.62
Health problems (2005, intensity weighted)	0.28	0.37	0.30	0.35	−0.48
Health problems (2006, intensity weighted)	0.33	0.39	0.30	0.39	0.67
Caseness score (ordinal)	22.62*	4.83	21.31	4.55	2.03
Overall health (ordinal; very good health = 5)	3.78	0.64	3.72	0.59	0.60
Health awareness (max. = 1)	0.35**	0.11	0.38	0.10	−1.66
Piped water (1/0)	0.83	0.38	0.81	0.40	0.34
Spraying (1/0)	0.60***	0.49	0.43	0.50	2.62
Duration spraying (h)	1.36***	1.73	0.58	1.07	3.81
Protective clothing (1/0)	0.37***	0.48	0.64	0.48	−3.41
Children (*N*)	1.73	1.95	1.83	1.70	−0.40
Hierarchy level (highest = 3)	1.37	0.67	1.30	0.64	0.82
Casual/seasonal (1/0)	0.51	0.50	0.61	0.49	−1.47
Piece rate (1/0)	0.17[a]	0.38	0.32	0.47	−2.51
Supervision (frequency/day of supervision)	1.17	1.42	0.99	0.84	1.07
Tenure (days)	1513.0	1938.0	1185.0	1550.0	1.34
Opportunity to find a new job (almost impossible = 3)[a]	1.68	0.75	1.81	0.74	−1.19
Risk of losing current job (very high risk = 5)[b]	1.86	1.04	2.03	0.94	−1.26

[a]This question was asked regarding the self-perceived chance of finding a job that is at least as good as the current one. Answer categories were 'Easy', 'Difficult' and 'Almost impossible'.
[b]Self-perceived chance to lose one's job within the next month. Answers were given on a five-point Likert scale.
*Significant difference at the 10% level; ** significant difference at the 5% level; *** significant difference at the 1% level.

for permanent workers. Tenure among workers who applied pesticides was significantly higher than among those who did not. The health problems within these 3 years were then aggregated using a frequency weight of their occurrence. As these aggregates for the 3 years increase with the frequency and number of health problems, the three variables were multiplied by −1 to measure good physical health.

Finally, for the functioning of mental health, a caseness score on the basis of the General Health Questionnaire (GHQ) (a screening device for mental disorders) was used. Respondents answered 12 questions by choosing one of four answer categories that indicate the frequency with which the described problem occurred over the previous four weeks (a problem occurred, for instance, 'not at all', 'no more than usual', 'rather more than usual' or 'much more than usual'. From these responses, the caseness score is calculated, which assigns 0 to every response that falls within the first two categories and 1 for answers that fall into the third or fourth categories. The validity of the GHQ in a developing-country context is confirmed by various studies (see Shankar and Famuyiwa (1991) for Nigeria, Montazeri *et al.* (2003) for Iran and Araya *et al.* (1992) for Chile). For the calculation of the caseness score from GHQ responses, see Banks *et al.* (1980).

In addition to the caseness score as an indicator for depression, self-assessed overall health was added, because there is strong evidence that these mutually influence each other (Evans *et al.*, 1994; Cassano and Fava, 2002).

Before moving on to the results of the model, Table 6.1 (above) introduces some characteristics of the sample, in which males and females represent almost equal proportions. The mean age in the sample was 30.7 years, with only small differences between farm types. Only a few workers had no primary schooling, which is in line with the comparatively high adult literacy rate in Kenya. Farm workers on small certified farms received more training than workers on small non-certified farms, although the intensity of training did not vary between the farm types.

Farm workers on both types of small farms reports the same frequency of health problems for the 3-year recall, which is reflected in the self-assessment of overall health. Overall health awareness was established by asking respondents how important health was perceived to be among other factors such as income and education. The sum of the items on the list was required to total 100, which was later normalized to 1. At the same time, a significantly higher proportion of workers on the non-certified farms are involved in pesticide application, while a lower proportion of workers on the non-certified farms use protective clothing. This could be a problem, because the duration of spraying is significantly higher on non-certified, compared with certified, farms. Table 6.2 further investigates this question by looking at the frequency of work-related health problems for a sub-sample of persons applying pesticides.

The occurrence of work-related health problems is similar between the two farm types except for skin rash, which occurs significantly more often among pesticide

Table 6.2. Work-related health problems in 2006 by farm type for pesticide appliers.

	Farm type				
	Small non-certified (n = 73)		Small certified (n = 40)		
Variable	Mean	SD	Mean	SD	t-value
Headache	0.21	0.22	0.16	0.18	1.15
Sneezing	0.22	0.22	0.19	0.23	0.64
Vomiting	0.03	0.11	0.03	0.10	−0.29
Stomach ache	0.08	0.18	0.06	0.17	0.68
Backache	0.14	0.21	0.09	0.18	1.24
Skin rash	0.08[a]	0.17	0.03	0.13	1.43
Dizziness	0.05	0.12	0.03	0.10	0.93
Blurred vision	0.05	0.17	0.05	0.14	0.15
Diarrhoea	0.00	0.04	0.02	0.07	−1.14
Eye irritation	0.06	0.17	0.05	0.14	0.30

SD, standard deviation.

[a] Significant difference at the 5% level.

appliers on non-certified farms. This may indicate sounder pesticide handling on small certified farms.

Results and discussion

The effect of GlobalGAP certification on the achievement of the four functionings is estimated by using the final model specifications shown in Tables 6.3 and 6.5. Additional models with alternative specifications to control other effects were estimated during the specification search of the models (not depicted here, but referred to in the text). Table 6.3 shows the estimation results of the measurement models.

For the functioning of being trained, days of training enter with a relatively high loading and a high value for r^2, indicating that 'being trained' consists mainly of the total days of training received in 2005 and 2006 and, to a lesser degree, of intense training.

For the functioning of earning a reasonable income, the hourly wage enters with a high factor loading. The corresponding r^2 indicates that the latent functioning explains a high degree of hourly wage variation. Besides the hourly wage, the total weekly income of an individual enters the latent functioning with a significantly lower loading and r^2, indicating that the functioning of earning a reasonable income is determined mainly by the hourly wage and not fringe and other in-kind benefits.

The physical health functioning is measured by the three indicator variables for health problems in the 3 years 2004–2006. As Table 6.3 shows, all factor loadings are high and the unobserved latent variable explains a high degree in their variation, ranging from 58% to > 90%. For mental health, the overall health enters with the highest factor loading and r^2. Therefore, the mental health latent variable correlates more highly with the overall health than with the caseness score response.

The indicators of model fit are presented in Table 6.4. The variables in the structural models explain 35.1–50.7% of the variation in the functionings. In addition to r^2, all other indicators point to a good model fit for all models.[8] The good model fit shows that multicollinearity is not a problem. This was also tested in various model specifications.

As shown by the structural model estimates in Table 6.5, GlobalGAP certification has a significant influence on training, where the significant dummy indicates that workers on GlobalGAP farms receive more training even after taking into consideration the type of labour. This finding corresponds well with the strict GlobalGAP protocol for working procedures, requiring better-skilled labourers. GlobalGAP certification

Table 6.3. Measurement models.[a]

Functioning	Variable	Loading	*p*-value	r^2
Being trained	Days of training	0.950	0.001	0.903
	Intense training	0.683	0.001	0.466
Reasonable income	Hourly wage	0.947	0.001	0.896
	Total income	0.455	0.001	0.207
Physical health	Health problems (2004)	0.810	0.001	0.656
	Health problems (2005)	0.975	0.001	0.951
	Health problems (2006)	0.777	0.001	0.604
Mental health	Overall health	0.639	0.001	0.408
	Caseness score	0.360	0.001	0.130

[a]Loadings are similar if models are run separately for certified and non-certified farms.

Table 6.4. Model fit.

	Being trained	p-value	Reason-able income	p-value	Physical health	p-value	Mental health	p-value
r^2	0.315		0.393		0.354		0.507	
Chi-square	56.517	0.007	58.202	0.004	121.130	0.004	70.303	0.001
Degree of freedom	33.0		33.0		83.0		36.0	
NFI	0.935		0.938		0.924		0.924	
RMSEA	0.058	0.289	0.060	0.246	0.046	0.617	0.067	0.114

NFI, normalized fit index; RMSEA, root mean square error of approximation.

Table 6.5. Structural models.[a]

Variable	Being trained	Reasonable income	Physical health	Mental health
Age	–0.247*	–0.085	0.209	–0.370**
Number of children	–0.005	0.198*	–0.178	–0.147
Detrimental health behaviour			–0.149	
Total income on farm	0.180**		–0.263***	0.202**
Effort	–0.021	0.216*	0.118	0.678***
Overtime (0/1)	–0.101	–0.052	0.032	–0.125
Piece rate (0/1)	–0.006	0.184**	0.131*	–0.053
Chance to find a new job	0.184**	0.155	–0.237***	–0.258**
Risk of losing current job	–0.103	–0.060	0.054	0.086
Monitoring intensity by supervisor	–0.023	–0.095	0.184**	–0.069
Piped water (0/1)	–0.071	0.150**	0.211***	0.035
Seasonal or casual (0/1)	0.061	0.160*	–0.100	–0.132
Tenure	0.312***	–0.004	–0.264***	0.053
Female	–0.170*	–0.008	0.127	–0.132
GlobalGAP certified (0/1)	0.256***	–0.006	–0.024	0.056
Hierarchy level: supervisor (0/1)	–0.132	0.402***	0.030	–0.127
Hierarchy level: team leader (0/1)	0.067	0.014	–0.048	0.151
Days of training		–0.100	0.292*	0.056
Health awareness			–0.249***	
Years of schooling	0.136	0.090	–0.083	0.118

[a]The data are free of outliers; variables were transformed in order to satisfy the normal distribution assumption. Some of the variables in the model fulfil the assumption of being normally distributed only approximately, but the divergence from the normal distribution was below a kurtosis of 7 and a skewness of 2, which is acceptable according to Bühner (2004).
*Significant at the 10% level; ** significant at the 5% level; *** significant at the 1% level.

does not have a significant influence on the other functionings. Training and human capital, respectively, represent a higher stability of employment relationships and, therefore, higher job security.[9]

Being certified could have an effect on earning a reasonable income through two different mechanisms: certification might have a positive impact on the marginal productivity of labour through improved crop management practices, or certification might have an impact on efficiency wages. However, results show that being certified does not lead to a better income in this sample after considering demographic factors, effort and piece rate. Effort was measured by a set of attitude questions (regarding the employer and the job) that were aggregated within the MIMIC model. Its positive and significant relation to reasonable income might be explained by the efficiency wage theory, suggesting that higher wages induce higher effort. The payment of piece rates is prevalent on all farm types, and therefore it does not influence the achievement of the functioning of earning a reasonable income differently among the different farm types (which was verified by additional estimates).

Even a second specification for reasonable income, in which the GlobalGAP variable could have captured the effects of more training or more effort on wages, does not show a significant GlobalGAP variable. The reason for no better achievement of the functioning of earning a reasonable income for workers on GlobalGAP farms might be the high financial burden that the standard imposes, especially on smallholders (see Hatanaka *et al.*, 2005; Humphrey, 2006). In this case, the extra margin of GlobalGAP-produced food remains with the party (exporter or farm owner) who invests in GlobalGAP-related infrastructure and provides the extensive training.

Similar to the functioning of earning a decent income, the GlobalGAP coefficient is insignificant in the models of physical and mental health, which could be seen positively as an indicator for overall good practices on the surveyed smallholder farms, even when omitting variables that are potentially influenced by GlobalGAP certification in a second estimation (not shown here).

The result for physical health does not change when only people applying pesticides are taken into consideration. Estimates for this subgroup showed no significant coefficient for the GlobalGAP variable, which is in contrast to the findings of Asfaw *et al.* (2009) who showed a significant reduction in pesticide-related health symptoms for small-scale farmers who had adopted GlobalGAP. This difference arises most probably due to the composition of the latent variable, which contains not only pesticide-related health problems but also a composite of more diverse symptoms weighted by their frequency. To elaborate further on this issue, the symptoms that were reported from 2004 to 2006 were isolated and compared between the groups of pesticide appliers and non-appliers. Only backache, skin rash and dizziness showed significant differences (10% level) between the two groups. Backache occurred significantly more often in the group of non-appliers in 2005/2006, while skin rash (all years) and dizziness (2004/2005) occurred significantly more often in the group of appliers. The insignificant GlobalGAP coefficient implies that stress factors inducing mental distress are the same among farms and that the idea of better working practices on GlobalGAP farms facilitating health is not supported by the data after considering age, schooling, gender and the availability of piped water. The findings do not vary much with the type of contract, i.e. whether a worker is on a casual or permanent contract, except for casual workers, who earn more on small farms compared with permanent workers. A possible explanation is that, in harvest times when labour is scarce, casual labourers receive an income premium.

GlobalGAP certification improves the welfare of workers in terms of higher human capital accumulation, which translates into more capabilities for an average worker. Besides this, an increase in human capital, i.e. an increase in the set of functionings (such as being able to read or being able to avoid health risks), is in itself a value. Findings show that currently it does not pay off in terms of better mental or physical health. However, given that, at the time of the survey, smallholder farms had only been certified since 2004, the time lag between certification and such secondary effects might not be sufficient to show up in the data.

Regarding the other coefficients of the physical health model, workers with more training show a higher level of physical health. Health awareness is negatively associated with physical health: people with greater health awareness seem to be more likely to identify their health problems. The variable total income indicates a negative relationship between wage and health. This may be due to a compensating wage differential, which compensates for adverse working conditions (Liu *et al.*, 1997).

Although the model fit is good for all models, the estimates for the structural model variables may be biased due to endogeneity. Labour type and training received, for instance, could be influenced by the unobserved ability of an individual, using years of schooling as a proxy. This would lead to an upwardly biased indicator of the influence of labour type on training received. Although possible for the estimate of labour type, this objection seems invalid for the GlobalGAP variable, since the amount of training a worker received includes only training received at the current farm. Thus, regardless of the question of whether better-trained workers are more attracted by GlobalGAP farms, the estimate is unlikely to be biased due to restricting the training to the current farm. The same problem might be true for the functioning of earning a decent income, because people with higher ability may prefer to work on certified farms where they earn more in terms of a daily wage. The higher wages on certified farms would then be a result of unobserved ability and not of efficiency wages. To reduce such a bias, the model partly controls ability with years of schooling. Finally, the model for physical health might suffer a similar problem. It could be argued that workers on GlobalGAP farms do not show better physical health in the analysis because GlobalGAP creates awareness of health problems, making workers on those farms more likely to state health problems than people on other farms. To refute this kind of argument, the model controls for health awareness. The results, therefore, are unlikely to be subject to such a bias.

Summary and Conclusions

The application of Sen's capability approach, in combination with structural equation modelling, helped to gain a detailed insight into the four functionings of being trained, earning a decent income and being physically and mentally healthy. Welfare is compared between workers on GlobalGAP-certified and non-certified small-scale farms, based on a random sample of 216 farm workers in vegetable production in Kenya.

The welfare estimates in this study suggest that private food standards, based on the example of GlobalGAP in horticulture in Kenya, do not lead to increased welfare in terms of better income or improved physical or mental health. The former might be due to the high cost of compliance for farms to become certified so that liquidity constrains the payment of potential efficiency wages, or that the overall marginal product of labour does not increase through certification. On the latter two dimensions of health, the survey captured workers on farms of early adopters, which may imply that, at the time of the survey, such effects might not yet have shown up. However, welfare increased because workers on GlobalGAP-certified farms show a significantly higher level of human capital accumulation through the functioning of being trained, compared with workers on non-certified farms. This finding gives some weight in favour of public support to the implementation of private production standards on small farms, as reflected in high donor support to GlobalGAP schemes in Kenya (Humphrey, 2008). However, an increase in this functioning could be achieved by alternative training schemes independent of private food production standards such as GlobalGAP (Mithöfer, 2008). Further, as shown by Humphrey (2008), similar results could have been expected without public donor support.

Further research

Given the increasing spread of private production standards, the monitoring of medium- to long-term effects on farm worker welfare would be valuable; the robustness of results needs to be checked through triangulation of analytical methods. Embedded in such research, the possibilities of leveraging entry points that such private market governance mechanisms provide in enhancing social standards need to be assessed further (see, for instance, Riisgaard, 2009).

Of interest for further research would be to isolate the health effects of GlobalGAP on those workers who might gain the most from the GlobalGap standards in terms of changing work practices – especially for those using pesticide applicators.

Future research could extend the analysis in this chapter by moving from functionings to analysing capabilities. Macroeconomic data predicting capabilities (such as the labour market situation, provision of health services, etc.) could be matched with information at the individual level of health (not self-stated, but assessed by a health care professional), family income and skills. This would allow for an assessment of the discrepancy between what would be possible for an individual to achieve and what has already been achieved and would, therefore, add the full dimension of opportunities to the analysis.

Endnotes

[1]This study was conducted when EurepGAP, Version 2.1 (October, 2004) was relevant. Since then EurepGAP has changed its name and logo to GlobalGAP, arguing that its proclaimed role in promoting the harmonization of good agricultural practice schemes had moved beyond Europe. The name change was announced at the 8th EurepGAP Conference, the EurepGAP Asia Conference, held in Bangkok on 6–7 September 2007. Therefore, throughout this chapter the term GlobalGAP is used, and can be considered as synonymous with EurepGAP.

[2]The GlobalGAP standard was introduced by the Eurep working group, which was founded in 1997 by 13 of the largest retailers in Europe (Busch *et al.*, 2005). Apart from food safety issues, the GlobalGAP standard covers aspects that are relevant for workers' welfare, such as occupational health and safety.

[3]It is important to note that the available data are subject to many problems – one of those being that the composition of crops in the horticulture export data of the Kenyan Horticultural Crops Development Authority (HCDA) is inconsistent in different time periods (Jaffee, 2003).

[4]Dolan excludes coffee in her estimation.

[5]In this study, the relationship between latent and observed variables is estimated with a 'maximum likelihood' procedure, because it is scale independent and can be applied in the presence of missing data without requiring their imputation (Arbuckle, 1996; Bühner, 2004).

[6]Because of this preselection of a district, the findings for smallholders are, literally, representative of this district only. Note that none of the included non-certified farms had a GlobalGAP certification beforehand.

[7]In order to control for migration among workers from certified to non-certified farms or vice versa, the questionnaire contained a section that asked for the employment history for a minimum of the previous 12 months. Findings showed that the workers in the sample did not switch between certified and non-certified farms, at least in the previous 12 months.

[8]A χ^2 value that is lower than the degrees of freedom and insignificant indicates a very good model fit (for a discussion of χ^2, see Tanaka, 1993). χ^2 is significant in all cases, which does not pose a problem for the model fit because the χ^2 values are not too large relative to the degrees of freedom. Because the sole assessment of model fit in terms of χ^2 is questioned as a good indicator in the literature (Tanaka, 1993), Table 6.4 also contains the normalized fit index (NFI). The standardized model fit (range between 0 and 1, with 0 representing the lowest possible model fit), where all observed variables are assumed to be uncorrelated (Gerbing and Anderson, 1993) for all functionings, is very reliable, as all values are close to 1. The same applies to the root mean square error of approximation (RMSEA), which summarizes the errors of approximation between population and implied models' variance–covariance structure. Values below 0.05 are considered to indicate a good model fit (Browne and Cudeck, 1993, pp. 144–147). The null hypothesis that the RMSEA is ≤ 0.05 cannot be rejected at the 10% level for the four functionings, indicating a very good fit for all models.

[9]This is, in fact, backed up by the data: tenure is highly correlated with the incidence of intense training, as well as with a low subjective risk of losing one's job.

References

Antle, J.M. and Pingali, P.L. (1994) Pesticides, productivity, and farmer health: A Philippine case study. *American Journal of Agricultural Economics* 76(3), 418–430.

Araya, R., Wynn, R. and Lewis, G. (1992) Comparison of two self-administered psychiatric questionnaires (GHQ-12 and SRQ-20) in primary care in Chile. *Social Psychiatry and Psychiatric Epidemiology* 27, 168–173.

Arbuckle, J.L. (1996) Full information estimation in the presence of incomplete data. In: Marcoulides, G.A. and Schumacker, R.E. (eds) *Advanced Structural Equation Modeling: Issues and Techniques*. Erlbaum, Mahwah, New Jersey, pp. 243–278.

Asfaw, S., Mithöfer, D. and Waibel, H. (2009) Food-safety standards, pesticide use and farm level productivity: The case of high-value crops in Kenya. *Journal of Agricultural Economics* 60(3), 645–667.

Asfaw, S., Mithöfer, D. and Waibel, H. (2010) What impact are EU supermarket standards having on developing countries' export of high-value horticultural products? Evidence from Kenya. *Journal of International Food and Agribusiness Marketing* 22(3/4), 61–86.

Banks, M.H., Clegg, C.W., Jackson, P.R., Kemp, N.J., Stafford, E.M. and Wall, T.D. (1980) The use of the general health questionnaire as an indicator of mental health in occupational studies. *Journal of Occupational Psychology* 53, 187–194.

Browne, M.W. and Cudeck, R. (1993) Alternative ways of assessing model fit. In: Bollen, K.A. and Long, J.S. (eds) *Testing Structural Equation Models*. Sage, Newbury Park, California and London, pp. 136–162.

Bühner, M. (2004) *Einführung in die Test- und Fragebogenkonstruktion*. Pearson Studium, Munich, Germany and Boston, Massachusetts.

Busch, L., Thiagarajan, D., Hatanaka, M., Bain, C., Flores, L. and Frahm, M. (2005) *The Relationship of Third-party Certification to Sanitary/Phytosanitary (SPS) Measures and the International Agri-food Trade: Final Report*. RAISE SPS Global Analytical Report 9, USAID, Washington, DC.

Cassano, P. and Fava, M. (2002) Depression and public health – An overview. *Journal of Psychosomatic Research* 53, 849–857.

Dolan, C.S. (2005) Fields of obligation: Rooting ethical sourcing in Kenyan horticulture. *Journal of Consumer Culture* 5, 365–389.

EurepGAP (2004) *Control Points and Compliance Criteria – Fruit and Vegetables* http://www.globalgap.org/documents/webdocs/UREPGAP_CPCC_FP_V2-1 Oct04_update_01July05.pdf, 13.03.2007

European Union (2006) *EU25 Import – Analysis by Country* http://trade.ec.europa.eu/doclib/docs/2006/april/tradoc_128152.pdf, 21.02.2007

Evans, R.G., Hodge, M. and Pless, I.B. (1994) If not genetics, then what? Biological pathways and population health. In: Evans, R.G., Barer, M.L. and Marmor, T.R. (eds) *Why are Some People Healthy and Others Not? – The Determinants of Health of Populations*. Walter de Gruyter, Berlin, Germany and New York, pp. 161–188.

FPEAK (2007) Grown under the sun. *Horticultural Insight* [Quarterly Publication of Fresh Produce Exporters Association Kenya] 6.

Gerbing, D.W. and Anderson, J.C. (1993) Monte Carlo evaluations of goodness-of-fit indices for structural equation models. In: Bollen, K.A. and Long, J.S. (eds) *Testing Structural Equation Models*. Sage, Newbury Park, California and London, pp. 40–65.

Hatanaka, M., Bain, C. and Busch, L. (2005) Third-party certification and the global agrifood system. *Food Policy* 30(3), 354–369.

Humphrey, J. (2006) Policy implications of trends in agribusiness value chains. *The European Journal of Development Research* 18(4), 572–592.

Humphrey, J. (2008) Private standards, small farmers and donor policy: EurepGAP in Kenya. Working Paper 308, Institute of Development Studies, University of Sussex, Brighton, UK.

Jaffee, S. (2003) From challenge to opportunity – transforming Kenya's fresh vegetable trade in the context of emerging food safety and other standards in Europe. Agriculture and Rural Development Discussion Paper 2, The World Bank, Washington, DC.

Kuklys, W. (2005) *Armatya Sen's Capability Approach*. Springer, Berlin and Heidelberg, Germany.

Liu, J.-T., Hammitt, J.K. and Liu, J.-L. (1997) Estimated hedonic wage function and value of life in a developing country. *Economics Letters* 57, 353–358.

Maertens, M. and Swinnen, J.F.M. (2009) Trade, standards and poverty: Evidence from Senegal, *World Development* 37(1), 161–178.

Mithöfer, D. (2008) Linking smallholders to high value crop markets: How does the group approach work? IIED Fresh Perspectives 25, Agrifood standards and pro-poor growth in Africa (http://www.agrifoodstandards.net/en/global/fresh_perspectives.html).

Montazeri, A., Harirchi, A.M., Shariati, M., Garmaroudi, G., Ebadi, M. and Fateh, A. (2003) The 12-item General Health Questionnaire (GHQ-12): translation and validation study of the Iranian version. *Health and Quality of Life Outcomes* 1(66) (http://www.hqlo.com/content/1/1/66, 27.03.2007).

Okello, J.J. and Swinton, S. (2007) Compliance with international food safety standards in Kenya's green bean industry: Comparison of a small- and a large-scale farm producing for export. *Review of Agricultural Economics* 29(2), 269–285.

Repetto, R. and Sanjay, B.S. (1996) *Pesticides and the Immune System: the Public Health Risks*. World Resources Institute, Washington, DC.

Republic of Kenya: Central Bureau of Statistics (2006) *Statistical Abstract 2006*. The Government Printer, Nairobi, Kenya.

Riisgaard, L. (2009) Global value chains, labour organization and private social standards: Lessons from East African cut flower industries. *World Development* 37(2), 326–340.

Sen, A. (1993) Capabilities and well-being. In: Nussbaum, M. and Sen, A. (eds) *The Quality of Life*. Oxford University Press, Oxford, UK and New York, pp. 30–53.

Sen, A. (1999) *Commodities and Capabilities*, 2nd edn. Oxford University Press, Oxford, UK and New Delhi, India.

Shankar, J. and Famuyiwa, O.O. (1991) Stress among factory workers in a developing country. *Journal of Psychosomatic Research* 35(2/3), 163–171.

Tanaka, J.S. (1993) Multifaceted conceptions of fit in structural equation models. In: Bollen, K.A. and Long, J.S. (eds) *Testing Structural Equation Models*. Sage, Newbury Park, California and London, pp. 1–9.

7 Group Culture and Smallholder Participation in Value Chains: French Beans in Kenya

Janneke Paalhaar and Kees Jansen

Introduction

Small-scale horticultural farmers in Kenya have formed associations[1] over the last two decades in order to gain a market share in the export market. There is considerable variation in access to the international market and the level of integration in the supply chain. Some associations seem to be more successful than others, even though factor endowments (e.g. access to land) do not differ substantially. The objective of this chapter is to examine whether group-related factors, such as structure and social cohesion within associations, shape collective action for the production and marketing of French beans for the export market.

At the start of this century horticultural produce (which includes vegetables, flowers and fruit) represented 13% of Kenya's export value and was one of the top four foreign exchange-earning industries (McCulloch and Ota, 2002). Due to the rapid and sustained growth of its exports, Kenya is among the largest exporters of vegetables to the European Union (EU) (Dolan and Humphrey, 2000). A handful of large exporters dominate the industry nowadays, having contracts with large UK and European retailers. They have established processing factories near Nairobi's international airport where horticultural products, delivered from rural areas, are sorted, washed, weighed, processed and packed. Recently there has been an increase in the export of processed vegetables, which have a higher value per weight than raw vegetables (McCulloch and Ota, 2002). Unlike the tea and coffee sectors, government interventions have been limited to the vegetable sector. Export of horticultural produce to European wholesale markets, South African retail and Middle East entrepôts is expanding. Outside the export market there is a functioning domestic horticulture market, generally with lower farm prices. French beans are the most important export vegetable crop in terms of volume, with more than 40,000 metric tons exported in 2005 (Mithöfer *et al.*, 2008). Predominantly shipped to European supermarkets in a pre-packed and pre-cut form, French beans require large investments in the coordination of supply and the creation of hygienic conditions at the farm/packing house (Narrod *et al.*, 2009).

French bean production is labour intensive, and a substantial part of the production costs is represented by non-family employment (Graffham *et al.*, 2006). Initially, most vegetable production for export was carried out by smallholders (McCulloch and Ota, 2002), but by the late 1990s 40% came from exporters' own farms or leased land, 42% came from large commercial farms and only

18% from smallholders (Dolan and Humphrey, 2000). This is partly the result of the tightening of European regulations on pesticide use and the costs of collecting output from multiple smallholders. On large farms, growing conditions can be more carefully controlled (McCulloch and Ota, 2002). Another reason for the decrease in smallholder participation in the export market is a lack of trust that the farmers will make specific investments and that the exporters will purchase the produce of the small-scale farmers (Ashraf *et al.*, 2009). To comply with Global Good Agricultural Practices (GlobalGAP) regulations, some of these associations have been trained by export companies and governmental extension service providers. However, many associations did not obtain GlobalGAP certification and exporters stopped sourcing French beans from these associations (Ashraf *et al.*, 2009). Although the smallholders' share in fresh bean export has declined, smallholders increasingly produce French beans for the domestic canning industry, which applies fewer food safety standards (only the pesticide residue limits remain) and mainly sources from smallholders (Narrod *et al.*, 2009). Those who produce for the export market benefit not only from higher farm prices but also from access to credit and extension services provided by export companies. Smallholders may grow vegetables for both export and local markets (Ashraf *et al.*, 2009).

In many African countries, producer organizations have gained importance over the last few decades (Bernard *et al.*, 2008; Bernard and Spielman, 2009). In Kenya, smallholders producing for the export market are usually organized in producer associations, which have to be registered as self-help groups with the Ministry of Culture and Social Services as required by law. It is impossible for small-scale farmers to sign individual contracts with exporters. Collective action is undertaken in order to pool together volumes of beans to attain economies of scale and to invest jointly in facilities needed to meet production standards. A management committee with several members leads the association. Associations may have their own technical assistants or trained leaders to help members meet the standards. Some exporters advance seeds on loan, supervise association members individually and penalize individuals who deviate from agricultural practices and standards demanded in the contract.

The above-mentioned reduction in the smallholders' share of the French bean export market was partly a result of changing standards. Even though many associations have been trained by export companies and governmental extension service providers on compliance with the GlobalGAP regulations, they were in many cases not able to make the necessary investments to comply with strengthening food standards. It has been argued that associations that could adapt their collective action to new standards and establish public–private partnerships were better able to survive (Narrod *et al.*, 2009). This raises the question as to why some associations are able to adapt their collective action and others not. There is a growing research interest in how local conditions and the organizational characteristics of producer associations influence changes in household income (Ashraf *et al.*, 2009). A turn to anthropology could help to broaden the concept of the person in social life. Anthropology has a long tradition of studying the role of social organizations and issues of trust underlying economic behaviour, themes that increasingly have attracted the interest of economists, particularly within new institutional economics.

Conceptual Framework

One interesting element of the anthropological view is that it tries to develop a broad perspective on collective action, in which it is not just a means to reduce transaction costs or to reach economies of scale. Group organization, from the anthropological point of view, is cultural, i.e. the result of people getting together, of mutual encouragement, cooperation and coercion (Douglas and Ney, 1998). Society not only is composed of individuals who order preferences and maximize satisfaction (as in the rational-choice model), but it also includes situations in which persons have a reciprocal stance to other intentional persons and social interactions (as in cultural theory) (Jansen, 2009). This requires an understanding of what culture is. Culture is not so much a pre-given thing or some extra context that limits human action, or a synonym for tradition, as it is inherent in human activity, creativity and thought (Douglas and Ney, 1998), and to most situations involving several competing cultures (Schwarz and Thompson, 1990; Richards, 2004). This means that the performance of a farmer association cannot be fully captured by measuring only how individual farmers succeed in producing commodities through the association. The focus should be broader and include the patterns of alliance and authority that hold between association members.

This study applies the so-called grid group cultural theory which offers an alternative to methodological individualism. This theory aims to capture the diversity of human preferences about 'ways of life' and relate this diversity to various possible forms of social organization. The theory identifies four basic forms of social organization (Fig. 7.1), each with its own advantages and disadvantages (Hood, 1998). The distinction between these different forms of social organization brings out some of the varying and contradictory attitudes within farmer associations about whom to blame, how to assess risks or how to judge actions (Vellema, 2005; Vellema *et al.*, 2006). These

Grid	Group: *Weak*	Group: *Strong*
High	**Fatalist culture** Farmers isolated in complex structures • Low cooperation, low trust, rule-bound approaches to organization • Minimum anticipation and lack of disposition to take responsibility; ad hoc responses to events	**Hierarchist culture** Strongly incorporated groups with complex structure • Socially cohesive, rule-bound approaches to organization • Elaborate organizational apparatus of controllers and overseers at all levels of the organization; greater managerial grip
Low	**Individualist culture** Competitive individualism • Atomized approaches to organization stressing negotiation and bargaining • Freedom to enter and exit transactions • Unchecked private gain at the expense of what is supposed to be a collective enterprise	**Egalitarian culture** Strongly incorporated groups with weak structure • High-participation structures in which every decision is 'up for grabs' • Shared opposition to outside world keeps group bound together • Failing negotiation and unwillingness to accept higher authority to break deadlocks; lack of ability to resolve disputes and feuds

Fig. 7.1. Four forms of social organization (from Douglas and Ney, 1998; Hood, 1998; and Vellema, 2005).

formations are constructed by measuring different types of individual and collective responses to imposition (grid) and incorporation (group). The 'grid dimension' explains the degree to which people's lives are circumscribed by conventions or rules, reducing the area of life that is open to negotiation. The 'group dimension' denotes the extent to which individual choice is constrained by group choice, by binding the individual into a collective body (Hood, 1998). Ostrander (cited by Tansey, 2004) defines the two axes by arguing that social order limits the freedom of individuals in two ways: with whom one interacts (group) and how one interacts with them (grid). Within cultural theory, it has been argued that collective action may emerge in groups as a result of two distinctive situations: smallness of scale (individuals are tightly bound by conditions of mutual reciprocation and trust) or coercion (individuals have no choice) (Tansey, 2004).

This chapter uses the idea of a relative grid and group dimension as a heuristic device in order to identify some major mechanisms that determine the organization of farmer associations in Kenya. Based on the framework in Figure 7.1, the major research question can now be defined: Which factors influence the group dimension and/or the grid dimension within an association?

Methodology and Data Collection

Two horticultural export companies (exporter A and exporter B) and seven associations with business relations with these exporters were selected (Table 7.1). These two export companies were chosen because they played an important role in the export market for smallholders, as well as in Europe. Exporter A exports mainly to Belgium, Germany, the Netherlands and other EU countries, and is currently switching sourcing from smallholders to large-scale farmers and would prefer to source exclusively from the latter. Export company B has a long history of working with smallholders, and their main market is the UK supermarket chain. Among the Kenyan exporters, exporter B seems the most loyal to small-scale farmers and the management states that it will continue to source from smallholders in the future, and it profiles its company as such. To keep the geographical factor constant, all the farmer associations, except for one, were selected from the same area and had access to an irrigation scheme in the Central Highlands of Kenya, about a two-hour drive from Nairobi. A guaranteed supply of water makes the production of French beans easier. The association without access to an irrigation scheme was nearer to Nairobi. In the study region exporter B has nine associations producing French beans, of which seven associations are certified for GlobalGAP. Exporter A also has nine associations producing French beans in this area, but none of these associations is GlobalGAP certified.[2]

Table 7.1. Association characteristics.

Association	Characteristics
1-A and 1-B	'Successful'
2-A and 2-B	'Not successful'
3-A	Rejected and signed contract with other exporter
4-A and 4-B	Rejected and no longer producing collectively

A, exporter A; B, exporter B.

To keep the factor 'training' constant, all selected associations had obtained the manual on Quality Management System for smallholders of the International Centre of Insect Physiology and Ecology (Nairobi) and had been trained in its application. For each export company, two associations of small-scale farmers were selected, one considered 'successful' and the other as 'not successful' by the field coordinators of the respective export company. Successful associations are those with which the exporters intend to continue to work. Furthermore, two associations that had transferred to other exporters would be included. However, since export company B did not have such an association, this resulted in only one such association, related to exporter A, being included. Finally, two associations were selected that were rejected by the export companies and had failed either to re-enter the export market or to set up alternatives. Of the seven studied associations only association 1-B was certified for GlobalGAP (Table 7.1).

Semi-structured interviews were held with six members of each association and dealt with the visions, organization, goals and interests of each of the associations. The chairman of each association and five other active members were selected using a simple random sampling technique: numbers 1 (who is the chairman), 5, 6, 10, 11 and 15 of the registered member list; 'active members' are those producing French beans throughout the year with a minimum demand of 20,000 seeds of French beans every 2/3 weeks. If it was not possible to speak with selected farmers at the time of visiting, a neighbour member was interviewed. Data were cross-checked with field observations, farm visits and observations at meetings and at the local marketplace. Additional data were collected by interviewing managers and the field coordinators of several export companies and experts in Kenya. Based on the gathered data, the ten most important indicators reflecting differences in grid and group dimensions of these associations were defined (Tables 7.2 and 7.3). The indicators reflect strengthening or weakening grid or group dimensions and are ranked from zero to two: zero if the indicator weakens the dimensions, one for an average indicator in comparison with the other associations and two if it relatively strengthens the dimension.

Results

The interviewed farmers expressed many different, interrelated reasons as to why they prefer to produce French beans rather than other crops, including:

- 'There is a continuous and good market for French beans';
- 'The temperature and climate are excellent for growing beans';
- 'This is the only possible market and there are no other possibilities for us';
- 'There are many exporters of French beans';
- 'It is possible to sell directly to exporters (no middleman)';
- 'We get a high price';
- 'We can sell to one person';
- 'They come and pick up the beans from the shed';
- 'French beans always come up';
- 'The crop needs fewer chemicals than other crops';
- 'The crop grows quickly, so we can get money quickly (in only 45 days)'; and
- 'They are easy to grow; farming French beans can be done by everybody, even those who are not educated'.

The last two responses were heard most frequently.

A preference for the export market rather than the domestic market is reflected in the following expressions:

- 'a good or higher price on the export market';
- 'the local market is too poor';
- 'to get rid of brokers';
- 'they pay every two weeks';
- 'the exporter picks up the product'; and
- 'the exporter is a stable buyer'.

Growing French beans for brokers is not popular, which is reflected in the following quotation from an interview with one farmer:

> There are always brokers, for every product! But you can refuse to sell maize, beans, bananas or tomatoes and keep them for some days, if you don't agree with the brokers' price, because these crops are tenable. French beans you cannot keep, so you have to sell to brokers on the day of harvesting. Maize, beans, bananas and tomatoes you can eat yourself, but French beans you cannot! Therefore, we only plant French beans when we have a contract with an exporter.

These considerations about brokers and the spot market are conducive to the formation of associations.

Interviewees not only referred to economic costs and benefits, such as better prices, as reasons for preferring to work collectively, but also used concepts such as 'unity', 'not being an individual', 'helping the community', 'borrowing money from the group', 'stand/negotiate as a group' and 'cooperation'.

Grid dimension

The number of rules, the kind of structure, the possibility of negotiation by individuals and the organizational structure within farmer associations tell us something about the grid dimension or imposition by the group (see Table 7.2). Based on a qualitative analysis of our interview data, we identified the ten indicators that seemed most strongly to influence the grid dimension, and these are described below.

1. The supply of seeds, fertilizer and pesticides. Seeds are always supplied by the exporter and payment deferred until harvesting, while fertilizers and pesticides are only sometimes supplied by the export company. Supply by the exporter and delivery on credit involve a high grid dimension for these associations, since it means external regulation (only association 3-A scores 2 points). When the associations deliver the pesticides and fertilizers to its members, this means less regulation (score 1 point),

Table 7.2. Grid dimension.

Indicator	Scale	1-A	2-A	3-A	4-A	1-B	2-B	4-B
Supply of inputs	2 = by export company 1 = by the association 0 = individual	1	1	2	0	1	0	0
Control over collective assets	2 = +commercial plots 1 = own chemical store 0 = scarce facilities	2	2	1	0	2	1	0
Plot for shed	2 = bought collective plot 1 = plot from member 0 = rent plot or no shed	2	2	1	0	2	1	0
Employees	2 = 4 employees 1 = 2 employees 0 = no employees	2	2	1	0	2	1	0
Spraying and control	2 = employ spray team 1 = by group of members 0 = individually	2	2	1	0	2	0	0
Elections and composition of MC	2 = +11 MC members 1 = elections every year 0 = elections not every year	1	1	0	0	2	0	1
Meeting MC and allowance	2 = +MC gets allowance 1 = meet once a month 0 = do not meet	2	1	1	1	2	1	0
Contribution	2 = save 8 KSh/kg 1 = save 5 KSh/kg 0 = no contribution	2	1	1	0	1	0	1
Control of side-selling	2 = 3 levels of control 1 = 2 levels of control 0 = no control	2	1	0	0	1	0	0
Access to credit	2 = from group account 1 = from external source 0 = no possibility	2	1	0	0	2	1	0
Total		18	14	8	1	17	5	2

KSh, Kenyan shilling; MC, management committee. Ranking: 0 = low, 1 = average, 2 = high.

and when association members buy their products individually from the shop, in cash, it means the least regulation because farmers can select individually which pesticides to use (score 0 points).

2. The control over collective assets makes interaction with the export market easier, but it requires defined contributions from members. There are two associations with no or scarce facilities, which may contribute to their failure to access the export market (4-A and 4-B). The other associations have better facilities and at least a grading shed and a chemical store of their own (1 point). Three of them share commercial plots, and thus have more control over collective production (1-B, 1-A and 2-A: 2 points).

3. A higher grid dimension can be attributed to associations with their own plot for their shed, because this requires commitment from the members in the form of payment, sharing and responsibility. Associations 1-B, 1-A and 2-A bought a plot for building the shed (2 points); 2-B and 3-A built a shed on the plot of a member (1 point due to relatively higher risks); 4-A and 4-B get 0 points because they rent a plot and do not have a shed, respectively.

4. More employees implies higher levels of monitoring of labour, suggesting a higher grid dimension.

5. If spraying has to be done by the members individually, the grid dimension is low because of high individual freedom (2-B, 4-B and 4-A: 0 points). Control may be difficult, particularly since association members often live at some distance from each other. The risk of exceeding the maximum pesticide residue limits increases. One association (3-A) uses a spraying team composed of members. Since these members are not educated in spraying, the risk of exceeding the maximum pesticide residue limits is still there, so this association gets 1 point. Other associations contract specialized spraying teams (1-B, 1-A and 2-A: 2 points).

6. Regular elections of the management committee and a higher number of committee members tend to increase the levels of control and limit space for patron–client relationships; therefore, 0 points are allotted when elections are not held every year (2-B, 3-A and 4-A). Associations that hold elections every year get 1 point. One association (1-B) scores 1 extra point, because its higher number of members on the management committee (11 rather than the normal 9 members) leads to a better division of tasks, which tends to strengthen the grid dimension.

7. Management committees that meet regularly will discuss problems better, share values, agree on objectives and define responsibilities. Zero points are allotted for associations that do not meet. The management committees of most associations meet at least once a month (1 point). Committee members who get an allowance (1-A and 1-B) will have to perform according to defined rules, and the expectations of other members will be higher (2 points).

A higher contribution to the association's account increases the grid dimension, because the individual freedom of the members decreases (0 points for no contribution: 2-B and 4-A). Some associations deduct some money from every sale of beans (KSh5/kg (about €0.05/kg)), 1 point; 2 points for the highest contribution of KSh8/kg (about €0.09/kg).

8. The number of levels at which side-selling is controlled is another element of the grid dimension. The first level of control is by the members of the group themselves. The second level is having a specific member of the group check the whole group, like a policeman. The third level of control is a check by the export company. These three levels of control score 2 points; two levels of control score 1 point. The other four associations score 0 points, because side-selling takes place and is not controlled. In three associations side-selling is strongly prohibited (1-A, 2-A and 1-B).

9. The grid dimension may increase when associations lend money to their members, because this often leads to more conventions and rules for controlling these loans (1-A and 1-B: 2 points). The regulatory effect of credit on associations is somewhat less when members borrow money from external sources, although the external prescriptions on its members will increase

(2-B and 2-A: 1 point). Associations with no internal or external credit score 0 points.

Group dimension

The group dimension concerns the extent to which farming and individual choice is constrained by the group (see Table 7.3). The ten indicators associated with the group dimension are listed below.

1. The age of the association tends to reflect the durability and strength of relationships within it; older associations have a higher group dimension. Associations 1-A, 1-B and 2-A have existed for 6 years or longer and score 2 points; 4-B and 3-A are 3 years old and score 1 point; the youngest associations no older than 1.5 years score 0 points.

2. It has been argued that the origin of members and their social and cultural background influences the definition of shared values and objectives, procedures and mutual

Table 7.3. Group dimension.

Indicator	Scale	1-A	2-A	3-A	4-A	1-B	2-B	4-B
Age of the association	2 = > 6 years 1 = 3 years 0 = < 1½ years	2	2	1	0	2	0	1
Origin of members	2 = all born in same village 1 = some members from elsewhere 0 = all settlers from elsewhere	1	1	1	2	1	0	1
Number of dropouts	2 = max. 1 dropout 1 = 10–15 dropouts 0 = 40 dropouts	1	1	2	1	2	0	1
Active members (%)	2 = 85–90% active 1 = 60% active 0 = < 50% active	2	2	0	0	1	0	0
Regularity of meetings	2 = twice a month 1 = once a month 0 = twice a year	0	1	1	1	0	1	2
Relative isolation from other associations	2 = isolated from other associations 1 = isolated but near Nairobi 0 = mixed with other groups	0	0	2	0	0	1	2
Recognition of merits	2 = reward for best farmer 1 = used to give a reward 0 = no reward	1	1	0	0	2	0	0
Trust and mutual support	2 = overall trust and help 1 = better relations with members 0 = do not help each other	0	1	1	0	2	1	1
Presented problems within the association	2 = no problems 1 = few problems 0 = many problems	0	0	1	1	2	0	1
Builds on participation in other groups	1 = all in other groups 0 = not all in other groups	0	1	0	1	1	0	1
Total		7	10	9	6	13	3	10

Ranking: 0 = low, 1 = average, 2 = high.

trust in producer associations (Santandreu and Castro, 2007). This suggests that the group dimension tends to be stronger in associations with more homogeneity between the members in regard to their origin and background. The association whose members are all settlers (coming from a different region) and belong to different tribes scores 0 points. The members of the other associations all belong to the Kikuyu tribe and score 1 point. The only association of which all the interviewees were born in the same village (4-A) scores 2 points.

3. The number of dropouts is important for the group dimension, as dropouts may be a result of, or a cause of, friction within an association and reflect a lack of social cohesion. From this point of view, no dropouts or only a few dropouts represent a high group dimension. In associations 1-B and 3-A only one member has left the association (2 points). The number of dropouts is higher for the other associations (10–15 members, 1 point), especially association 2-B, where more than 40 farmers quit (0 points).

4. Associations with a high percentage of *active* members function better and are more coherent. Associations with more than 85% of active members score 2 points (1-A and 2-A). The association with the second highest share of active members (1-B, 60%) scores 1 point, and the others with 50% of active members or fewer score 0 points.

5. Transparent and regular communication within the association on its vision, objectives and decisions sustains the involvement of the members and helps avoid disagreements, mistrust and misunderstanding (Dubbeling, 2007). So, to strengthen the group dimension and become a successful association, it is clearly important to meet regularly, as does association 4-B, whose members meet twice a month (2 points). The other associations meet less often, i.e. once a month (1 point) except for 1-A and 1-B, who meet only three and two times a year, respectively (0 points). Because these last two associations are fairly successful and score highly on other factors, it may be the case that more experienced, older associations need fewer meetings.

6. The relative geographical isolation of an association makes it more difficult for group members to interact with outsiders, and may strengthen group boundaries. The closer a particular association is to Nairobi or other associations, the more contact it can have with brokers, other farmers and other export companies. Geographically isolated associations seem to perform better, because the temptation to side-sell the French beans or to be double-contracted is lower for these associations. In such associations trust and social cohesion tend to be higher and, therefore, the group dimension stronger. Associations 4-B and 3-A are isolated (2 points) while 2-B is isolated although located near Nairobi, which probably creates a greater opportunity for contact with brokers (1 point).

7. The recognition of merits within associations may increase the social pressure between members. One association rewards those members with the highest production levels by making a present in the form of a farming tool (1-B: 2 points). Two associations (1-A and 2-A) used to reward the most successful member of the association, although new management committees have terminated this practice (1 point). The other associations never rewarded the best farmer (0 points).

8. The extent to which associations perceive that members help each other, work together, trust the other members and know each other is in itself an indicator for the group dimension. Associations whose members are used to working together and trust each other have a higher group dimension. In two associations people consider it uncommon for members to help each other on the farm or in social life (1-A and 4-A: 0 points). Then there are associations in which it is more common to help each other when necessary and whose interviewees say they have a better relationship with the members of the association than with other farmers (1 point). Finally, there is only one association whose members say they trust each other and help each other (1-B: 2 points).

9. Group cohesiveness is higher when inherent conflicts are controlled. In a relatively short period of research, it is difficult to identify and understand most of the internal problems since these tend to be too sensitive to discuss with outsiders. Furthermore, problems that are disclosed tend to be epiphenomenal for other, underlying problems. However, if group cohesiveness and levels of control decrease, internal disagreements and discontent may be more easily exposed to outsiders and, therefore, when members reveal problems this can be taken as an indicator of group cohesiveness. Associations that presented a variety of problems about money, linkages with brokers, double-contracting or the shed have a low group dimension (0 points) and, according to the export companies, these associations are less successful. However, some associations that scored quite high for the other indicators may present problems (1-A and 2-A). Other associations revealed some, but smaller, problems (1 point), while only in one association did the members say they never had any internal problems (1-B: 2 points).

10. When an association builds on relationships between members developed in other groups or networks (e.g. money groups), this may strengthen the group's internal relationships and increase levels of trust. Associations that did not do so score 0 points, the others 1.

Grid group map of farmer associations

Different factors may reinforce each other. For example, higher contributions by association members may make it possible to buy a plot for the shed or employ a spray team and 'the age of a group' influences the number and quality of assets, as older associations tend to have a higher budget, more experience and better facilities. Another example is the interaction between 'access to credit' and 'the percentage of active members'. Interviewees indicated that a lack of credit (no money to buy land, inputs or equipment) is a cause for the issue of inactive members. Given the low number of associations, it is not possible to calculate precise correlations, but the data nevertheless allow for investigation into what extent associations act with a similar rationality or whether they follow different trajectories and different forms of group organization to reach the same objective, namely selling beans to the export market. If we fit the sums for the different indicators in a single grid group map we find the dispersion displayed in Fig. 7.2.

The association with high scores for both the grid and group dimension is placed in the upper right-hand corner. This association (1-B) is the only one in the study that has been certified for GlobalGAP. The different types of associations have different numbers.

The two associations that have been defined as 'successful' by the export companies are the ones with the highest grid dimension (1-A and 1-B), but not necessarily the highest group dimension; one of these (1-A) has a lower group dimension than several other associations. This supports the proposition that export companies prefer, in particular, a high grid rating: clear rules, strict penalties and high levels of hierarchical control within the association rather than being concerned about the group dimension. The argument that grid is a substitute for group might work in the short term, but may not serve in the long term in case both dimensions turn out to be indispensable for the success of the association.

Although only one association has been certified for GlobalGAP, other associations too have implemented a broad range of the requirements for certification. This may have led to a more prudent pesticide and fertilizer application and increased quality of seed. According to the interviewees, the required record keeping has enabled them to evaluate the profitability of their operations. GlobalGAP regulations, even though they have led only to certification in one case, seem to have influenced the culture of various associations, particularly through the grid dimension, e.g. more and stricter rules regarding spraying. The case of association 1-B suggests that a high group dimension may be conducive to qualification for GlobalGAP certification, since high

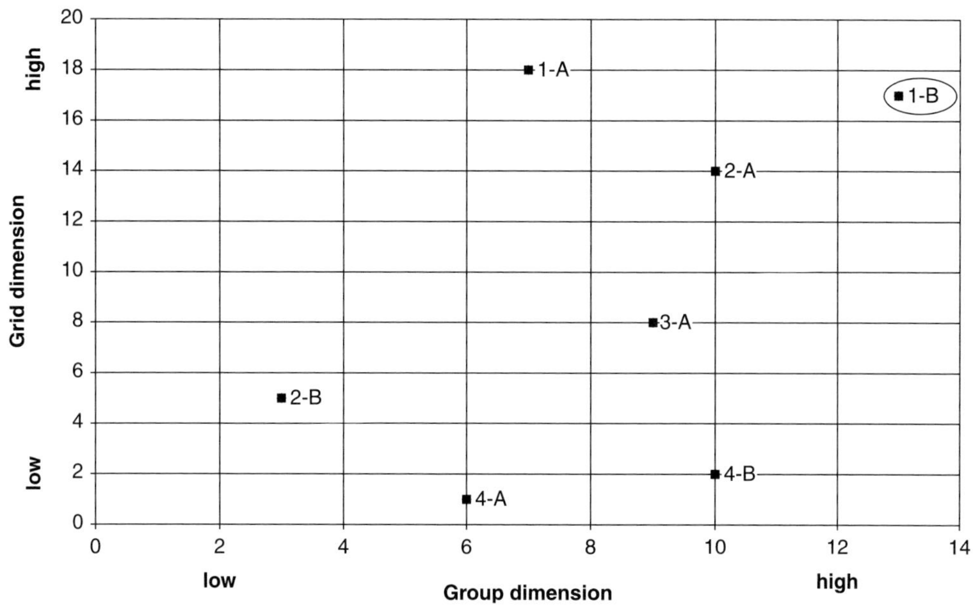

Fig. 7.2. Grid and group dimensions of associations.

levels of mutual trust stimulate the functioning of an association. In turn, becoming certified can further increase the social cohesion within an association. Association 1-A applied to become certified for GlobalGAP but failed, because some farmers were double-contracted with two export companies, which is against the rules. Our interviews suggest that this failure has lowered its group dimension due to internal disagreement about the expenditures for the certification. Due to this and other reasons 12 to 15 farmers (of the initial 33 members) quit the association. Nevertheless, the association has maintained a good relationship with the export company.

The variation in relationships between associations and export company A seem to be correlated to the grid dimension; association 2-A is considered more problematic than 1-A by the exporter, and internal problems with the management of funds caused the exporter to label this association unsuccessful. These problems have recently been solved and the offenders were punished for their activities, according to the chairman. Association 3-A, formed in 2004, was dropped after 2 years by exporter A for reasons unknown to its members. Some time later, the association was able to sign a contract with exporter C, and its membership of 30 farmers is stable. Association 4-A was dropped by the export company when their French beans exceeded the maximum pesticide residue limits. Collective production of French beans halted, but individual members sell French beans to brokers. They still hire facilities collectively and meet to discuss future activities.

The variation in the grid and group dimensions is high for associations that have interacted with exporter B. Association 1-B is one of the most stable associations in this study. This association had been selling French beans to exporter A until 2003, when it switched to exporter B. The association was 'not satisfied with the contract' with exporter A and indicated that it was pleased with the existing relationship with exporter B. Association 2-B is a young association with 70 members, and signed a six-month contract with exporter B in September 2006. According to the export company, this association is 'not successful', because the farmers are in debt to the company and do not produce the desired quantity of French beans. There is a difference of opinion about the delivery of seed and, therefore, association 2-B temporarily stopped producing French beans for exporter B. The reduction of association membership to 30 farmers may have caused this conflict, since the exporter delivers seed according to the original contract while the smaller number of farmers cannot grow all this seed. Association 4-B never succeeded in putting into place a system to control its member contributions and practices; the NGO that helped the association to start in 2004 with 22 members eventually dropped it. The exporter left and some members now produce individually for brokers. The management committee is still active and intends to renew relationships with exporter B.

Summary and Conclusions

This chapter has identified a series of factors in both the grid and group dimension that co-determine the participation of Kenyan smallholders in the French bean export market. This exploratory study suggests that the divergent forms of social organization lead to more than one way in which associations can become embedded in export value chains. No single factor explains success or failure. Each of the configurations of the grid and group dimension provides the basis of a more-or-less viable way of acting. The issue is not so much which configuration is right, but rather which one is more appropriate to the task at hand.

This chapter elucidates some problems when farmer associations are conceptualized as a sort of collective of individual actors (groups without group characteristics, without group culture), as rational, self-conscious actors seeking only to fulfil materialistic needs. Such a conceptualization, in fact, reasons from within the 'individualist culture' as described in the grid group map (Fig. 7.1). Within an 'individualist culture' an optimal environment for fostering entrepreneurial creativity would have limited rules, low social cohesiveness and market actors with business-oriented goals only. In this study, however, a low grid dimension, which goes together with a competitive individualism, was not a characteristic of successful associations. Instead, the associations that were classified as successful by the exporters displayed high levels in the grid dimension. It has been argued above that the smallness of scale (individuals are tightly bound by the conditions of mutual reciprocation and trust) or coercion (individuals have no choice) can be involved by collective action. The differences between the two most successful associations (1-A and 1-B) suggest that the group dimension is not unimportant, but the overall picture is that the current environment of new standards, such as GlobalGAP, mediated by the export companies, locates success predominantly in the grid dimension, supporting a more hierarchical way of social organization that puts the emphasis on a system of strict rules and penalties. Those development interventions that focus on strengthening an egalitarian culture in farmer associations may have to find new ways to deal with this context.

Further research

This analysis of the grid and group dimensions dovetails with literature that has taken the group dimension into consideration. Transparent and regular communication within farmer associations on their vision, objectives and decisions will sustain the involvement of the members and build up trust and mutual understanding (Dubbeling, 2007). Associations with shared values, democratic decision making and clear objectives will more easily achieve good results (Santandreu and Castro, 2007) and the high levels of the social cohesion of the association will increase the viability of certification. Social sanctions are most effectively applied within strong social groups (Miguel and Gugerty, 2005). This study confirms that the group dimension is important for the effective functioning of associations and successful participation in the export market, but argues that the grid dimension may even be more decisive. The structure, rules and penalties, and the internal organization of associations influence their functioning and success in the export market. This latter aspect needs more discussion, while attention to the difference between the grid and group dimension may help in better understanding why associations such as 4-B, with a high group but a low grid dimension, are unsuccessful.

Endnotes

[1]These associations are mostly called groups, but since 'group' has a special connotation in this chapter we use the term 'association' for a farmer group.

[2]GlobalGAP certification is a prerequisite for exports to Europe. However, since this is a private (voluntary) standard, non-certified vegetables are also exported to Europe depending on the requirements of the particular national market and/or marketing channel, such as a particular supermarket chain or wholesale market requirements.

References

Ashraf, N., Xavier, G. and Karlan, D. (2009) Finding missing markets (and a disturbing epilogue): Evidence from an export crop adoption and marketing intervention in Kenya. *American Journal of Agricultural Economics* 91(4), 973–990.

Bernard, T. and Spielman, D.J. (2009) Reaching the rural poor through rural producer organizations? A study of agricultural marketing cooperatives in Ethiopia. *Food Policy* 34(1), 60–69.

Bernard, T., Collion, M., de Janvry, A., Rondot, P. and Sadoulet, E. (2008) Do village organizations make a difference in African rural development? A study for Senegal and Burkina Faso. *World Development* 36(11), 2188–2204.

Dolan, C. and Humphrey, J. (2000) Governance and trade in fresh vegetables: The impact of UK supermarkets on the African horticulture industry. *Journal of Development Studies* 37(2), 147–176.

Douglas, M. and Ney, S. (1998) *Missing Persons: a Critique of the Social Sciences*. University of California Press, Berkeley, California.

Dubbeling, M. (2007) An inter-regional action-research agenda: Recommendations for strengthening social organisations of urban and peri-urban producers. *Urban Agriculture Magazine* 17, 7–8.

Graffham, A., Karehu, E. and MacGregor, J. (2006) Impact of EurepGAP on small-scale vegetable growers in Kenya (available at www.agrifoodstandards.net/en/filemanager/active?fid=83; accessed 16 December 2009).

Hood, C. (1998) *The Art of the State: Culture, Rhetoric, and Public Management*. Clarendon Press, Oxford, UK.

Jansen, K. (2009) Implicit sociology, interdisciplinarity and systems theories in agricultural science. *Sociologia Ruralis* 49(2), 172–188.

McCulloch, N. and Ota, M. (2002) Export horticulture and poverty in Kenya. IDS working paper 174, Institute of Development Studies, Brighton, UK.

Miguel, E. and Gugerty, M.K. (2005) Ethnical diversity, social sanctions, and public goods in Kenya. *Journal of Public Economics* 89(11/12), 2325–2368.

Mithöfer, D., Nang'ole, E. and Asfaw, S. (2008) Smallholder access to the export market: The case of vegetables in Kenya. *Outlook on Agriculture* 37(3), 203–211.

Narrod, C., Roy, D., Okello, J., Avendaño, B., Rich, K. and Thorat, A. (2009) Public–private partnerships and collective action in high value fruit and vegetable supply chains. *Food Policy* 34(1), 8–15.

Richards, P. (2004) Private versus public? Agenda setting in international agro-technologies. In: Jansen, K. and Vellema, S. (eds) *Agribusiness and Society: Corporate Responses to Environmentalism, Market Opportunities and Public Regulation*. Zed, London, pp. 261–288.

Santandreu, A. and Castro, C. (2007) Social organisations of agricultural producers in Latin America and Europe: Lessons learned and challenges. *Urban Agriculture Magazine* 17, 5–6.

Schwarz, M. and Thompson, M. (1990) *Divided We Stand: Redefining Politics, Technology and Social Choice*. Harvester Wheatsheaf, Hemel Hempstead, UK.
Tansey, J.A. (2004) Risk as politics, culture as power. *Journal of Risk Research* 7(1), 17–32.
Vellema, S. (2005) Regional cultures and global sourcing of fresh asparagus. In: Fold, N. and Pritchard, B. (eds) *Cross-Continental Agro-Food Chains: Structures, Actors and Dynamics in the Global Food System*. Routledge, London, pp. 124–136.
Vellema S., Loorbach, D. and van Notten, P. (2006) Strategic transparency between food chain and society: Cultural perspective images on the future of farmed salmon. *Production Planning and Control* 19(6), 624–632.

8 Export Vegetable Supply Chains and Rural Households in Senegal

Miet Maertens, Liesbeth Colen and Jo Swinnen

Introduction

International trade in fresh and processed vegetables has been increasing sharply over the past two decades (Aksoy and Beghin, 2005). Developing countries have become the main exporters of fresh and processed vegetables and high-income countries are among the major vegetable importers – for example, the value of vegetable imports in the European Union (EU) has surpassed all other categories of agricultural imports since the early 1990s, while the USA is the single largest importer of vegetables worldwide (Weinberger and Lumpkin, 2005). An important share of the global trade volume comes from Latin America and the Caribbean; and from China, where vegetable exports started growing strongly from the late 1970s (Weinberger and Lumpkin, 2005). During the 1990s and early 2000s, vegetable exports from sub-Saharan Africa (SSA) grew sharply. The total value of SSA vegetable exports increased from US$73 million in 1980 to US$159 million in 1990, and further to US$474 million in 2006 (FAOSTAT, 2009). Some SSA countries, such as Kenya and Senegal, have become important vegetable-exporting countries and Africa as a whole has become the main external supplier of vegetables to the EU (Weinberger and Lumpkin, 2005).

Diversification towards high-value export crops, such as vegetables and horticultural commodities in general, has often been promoted as a pro-poor development strategy because of the direct link with the rural economy, the relatively high value of horticulture produce and the labour intensity of vegetable production (Carter *et al.*, 1996; Anderson and Martin, 2005; Diop and Jaffee, 2005). Such export diversification strategies have been included explicitly in the poverty reduction policies of many developing countries, such as in the Poverty Reduction Strategy Paper (PRSP) of Senegal, for example.

Since the 1990s global food systems have experienced important structural changes, which have cast doubt on the beneficial development effects of increased high-value exports from developing countries. The main concern is that, due to increasing public and private food standards,[1] increasing foreign direct investment[2] and the resultant increases in the level of vertical coordination and consolidation in the agrifood sectors of developing countries, smallholder farmers have become increasingly marginalized in high-value export chains. It is argued that smallholders might not reap the benefits from the growth in high-value agricultural exports, either because they are excluded as primary producers or because their benefits are squeezed by more powerful actors in the chain, such as large multinational food companies. First, the proliferation and tightening of food standards, public regulations and private standards can lead to the exclusion of smallholder farmers because compliance costs are prohibitively high for these producers (Unnevehr, 2000; Garcia Martinez and Poole, 2004). Secondly, increased consolidation in food processing and distribution and increased foreign direct investment (FDI) in developing countries' agrifood sectors can lead to the exclusion of smallholder suppliers because they cannot provide the volumes and quality that large buyers demand, and also because transaction costs favour sourcing from large and more capitalized suppliers (Farina and Reardon, 2000; Weatherspoon and Reardon, 2003; Neven *et al.*, 2009). Thirdly, the increased consolidation in food distribution could decrease farmers' bargaining power vis-à-vis large companies, and reduce their potential benefits (Gibbon, 2003).

These arguments are, however, subject to debate. Empirical studies – including case studies on vegetable exports from different SSA countries – provide competing conclusions about the effects of increased high-value exports and structural changes in global high-value supply chains on rural incomes and poverty. Studies from Kenya, Côte d'Ivoire, Ghana and South Africa, for example, suggest that horticulture exports are increasingly grown on large estate farms, thereby excluding smallholders and increasing rural poverty (Jaffee, 2003; Minot and Ngigi, 2004; Danielou and Ravry, 2005; Neven *et al.*, 2009). In Kenya and Zambia, the introduction of Global Good Agricultural Practices (GlobalGAP) has led to the exclusion of a large number of smallholder farmers, while those farmers who do succeed in obtaining certification clearly benefit through technology transfers, increased efficiency and farmers' health (Graffham and MacGregor, 2007; Graffham *et al.*, 2007). On the other hand, in South Asia (Gulati *et al.*, 2007), China (Wang *et al.*, 2006) and Eastern Europe (Dries and Swinnen, 2004) studies have found that smallholder producers are included in, and benefit from, high-standard export supply chains. Evidence from Madagascar shows that smallholder farmers are able to supply high-quality vegetable exports through innovative contract farming schemes with intensive farm assistance programmes and on-farm monitoring, leading to important direct and indirect benefits for these farmers (Minten *et al.*, 2009).

Another key argument is that the important income and poverty effects of increased high-value exports might be realized through labour markets rather than through product markets. As vegetable production is labour intensive and stringent food standards increase the need for labour-intensive postharvesting and other activities, such as sorting, grading, washing, packing and labelling, increased vegetable exports might contribute to the creation of rural employment opportunities. Labour market effects have mostly been ignored in empirical studies, but some recent studies have shown that horticulture export sectors contribute importantly to the creation of rural employment (McCullough and Ota, 2002; Humphrey *et al.*, 2004; Maertens *et al.*, 2008; Maertens and Swinnen, 2009; Neven *et al.*, 2009).

In this chapter we describe the results of case studies on the welfare implications of green bean and tomato exports in Senegal – currently the second largest exporter of green beans and the largest exporter of tomatoes in SSA. Both case studies show that the growth in vegetable exports has contributed importantly to poverty reduction in the specific areas from where vegetables are sourced. In addition, we show that, while the inclusion of smallholder suppliers in high-value vegetable export chains certainly creates benefits for these farmers, the main poverty-reducing effects come through labour markets and the creation of employment accessible for the poorest rural households.

In the next section, we document the growth in vegetable exports in Senegal in the past decade. In the third section (Increasing Standards in Global Vegetable Chains), we give an overview of how the levels of food standards, public regulation and private standards in vegetable supply chains have increased in recent years, with the main focus on standards in the EU, which is the main destination for vegetable exports from Senegal, and from SSA in general. In the fourth section (Governance in Vegetable Export Supply Chains in Senegal), we describe the structure and organization of the green bean and tomato export supply chains in Senegal. In the fifth section (Rural Households in Export Supply Chains), we describe how and to what extent local households are involved in vegetable export production in the two case study regions. In the sixth section (Welfare Implications of Increased Export Production) we summarize the results of a detailed analysis of the income and poverty effects of increased vegetable exports for local households in the two case study regions. We draw some main conclusions and implications for policy making and further research in the final section.

Vegetable Export Growth in Senegal

After decades of dependency on groundnuts as the main agricultural export commodity throughout the 1960s, 1970s and 1980s,[3] Senegal has adopted a strategy of agricultural export diversification towards higher-value commodities, particularly since the devaluation of the franc CFA (Communauté Financière Africaine) in 1994. The vegetable sector has played a crucial role in this strategy. Vegetable exports from Senegal have increased tremendously since the start of the new millennium in terms of volume, from 1050 metric tons in 1996 to almost 16,500 metric tons in 2006 (Fig. 8.1). The two main vegetable export crops are green beans and tomatoes, accounting for 49% and 42% of the total volume of vegetables exported in 2006. Other vegetables that are exported in minor quantities include cucumbers, onions, mushrooms, lettuce and chilli peppers. In recent years, Senegal has developed into one of the main vegetable exporters within SSA. In fact, in SSA Senegal ranks first in tomato exports and second in green beans after Kenya.[4]

Apart from some small volumes exported to neighbouring countries, Senegal's vegetable exports are destined mainly for countries in the EU, particularly France, the Netherlands and Belgium. Currently, Senegal ranks fifth as an external supplier of fresh and chilled tomatoes to the EU (after

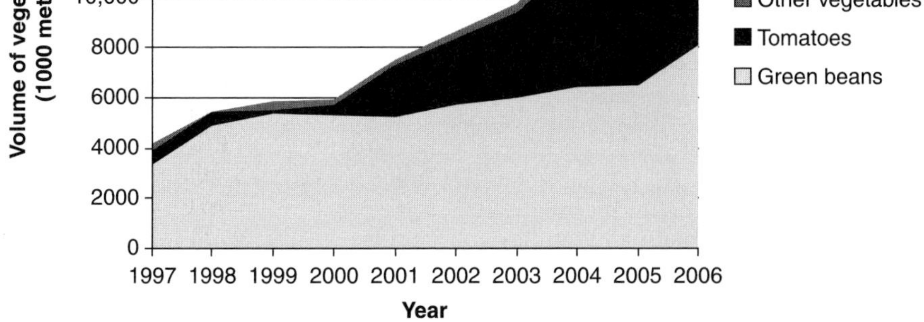

Fig. 8.1. Vegetable exports from Senegal, 1997–2006 (from ONAPES, 2007, unpublished data).

Morocco, Turkey, Israel and Syria), and fourth as an external supplier of green beans to the EU (after Morocco, Egypt and Kenya).[5]

Increasing Standards in Global Vegetable Chains

During the 1990s and 2000s food quality and safety requirements increased rapidly, especially in the EU – the main destination of vegetables exported from Senegal – and other high-income regions. Public authorities have imposed a series of stringent public regulations; in addition, many large trading and retailing companies have established private food standards that are often even more demanding. Fresh food imports in particular, such as vegetables, are subject to strict standards because fresh food is prone to food safety risks, and because such products are often used as strategic products by large retail chains (Dolan and Humphrey, 2000; Jaffee and Henson, 2005).

EU public regulations

European Union legislation puts forward specific public standards concerning food quality and food safety for vegetables. First, vegetable exports need to comply with the general regulations on the health control of foodstuffs of non-animal origin. These specify (i) traceability rules; (ii) the responsibilities of food importers; (iii) hygiene rules (based on the Hazard Analysis and Critical Control Points (HACCP) principle); (iv) conditions concerning contaminants in food (such as maximum pesticide residue levels); and (v) conditions on the use of genetically modified crops, etc. Secondly, EU general labelling rules must be fulfilled, with labels indicating the place of origin of the food produced, the net quantity and expiry date, etc. Thirdly, the EU requires compliance with marketing standards specific to fresh fruit and vegetables. These marketing standards describe the minimum requirements of quality, size, presentation and marking for each type of fruit or vegetable, and specify different quality classes for each product.[6]

Regular controls on the safety, labelling and marketing standards of imported vegetables are carried out by authorities in the EU member states at the point of entry into the EU, to ensure that they comply with all EU requirements. However, certain exporting countries are permitted to perform the conformity-checking operations themselves, prior to entry into the EU. Since 2006 Senegal has been among the few African countries, along with Kenya, South Africa and Morocco, that is accredited by the EU to perform its own conformity assessments and controls prior to shipment to the EU. This accreditation reduces exporters' risks and costs related to non-conformity and the entry denial of produce.

Private food standards

There exists a large variety of private standards and certification schemes. Some of these apply to one single retailer (e.g. Carrefour Filière Qualité, Tesco Nature's Choice), while others are national (e.g. the British Retail Consortium (BRC) Global Standard) or international initiatives (e.g. GlobalGAP). To some extent, private companies engage in standard setting to reduce reputation or commercial risks associated with food safety hazards, especially when the public food safety capacity is weak. On the other hand, private standards are used as a competitive strategy for product differentiation (Henson, 2006; Henson and Humphrey, 2008).

The most important private standard for vegetable exports from Senegal to the EU is GlobalGAP; other, more stringent, private standards, such as BRC and Tesco Nature's Choice, have been adopted by only a few large-scale producers. Currently GlobalGAP includes many more products, but the initial focus was on fresh fruit and vegetables. GlobalGAP standards consist of a series of protocols for the application of good agricultural practices to the production of fruit and vegetables, including requirements in relation to site and soil management, irrigation, waste and pollution management, harvesting, the handling of produce, personal hygiene, packing and storage, quality control, etc., and the provision of guidelines for producers' compliance with these standards (GlobalGAP, 2009).

Governance in Vegetable Export Supply Chains in Senegal

In the remainder of this chapter, we discuss two specific case studies of vegetable exports from Senegal: green beans from les Niayes, a region stretching along the coast north of Dakar, and tomato from the Senegal River Delta area in the north of the country (Fig. 8.2). Both vegetables are typically produced for export, and domestic trade is limited to small volumes that, in the main, do not conform to export standards and requirements and are traded in urban markets for consumption in restaurants and hotels. To reveal how the structure and governance system of the vegetable export chains has adapted to the changing requirements in EU markets, we interviewed (i) 9 of the 20 green bean-exporting companies who, *in toto*, represented one-third of the total green bean export volume (2005); and (ii) in 2006, the dominant tomato-exporting company, accounting for almost the entire volume of tomato exports. We also interviewed the two main horticulture exporter organizations.[7] A more detailed description of data collection can be found in Maertens *et al.* (2008) and Maertens and Swinnen (2009). These company-level interviews show that the increasingly stringent EU standards have major consequences for the organization and structure of the vegetable export sector in Senegal.

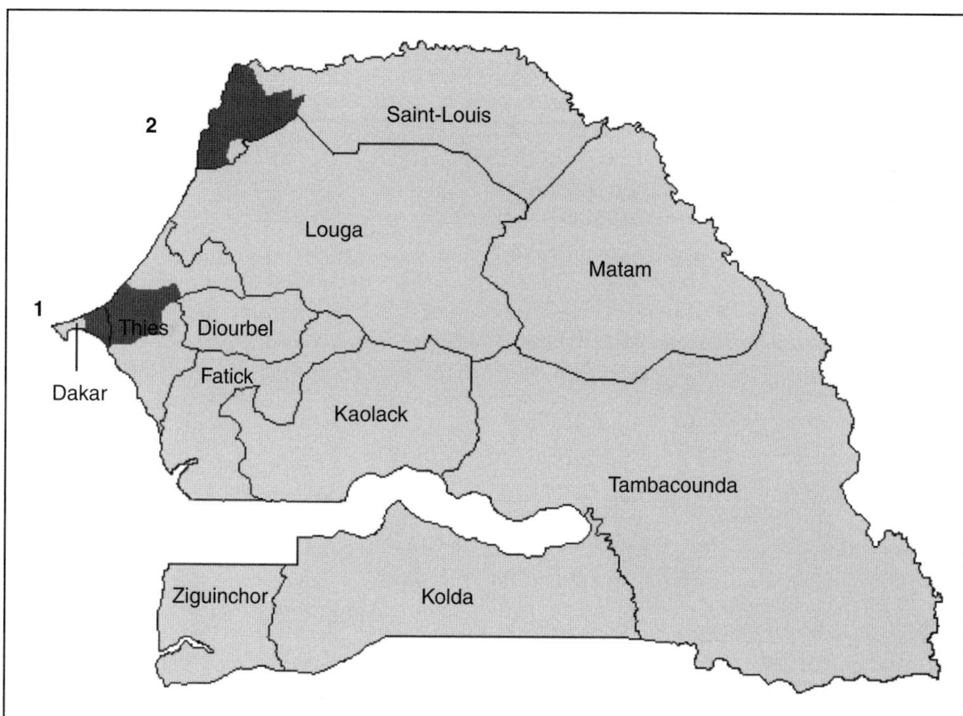

Fig. 8.2. Map of the research areas in (1) les Niayes and (2) Senegal River Delta, Senegal (from Atlas du Sénégal – IRD – Cartographie A. LE FUR – AFDEC).

Standards compliance and certification

Exporters are forced to keep up to date with the changing legislation and make additional investments in order to comply with food standards. The target of the largest vegetable exporters is to obtain GlobalGAP certificates. In the past couple of years these firms have made substantial investments to comply with these private requirements, including investments in cold storage facilities and transport capacity, facilities for selection and packaging, control mechanisms, improvements in sanitary conditions at the conditioning station, etc. In our sample of bean and tomato exporters, only one out of nine bean-exporting companies had been GlobalGAP and HACCP certified since 2004, while three others were in the process of certification. The tomato-exporting company in our sample has been GlobalGAP and HACCP certified since 2003, and became certified more recently by several other private certification schemes, including the International Organization for Standardization (ISO) and Tesco Nature's Choice, and is in the process of certification for the British Retail Consortium (BRC).

Consolidation and foreign direct investment

Many exporters face financial constraints when investing in upgrading their supply capacity in conformity with EU regulations and private standards. Only larger companies, and in particular (partially) foreign-owned companies with access to foreign capital, are able to make the necessary investments. Many smaller exporters have left the market in recent years because they cannot incur the high costs related to the compliance with stringent standards. Once GlobalGAP was initiated in 1997 and since the new EU Food Law of 2002, the vegetable export sector in Senegal has been consolidating. In the green bean sector, the number of exporting companies fell from 27 in 2002, to 20 in 2005 and to 14 in 2008, with mainly smaller exporters dropping out. Moreover, the market share of the three largest companies increased from less than half in 2002 to two-thirds in 2005. In the tomato export sector the consolidation is even more extreme; this sector is dominated by a single company, a foreign company that is part of a French multinational holding, which started exporting tomatoes to the EU in 2003 and achieved a market share of more than 95% in the 2006/2007 season.

Vertical coordination and ownership integration

Higher food standards increase the need for tighter coordination and lead to important changes in the governance system of vegetable export supply chains. This is apparent in the increasing degree of vertical coordination at different levels of the chains. First, in recent years many large exporters have shifted from loose agreements with downstream EU importers to more binding contracts, including the provision of price, the quantity and timing of delivery and sometimes even the pre-financing of delivery by buyers. In the tomato export sector, the dominant exporter is part of a multinational holding in which overseas transport and distribution within the EU are organized through affiliates of the holding.

Secondly, in order to guarantee product quality, safety and traceability throughout the supply chain and to ensure the accurate timing of production and harvesting, vegetable exporters have engaged in tighter vertical coordination with upstream suppliers. In the green bean sector, increased vertical coordination includes contract farming schemes with more sophisticated contracts and tighter contract coordination, and a shift to vertical integration. Large exporters increasingly engage in vertically integrated estate production as a strategy to comply with increasing food standards, and the growth in bean exports is related mainly to the expansion of production on these large-scale plantations. As a result, the share of produce that is sourced from smallholders on a contract basis decreased from 95% in

1999 to 52% in 2005. The tomato export supply chain in the Senegal River Delta area is an extreme case of vertical coordination; the dominant exporter relies completely on its own integrated agroindustrial production.

The described changes in the structure and organization of vegetable export chains, especially the ongoing shift from smallholder contract farming to large-scale agroindustrial production, have major implications for rural households in the producing areas. This is further described and analysed in the following two sections.

Rural Households in Export Supply Chains

To study the extent of involvement of rural households in the vegetable export supply chains and the welfare implications thereof, we implemented a household survey in both case study regions. These surveys covered 599 households in five rural communities: 300 households in three rural communities in les Niayes, from where the large majority of exported green beans originate, and 299 households in two rural communities in the Senegal River Delta region, from where practically the entire volume of export tomatoes originates. More details on sample selection and data collection can be found in Maertens *et al.* (2008) and Maertens and Swinnen (2009).

In this section we first describe the extent of local household involvement in vegetable export sectors for the two case studies, and then discuss the characteristics of households that are involved in export supply chains through product and labour markets. In the following section we give more details on the income and poverty effects.

Household participation

As exports of green beans and tomatoes have increased sharply in Senegal over the past decade, the participation of local rural households in green bean and tomato export production also increased sharply. In the sampled villages in les Niayes region, the participation of households in green bean export production increased from fewer than 15% of households in 1996 to 40% in 2005 (Fig. 8.3). As a result of changes in the sourcing strategies of large exporting companies in response to higher standards, the share of local households supplying green beans to the export industry in contract farming arrangements dropped sharply after 2000. In our sample, 72% of contracted bean farmers lost their contract with the

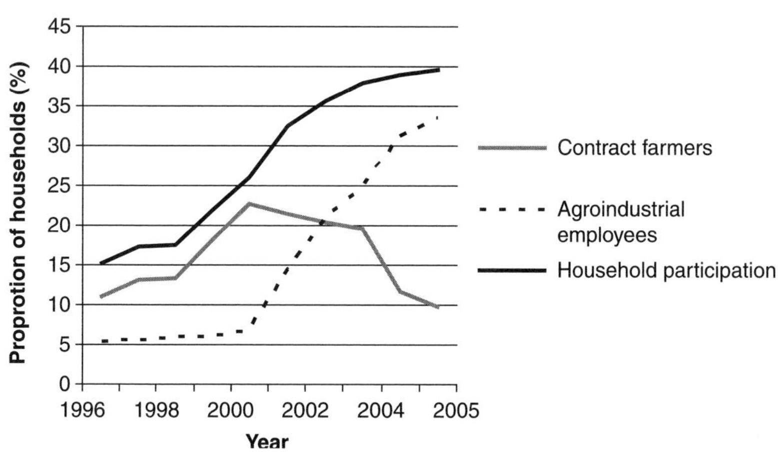

Fig. 8.3. Participation of local households as contracted suppliers and agroindustrial employees in green bean export production, 1996–2005 (les Niayes area; from Maertens and Swinnen, 2009).

export industry in the period 2000–2005, falling back on production for local markets and other income-generating activities. On the other hand, the share of local households with jobs in the fields and in the processing plants of agroexporting companies increased tremendously, from 6% in 2000 to almost 35% in 2005. Almost half of the previously contracted farmers who lost their production contract with the bean export industry engaged in this employment. Most employees are recruited on a day-to-day basis during the export season. The jobs consist of work in the fields, such as planting, fertilizer and pesticide application, harvesting and work in the processing plants, such as the sorting, packaging and labelling of produce. It was estimated that, in 2005, the green bean export sector contracted almost 1000 farmers for the supply of primary produce and employed some 12,000 workers. It must be noted that households in this area generally only allocate part of their land and/or labour resources to these export-related production and employment activities, and that they primarily remain independent smallholder farm households producing for the local market.

The growth in tomato exports also created employment in the Senegal River Delta area: in 2006 the tomato export industry employed more than 3000 workers – both in the fields of the estate farm and in the processing unit. Eighty per cent of these were temporary seasonal workers or day labourers, and were recruited from neighbouring villages. For permanent jobs, more highly skilled labourers are hired. In the sampled villages, over one-third of households had members working as employees in the tomato export industry (Fig. 8.4), and this proportion increased from 14% in 2003 to 39% in 2006. The proportion was highest – almost 50% of households – in the community of Gandon, which includes villages in the immediate surroundings of the production sites and processing unit of the tomato-exporting company. In the adjacent community, Ross Béthio, and including more distant villages, the proportion of households employed in the tomato-exporting industry was about 30%.

Household characteristics

The distributional implications and potential poverty-reducing effect of expanding vegetable exports critically depend on which type of households are included as either contracted suppliers or agroindustrial employees in the export chains. In

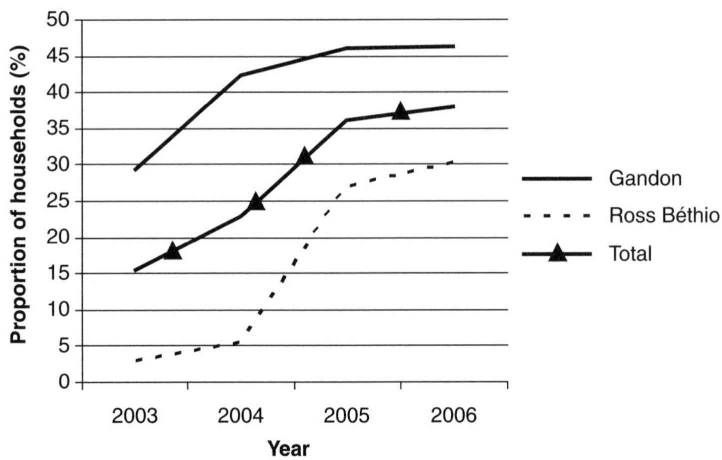

Fig. 8.4. Participation of local households as agroindustrial employees in tomato export production, 2003–2006 (Senegal River Delta area; from Maertens *et al.*, 2008).

Table 8.1, we compare the characteristics of three groups of household in the les Niayes area: (i) those working as employees in the green bean export industry; (ii) those supplying green beans to the export industry as contracted farmers; and (iii) those who do not participate at all in green bean export production. From this comparison, it is clear that those households working as agroindustrial employees are larger households, with more workers and less land, livestock and fewer non-land assets; they are relatively poorer than the other households. Households engaging in green bean contracting are relatively large households that are relatively better off in terms of land, livestock and non-land assets. At the same time, no difference could be found in the level of education across the different types of household.

With respect to tomato exports in the Senegal River Delta region, the same observations are made (Table 8.2). Households employed in the tomato export industry tend to be larger households with more workers, and relatively poorer households with less land, livestock and fewer non-land assets, while no difference in the education level was observed across the households.

These results indicate that participation in contract farming with exporting firms is biased towards relatively larger farms and better-off households, while participation in agroindustrial employment in the export vegetable sector is directed towards relatively poorer households. In addition, wage employment in the export agroindustry does not require certain levels of education, which contrasts with some

Table 8.1. Comparison of household characteristics (les Niayes area; from Maertens and Swinnen, 2009).

	Total sample	Non-participants	Employees in the green bean agroindustry	Green bean contract farmers
Households in sample (*n*)	300	158	109	59
Human and social capital				
Age of household head (years)	54.0	53.0	56.0	53.0
Labourers (*n*)[a]	6.9	6.4	7.7	7.7
Dependency ratio[b]	0.568	0.571	0.566	0.527
Female-headed households (%)	3.0	3.3	2.8	0.0
Household head with primary education (%)	17.6	16.5	18.8	19.4
Ethnicity (non-Wolof, %)[c]	27.0	31.0	17.0	32.0
Membership of an organization (%)	58.0	54.0	62.0	77.0
Physical capital				
Per adult-equivalent landholdings (ha)[d]	0.83	0.84	0.78	1.03
Livestock units[e]	2.64	2.87	1.84	4.14
Non-land assets (1000 FCFA)	270.7	320.9	176.9	308.8

[a] Labour is measured as the number of persons older than 15 and able to work.
[b] Dependency ratio is calculated as the number of dependants (children below the age of 15, students and those unable to work) over the total household size.
[c] In Senegal, non-Wolof households refer to ethnic minorities.
[d] Per adult-equivalent measures are calculated using the modified OECD adult equivalence scales. A weight of 1.0 is assigned to the first adult, 0.5 for all other adults and 0.3 for children.
[e] One livestock unit equals 1.0 cow/horse, 0.8 donkey or 0.2 sheep/goat/pig.
FCFA, Franc Communauté Financière Africaine.

Table 8.2. Comparison of household characteristics (Senegal River Delta area; from Maertens *et al.*, 2008).

	Total sample	Non-participants	Employees in the tomato agroindustry
Households in sample (*n*)	299	128	171
Human and social capital			
Age of household head (years)	56.0	54.0	59.0
Labourers (*n*)[a]	6.6	4.8	9.3
Dependency ratio[b]	0.511	0.528	0.484
Female-headed households (%)	3.8	2.7	5.5
Household head with primary education (%)	26.7	28.3	24.3
Ethnicity (non-Wolof, %)[c]	55.0	57.0	50.0
Membership of an organization (%)	50.0	48.0	53.0
Physical capital (per adult-equivalent)[d]			
Landholdings (ha)	0.18	0.2	0.13
Livestock units[e]	0.79	0.86	0.68
Non-land assets (1000 FCFA)	15.5	17.4	12.7

[a] Labour is measured as the number of persons who are older than 15 and able to work.
[b] Dependency ratio is calculated as the number of dependants (children below the age of 15, students and those unable to work) over the total household size.
[c] In Senegal, non-Wolof households refer to ethnic minorities.
[d] Per adult-equivalent measures are calculated using the modified OECD adult equivalence scales. A weight of 1.0 is assigned to the first adult, 0.5 for all other adults and 0.3 for children. Physical capital refers to the year 2001, before the tomato export agroindustry was established in this area.
[e] One livestock unit equals 1.0 cow/horse, 0.8 donkey or 0.2 sheep/goat/pig.
FCFA, Franc Communauté Financière Africaine.

earlier studies on rural off-farm employment (Dercon, 1998; Barrett *et al.*, 2001). These are important outcomes when trying to understand the overall welfare and poverty implications of increased vegetable exports and supply chain restructuring.

Welfare Implications of Increased Export Production

Vegetable exports in Senegal have had a key impact on the incomes of both contracted farmers and agroindustrial employees, and have contributed greatly to poverty reduction in the case study regions. In this section, we discuss some descriptive results of the income and poverty effects. These results are corroborated by more detailed econometric analyses in Maertens and Swinnen (2009) for the green bean case study and in Maertens *et al.* (2008) for the tomato case study.

Income effects

A comparison of mean incomes across households reveals that, in both case studies, the average incomes for households involved in vegetable export production are substantially larger than for other households (Figs 8.5 and 8.6). Figure 8.5 shows that green bean contract farmers, with larger land and capital endowments, have incomes that are more than three times higher than the incomes of non-participating households, while the average income of employees in the green bean export industry is 2.5 times higher than for non-participating households, despite the fact that these are households with less land and fewer capital endowments. In addition, an important share of income (about 30%) for agroindustrial employees comes from the wages earned through this employment.

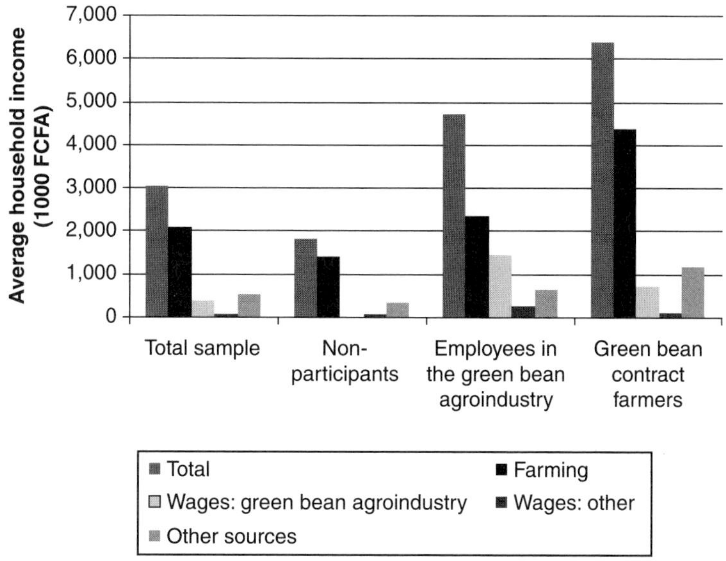

Fig. 8.5. Comparison of average household income from various sources (les Niayes area; from Maertens and Swinnen, 2009). FCFA, Franc Communauté Financière Africaine.

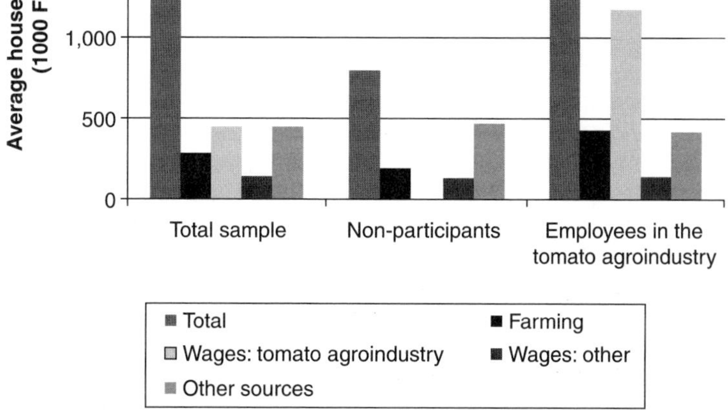

Fig. 8.6. Comparison of household income from different sources (Senegal River Delta area; from Maertens *et al.*, 2008). FCFA, Franc Communauté Financière Africaine.

With regard to the case study on tomato exports in the Senegal River Delta area, the differences in income are similar: employees in the tomato agroindustry have an average income that is 2.7 times higher than that for other households (Fig. 8.6). However, these households derive slightly more than half of their income from the wages earned in this sector.

In summary, we find not only that vegetable contract farmers with more labour, land and capital endowments have the highest incomes, but also that employees in the vegetable export industries have incomes that are substantially higher than for other households, despite their being relatively poorer households with less land and fewer capital endowments.

Poverty reduction

These income effects translate into important poverty reduction effects, as revealed by the proportion of households living on per adult-equivalent incomes that fall below the national rural poverty lines. The overall incidence of poverty in both research areas was estimated to be 42%, which is considerably below the national rural poverty rate of 58%. We found both poverty and extreme poverty to be substantially lower among households participating in vegetable export compared with non-participating households (Figs 8.7 and 8.8). Among contracted green bean farmers, poverty was only 13% and extreme poverty only 1%. Among households employed in the green bean export industry, the incidence of poverty was 40% and of extreme poverty, 5%, which is substantially lower than the rate of poverty (46%) and extreme poverty (17%) among non-participating households (Fig. 8.7). Similar results were found in the second case study on tomato exports: 35% of households employed in the tomato export industry were poor and 6% extremely poor, while the figures were 46% and 18%, respectively, for non-participating households (Fig. 8.8).

The large income effects and poverty figures suggest that the vegetable export industry creates possibilities for households with lower land and non-land assets to escape from poverty through the creation of employment opportunities for low-skilled labour.

Summary and Conclusions

The main conclusion from the case studies described in this chapter is that the sharp growth in vegetable exports from Senegal has made an important contribution to improving rural incomes and reducing rural poverty in the country. The case of Senegal documents that, despite increasing consolidation in global food supply chains,

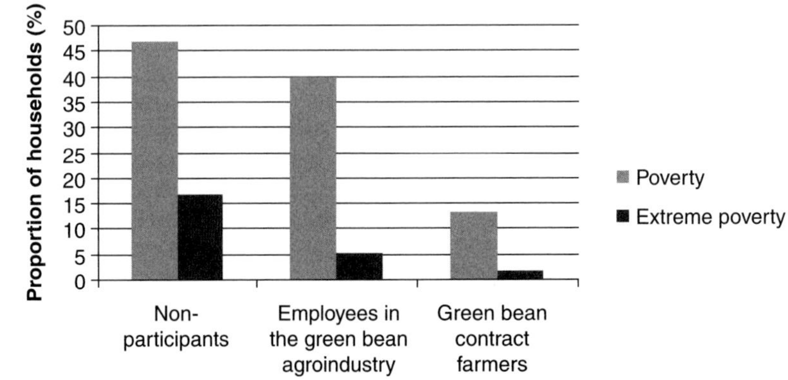

Fig. 8.7. Comparison of poverty levels across households (les Niayes area; from Maertens and Swinnen, 2009).

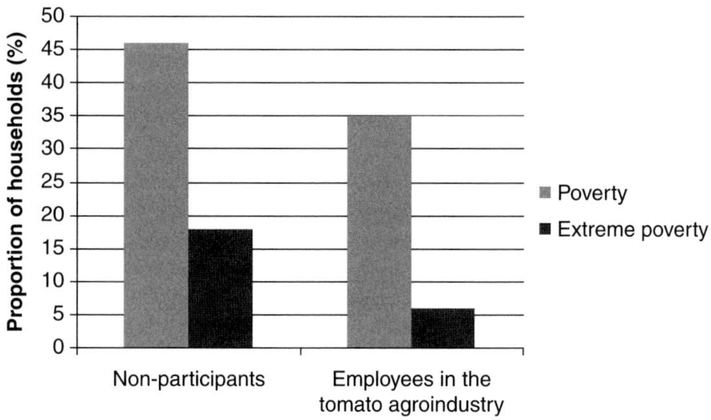

Fig. 8.8. Comparison of poverty levels across households (Senegal River Delta area; from Maertens *et al.*, 2008).

the increasing dominance of multinational holdings in these chains, the increasing standards for fresh produce imports in high-income markets and the expansion of both vegetable exports and high-value horticulture crops in general should remain an important agricultural and pro-poor development strategy in poor countries. Although the benefits from vegetable export growth in Senegal are concentrated in specific rural communities, predominantly in les Niayes and the Senegal River Delta regions, and not yet shared equally all over the country, there is scope for ongoing investments in other regions and communities of the country to result in expanded income benefits and poverty-reduction impacts. Likewise, there might be scope for pro-poor rural development through the development and expansion of high-value and high-standard export supply chains throughout sub-Saharan Africa.

A second key point made in this chapter is that the main poverty-reducing effects of vegetable export expansion come through labour markets and the creation of employment accessible to the poorest rural households, rather than through product market effects and the inclusion of smallholder producers in vegetable export chains through contract farming mechanisms. This challenges the implicit assumption underlying many studies that export supply chains need to integrate smallholder suppliers if export expansion is to benefit rural incomes and poverty reduction. In fact, the growth in high-value export production, even if realized on large-scale estate farms, should be seen as a source of employment for the rural poor and not merely interpreted as the exclusion of smallholder producers. This calls for a broader policy view and an expansion of the focus of policy makers, development agencies and donors on smallholder inclusion in lucrative horticulture, and also for other high-value export supply chains to work towards incorporating labour market issues.

Further research

Our findings, i.e. that high-value export sectors in developing countries benefit rural poverty reduction mainly through labour market effects, remain specific to the case studies we have described in this chapter. Other studies have come to different conclusions on the exclusion of smallholder producers and poverty impacts of high-value trade, but it is not yet well understood how local circumstances and supply chain factors drive these different outcomes and further comparative empirical research would be needed to resolve these issues. A conceptualization of different effects and different supply chain structures could contribute to a better understanding of the factors driving diverging outcomes.

In addition, the dynamic and long-term effects of high-value trade remain unclear. It is quite likely that the initial effects differ greatly from effects in the long term. With further increase in public regulations and the increased use of private standards the organization and structure of supply chains will continue to change, with implications for local rural households. Learning processes by companies, farmers and workers might induce further changes. Many questions remain on the sustainability of the observed income and poverty effects of high-value trade.

Finally, sustainability entails an environmental component. While the socio-economic implications of high-value trade and supply chain upgrading have been described and analysed in a large number of case studies – albeit often not comparable – the environmental implications remain relatively under-researched. Environmental outcomes of supply chain upgrading and high-value trade are likely to become important as standards, especially private standards, increasingly focus on environmental aspects of food production and trade.

Endnotes

[1]The increase and proliferation of public and private food standards and the implications for food supply chains are described, for example, in Farina and Reardon (2000), Jaffee and Henson (2005) and Maertens and Swinnen (2008).

[2]See, for example, Weatherspoon and Reardon (2003) and Colen *et al.* (2009) for a discussion on the increase of foreign direct investment in developing countries' agrifood sectors, especially in the food distribution sectors; and Dolan and Humphrey (2000) and Dries and Swinnen (2004) for a discussion on the implications of foreign direct investment for supply chains.

[3]For decades, Senegal has been dependent on groundnuts as the main agricultural export commodity, providing up to 80% of the total agricultural export revenue in the 1960s and 1970s, and leading to a collapse in export revenues and a major crisis in the Senegal groundnut sector as a result of sharply falling world groundnut prices in the 1980s (Badiane, 2001).

[4]In 2006, Kenya ranked first as the SSA exporter of green beans, exporting almost 60,000 metric tons, and Senegal second with 6500 metric tons. In the same year, Senegal was the number one exporter of tomatoes with almost 7000 metric tons, followed by South Africa with 6824 metric tons (FAOSTAT, 2009).

[5]In 2006 the EU imported 6000 metric tons of tomatoes from Senegal, which represented 1.86% of the EU's total external supply of tomatoes. In the same year, the EU imported 7000 metric tons of green beans from Senegal, which represented almost 3% of the EU's total external supply (Eurostat, 2009). It must be noted that this is a small (yet rapidly growing) share of overall EU imports and that Senegal could be susceptible to potential changes in EU demand.

[6]For example, the marketing standards for tomatoes specify the minimal requirements (tomatoes must be clean, pest-free, fresh, intact ...) and the classification of tomatoes into different quality classes according to their size, colour, uniformity and tolerance to deviation from quality and size requirements. Similar marketing standards exist for green beans.

[7]These organizations are SEPAS (Syndicat des Exportateurs des Produits Agricoles) and ONAPES (Organisation National des Producteurs Exportateurs de Fruits et Légumes de Sénégal).

References

Aksoy, M.A. and Beghin, J.C. (2005) *Global Agricultural Trade and Developing Countries*. The World Bank, Washington, DC.

Anderson, K. and Martin, W. (2005) Agricultural trade reform and the Doha Development Agenda. *World Economy* 28, 1301–1327.

Badiane, C. (2001) Senegal's trade in groundnuts: Economic, social and environmental implications. TED Case Studies 646, Trade Environment Database Project, American University, Washington, DC.

Barrett, C.B., Reardon, T. and Webb, P. (2001) Non-farm income diversification and household livelihood strategies in rural Africa: Concepts, dynamics and policy implications. *Food Policy* 26, 315–331.

Carter, M.R, Barham, B.L. and Mesbah, D. (1996) Agricultural export booms and the rural poor in Chile, Guatemala and Paraguay. *Latin American Research Review* 31(1), 33–65.

Colen, L., Maertens, M. and Swinnen, J.F.M. (2009) Foreign direct investment as an engine for economic growth and human development: A review of the arguments and empirical evidence. *Human Rights and International Legal Discourse* 3(2), 177–227.

Danielou, M. and Ravry, C. (2005) The rise of Ghana's pineapple industry. Africa Region Working Paper Series 93, The World Bank, Africa Region, Washington, DC.

Dercon, S. (1998) Wealth, risk and activity choice: Cattle in western Tanzania. *Journal of Development Economics* 55, 1–42.

Diop, N. and Jaffee, S. (2005) Fruits and vegetables: Global trade and competition in fresh and processed product markets. In: Aksoy, A.M. and Beghin, J.C. (eds) *Global Agricultural Trade and Developing Countries*. The World Bank, Washington, DC, pp. 237–257.

Dolan, C. and Humphrey, J. (2000) Governance and trade in fresh vegetables: The impact of UK supermarkets on the African horticulture industry. *Journal of Development Studies* 37(2), 147–176.

Dries, L. and Swinnen, J.F.M. (2004) Foreign direct investment, vertical integration and local suppliers: Evidence from the Polish dairy sector. *World Development* 32(9), 1525–1544.

Eurostat (2009) Eurostat statistical website, http://epp.eurostat.cec.eu.int (accessed 2009).

FAOSTAT (2009) www.faostat.fao.org (accessed 2009).

Farina, E.M.M.Q. and Reardon, T. (2000) Agrifood grades and standards in the extended Mercosur: Their role in the changing agrifood system. *American Journal of Agricultural Economics* 82(5), 1170–1176.

Garcia Martinez, M. and Poole, P. (2004) The development of private fresh produce safety standards: Implications for developing and Mediterranean exporting countries. *Food Policy* 29(3), 229–255.

Gibbon, P. (2003) Value-chain governance, public regulation and entry barriers in global fresh fruit and vegetable chains into the EU. *Development Policy Review* 21(5/6), 615–625.

GlobalGAP (2009) www.globalgap.org (accessed 2009).

Graffham, G. and MacGregor, A. (2007) Impact of EurepGAP on small-scale vegetable growers in Zambia. *Fresh Insights* 5, IIED-DFID-NRI, London.

Graffham, G., Karehu, E. and MacGregor, J. (2007) Impact of EurepGAP on small-scale vegetable growers in Kenya. *Fresh Insights* 6, IIED-DFID-NRI, London.

Gulati, A., Minot, N., Delgado, C. and Bora, S. (2007) Growth in high value agriculture in Asia and the emergence of vertical links with farmers. In: Swinnen, J.F.M. (ed.) *Global Supply Chains, Standards and the Poor*. CAB International, Wallingford, UK, pp. 91–108.

Henson, S. (2006) The role of public and private standards in regulating international food markets, Paper prepared for the *IATRC Summer Symposium*, Bonn, Germany.

Henson, S. and Humphrey, J. (2008) Understanding the complexities of private standards in global agrifood systems. Paper presented at the *International Workshop on 'Globalization, Global Governance and Private Standards'*, Leuven Centre for Global Governance Studies, 4–5 November 2008, Leuven, Belgium.

Humphrey, J., McCulloch, N. and Ota, M. (2004) The impact of European market changes on employment in the Kenyan horticulture sector. *Journal of International Development* 16(1), 63–80.

Jaffee, S. (2003) From challenge to opportunity: Transforming Kenya's fresh vegetable trade in the context of emerging food safety and other standards in Europe. Agricultural and Rural Development Discussion Paper, The World Bank, Washington, DC.

Jaffee, S. and Henson, S. (2005) Agrofood exports from developing countries: The challenges posed by standards. In: Aksoy, A.M. and Beghin, J. (eds) *Global Agricultural Trade and Developing Countries*. The World Bank, Washington, DC, pp. 91–114.

Maertens, M. and Swinnen, J.F.M. (2008) Standards as barriers and catalysts for trade, growth and poverty reduction. *Journal of International Agricultural Trade and Development* 4(1), 47–61.

Maertens, M. and Swinnen, J.F.M. (2009) Trade, standards and poverty: Evidence from Senegal. *World Development* 37(1), 161–178.

Maertens, M., Colen, L. and Swinnen, J.F.M. (2008) Globalization and poverty in Senegal: A worst case scenario? LICOS discussion paper 217/2008, Leuven, Belgium (www.econ.kuleuven.be/LICOS/DP/dp.htm).

McCullough, N. and Ota, M. (2002) Export horticulture and poverty in Kenya. IDS Working Paper 174, Institute for Development Studies, Brighton, UK.

Minot, N. and Ngigi, M. (2004) Are horticultural exports a replicable success story? Evidence from Kenya and Côte d'Ivoire. EPTD/MTID discussion paper, IFPRI, Washington, DC.

Minten, B., Randrianarison, L. and Swinnen, J.F.M. (2009) Global retail chains and poor farmers: Evidence from Madagascar. *World Development* 37(11), 1728–1741.

Neven, D., Odera, M.M., Reardon, T. and Wang, H. (2009) Kenyan supermarkets, emerging middle-class horticultural farmers and employment impacts on the rural poor. *World Development* 37(12), 1802–1811.

Unnevehr, L.J. (2000) Food safety issues and fresh food product exports from LDCs. *Agricultural Economics* 23(3), 231–240.

Wang, H., Dong, X. and Rozelle, S. (2006) Producing and procuring horticulture crops with Chinese characteristics: The case of Northern China. Contributed paper at the *International Association of Agricultural Economists' Conference*, 12–18 August, Gold Coast, Australia.

Weatherspoon, D.D. and Reardon, T. (2003) The rise of supermarkets in Africa: Implications for agrifood systems and the rural poor. *Development Policy Review* 21(3), 333–356.

Weinberger, K. and Lumpkin, T.A. (2005) Horticulture for poverty alleviation: The unfunded revolution. AVRDC Working Paper 15, The World Vegetable Center, Shanhua, Taiwan.

9 Comparative Assessment of the Marketing Structure and Price Behaviour of Three Staple Vegetables in Lusaka, Zambia

David Tschirley, Munguzwe Hichaambwa and Mukwiti Mwiinga

Introduction

Rapidly growing urban populations and renewed growth in per capita incomes in sub-Saharan Africa (SSA) are creating major opportunities for local farmers by driving rapid growth in the domestic market demand for food. At the same time, these trends put enormous stress on the supply chains on which these farmers rely to respond to this increasing demand. Africa has the highest urban growth rate of any developing area, currently 3.7% per year, and this is projected to remain above 3% through to 2030. Urban populations will grow by about 170% over the next 30 years, far outstripping rural growth and pushing the urban population share above 50% (United Nations, 2007). Rising incomes multiply the impact of growing populations on marketing systems. After essentially no growth in per capita incomes during the 1990s, SSA achieved total per capita GDP growth of 15% from 2000 to 2006, higher than in Latin America; six SSA countries ranked among the top 30 worldwide in per capita GDP growth during this period. Together, these two trends could fuel increases in demand for marketed food of more than 5% per year, doubling marketed volumes in 12–14 years.

Currently, fresh produce marketing systems are the biggest users of the 'public' marketing infrastructure, and have been most affected by the widespread underinvestment in these systems. As formal public marketplaces have been overwhelmed by rising volumes of commodities and numbers of traders, the informal marketing sector has exploded, raising major concerns about congestion and hygiene among city planners. The chaotic state of many of these markets means that few, if any, comprehensive programmes have been put in place actively to link farmers to them.

Since about 2000, great attention has been paid to the 'supermarket revolution' in developing countries.[1] The term may be appropriate to describe what has happened in some countries of Latin America and East Asia, as income growth and openness to foreign direct investment drove widespread growth in the market share of supermarkets, with important implications for farmers, traditional traders, consumers and even municipal finances. Among many funding agencies this phenomenon was viewed with concern for its potential to exclude small farmers, but also with hope: as a way to deal with the distressingly complex challenge of how to modernize the chaotic and increasingly inadequate supply chains serving urban areas.

Nevertheless, after the initial burst of enthusiasm through the middle part of this decade, there now exists a broad consensus that this phenomenon is likely to proceed much more slowly than once thought in SSA (Humphrey, 2006; Traill, 2006; Minten, 2008; Tschirley *et al.*, 2009). This is due in large measure to much lower incomes: per capita purchasing power parity income across *all* of Africa was only 40% that of the 15 *poorest* Latin American countries in 2006, and one-quarter that of Latin America as a whole (Maddison, 2009). This anticipated slow growth includes fresh produce supply chains, where both the promise and the perils of supermarket expansion have received the greatest attention. Throughout nearly the entire continent, the so-called 'traditional' marketing sector – open-air markets, dispersed informal vendors and traditional shops – is expected to play a dominant role in fresh produce marketing for the next two to three decades.

If correct, this emerging consensus has profound policy implications. It suggests that private investment in modern, integrated supply chains cannot be relied upon to solve the multitude of problems – logistical inefficiencies, deteriorating infrastructure, high product wastage, urban congestion and food safety concerns – that increasingly plague traditional production and marketing systems over a time frame acceptable to most policy makers and donors. In addition, Africa's high rate of urban population growth means that a rapidly rising share of the population will be subjected to these problems over time. Public engagement will be central to any improvement in these areas.

This public engagement must be based on a solid understanding of these systems and on new approaches to public–private sector collaboration to improve them. Nonetheless, while there is wide appreciation of the poor performance of many of these systems, little comparative knowledge has been generated to quantify the range of observed performance. We begin filling this gap by examining the marketing structure and price behaviour of tomatoes, rape and onions serving Lusaka, Zambia. These crops are perhaps the three main 'staple vegetables' in East and southern Africa, eaten on a daily basis by most people; in Lusaka, they account for more than half of all vegetable consumption. They show great variability in production and perishability characteristics and, therefore, are likely to illustrate a substantial portion of the range of marketing structures and price behaviours seen in these traditional systems.

The next section discusses data and methods. The third section (Marketing Structure) generates a quantitative description of the system serving Lusaka, including the geographical distribution of marketed production reaching the city, the structure of the city's fresh produce marketing channels and the role of brokers in the wholesale trade. The fourth section (Price Behaviour) examines the price behaviour of the three crops, including seasonality, price predictability, estimates of short-term price flexibility and its implications, and the level and behaviour of wholesale–retail marketing margins. The fifth section (Summary and Conclusions) includes initial thoughts regarding key challenges facing these systems and possible priorities for investment.

Data and Methods

Primary data for this study come from three sources. Michigan State University's (USA) Food Security Research Project (FSRP) has collaborated with the Zambia National Farmers' Union (ZNFU) since January 2007 to collect detailed information on prices and quantities of tomatoes, rape and onions in Lusaka's dominant wholesale market (Soweto). On Monday, Wednesday and Friday of each week market reporters randomly collect three prices paid by buyers (primarily small-scale retailers) for each product during each hour from 06.00 to 11.00 hours They collect basic information on all trucks entering the market with the three products, including time of arrival, quantity, geographical origin of the product and whether the seller will be working through a broker or selling directly. Retail prices are collected on these same three days in Chilenje Market, one of the city's open-air retail markets (hereafter referred to as Chilenje) and on Tuesdays and Thursdays in three supermarkets (Shoprite, Spar and Melisa). Chilenje is one of the main open-air retail markets in the city serving lower- and middle-income Zambians.[2] These data allow computation of total volumes and values flowing through Soweto and detailed assessment of seasonality, intra- and inter-day price variability and marketing margins. Building on their market knowledge and rapport with traders, market reporters interviewed traders during early 2010 on the destination of products flowing out of Soweto.

Wide variation in quality and a lack of formal grades and standards pose difficulties for the price analysis of fresh produce markets in most developing countries, and Zambia is no different in this regard. Nonetheless, analysis in this study is facilitated by the fact that FSRP/ZNFU have worked with traders to develop a (still informal) set of standards for a low, medium and high 'standard' quality for all three crops being monitored. All price collection focuses on 'medium standard' quality, which should reduce the artificial price variation in our data.

Additional primary data come from the FSRP Urban Consumption Survey (UCS). This survey interviewed over 1800 households in four urban centres in Zambia, including over 600 in Lusaka, over two rounds in August 2007 and February 2008. Detailed data on household food consumption and places of purchase allow us to estimate the total size of the Lusaka market for these three products, and the market share of various types of retail outlets (open-air markets, street vendors, supermarkets and others). In combination, Soweto monitoring data and the UCS provide most of the data for the detailed market maps in the following section. These data were complemented with a retail 'market-mapping' exercise to estimate the share of Soweto in retailer purchases of these three crops. This mapping exercise involved counts of the total number of fresh produce traders for each crop in the top ten retail markets of the city, along with Likert scale questions to a sample of traders on the relative volumes purchased in Soweto, in other wholesale markets and directly from farmers in production areas.

Marketing Structure

We summarized the results of the geographical distribution of marketed production entering Lusaka, and of the structure of the marketing channels for the three products, before turning to an assessment of the role of brokers in Soweto.

Lusaka's 'market sheds': the geographical distribution of marketed production

By 2000, Lusaka's population (about 1.1m) had doubled over the previous 20 years. We conceive the city's market shed as the geographical extent over which products flow into Lusaka. Figure 9.1 maps these market sheds, based on the districts that provided 80% of Lusaka's supply of each crop from mid-January 2007 through to mid-January 2009.[3] The geographical extent of these market sheds follows the perishability characteristics of the crops: the city draws nearly 60% of its onion supply from imports, with half of this coming from Johannesburg, about 1200 km away. The city draws on five nearby districts for 87% of its tomato supply, while it needs only the two closest districts to assemble 83% of its rape supply. Rough estimates of the mean distance to market are 44 km for rape, 69 km for tomatoes and 539 km for onions.[4] Very little of the supply reaching Soweto comes from peri-urban areas. Even for rape, the most perishable of the three, production areas in the two main supply districts, though close to Lusaka, are rural in nature and not peri-urban.

Within each district, smaller production areas were identified based on definitions used by farmers and traders selling in Soweto; typically these areas follow existing local boundaries, which could include one

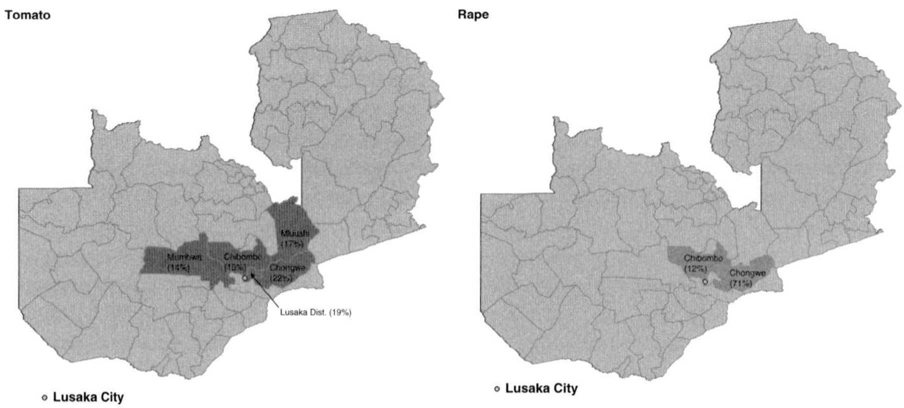

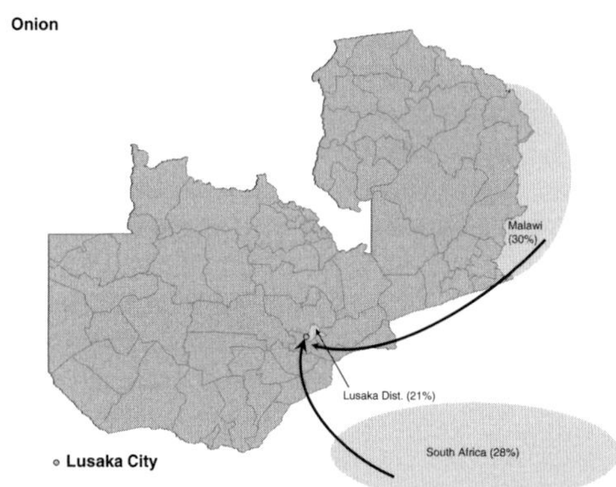

Weighted average distance to market

Crop	Km
Tomato	69
Rape	44
Onion	593

Straight line (air) distance from district town to central Lusaka (http://distancecalculator.globefeed.com/country_distance_calculator.asp)

Fig. 9.1. Shares of tomato, rape and onion supplied to Soweto, Lusaka, by district.

or several villages. This measure reinforces the finding that the production of rape is substantially more concentrated than that of tomatoes: 53% of the total rape supply to Lusaka comes from the top three areas, while only 28% of tomatoes comes from the top three (Table 9.1). Figures for onions are misleading in the sense that Malawi and Johannesburg were each classified as a single 'area', although each undoubtedly draws on many production areas within those countries for the supplies that reach Lusaka.

Table 9.1. Geographical concentration of marketed production flowing through Lusaka.

Concentration measure	Crop		
	Tomato	Rape	Onion
Total number of identified production 'areas'	115	93	42
Share of total supply from:			
Top 3 areas (%)	28	53	65
Top 5 areas (%)	39	59	75
Top 10 areas (%)	61	73	93

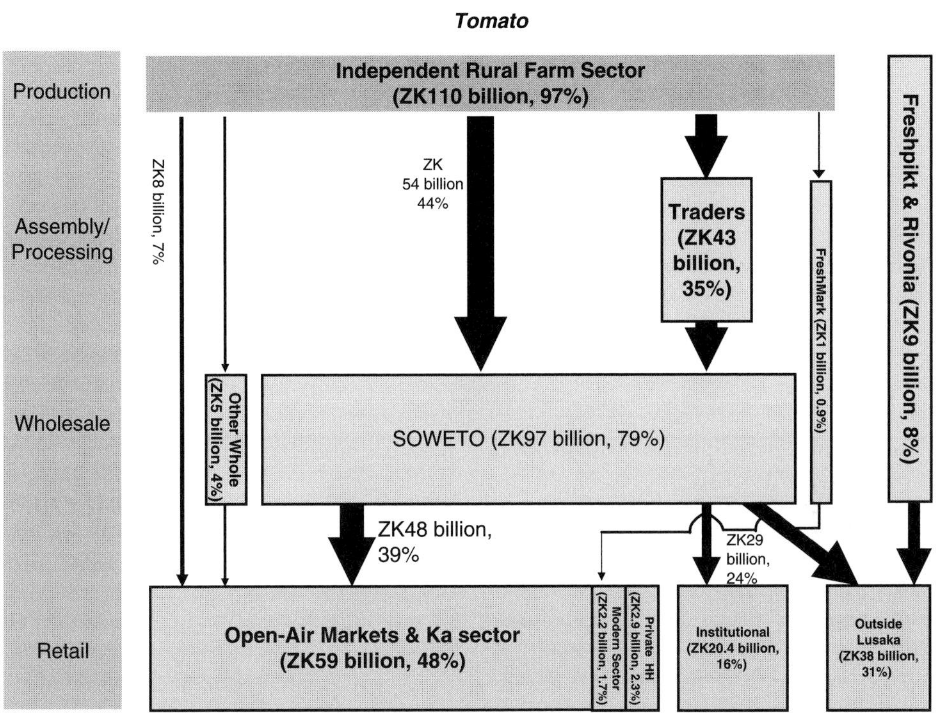

Fig. 9.2. Simplified channel map for tomato in Lusaka. ZK, Zambian Kwacha (currency). *Ka* is the diminutive in Njala; thus *kashop* is a small, rudimentary shop, *katable* is a small table on which a vendor sells his/her wares and *kantemba* is a small *ntemba* or kiosk. Rivonia is a suburb of Johannesburg.

Channel maps

We used data from the price and quantity collection system in Soweto, along with urban consumption survey results and interviews with sellers in Soweto, to construct channel maps for the three crops (Figs 9.2, 9.3 and 9.4). Key aspects of each map are summarized in Table 9.2.

Soweto is a sprawling retail market – by far the largest in Lusaka – that also serves as the dominant wholesale market in the city. It carries a wide range of food and non-food items, with over 1000 sellers of fresh

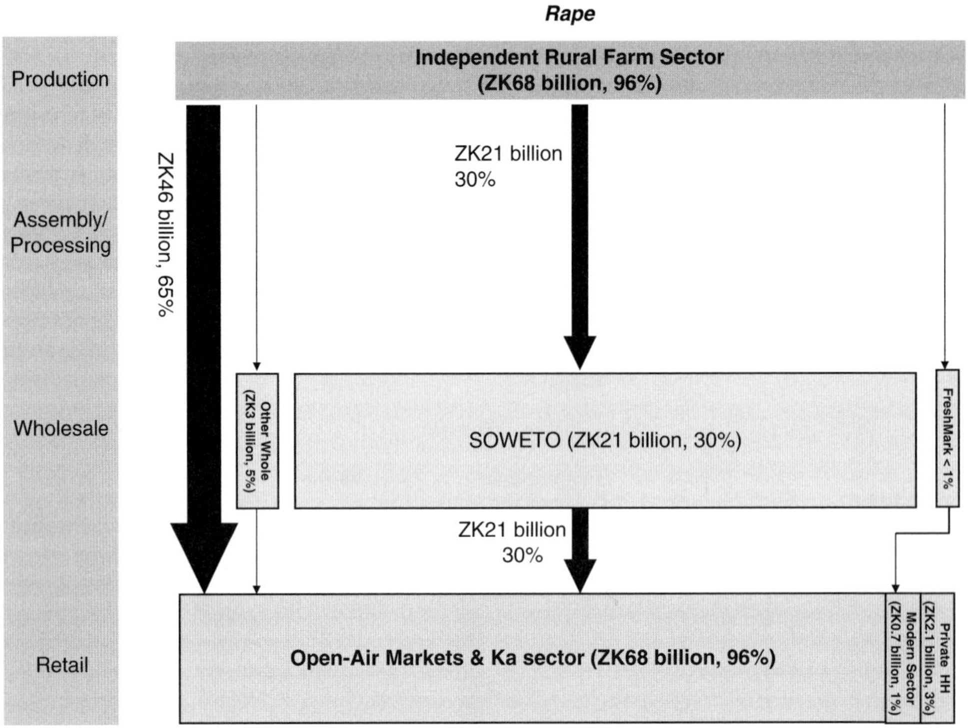

Fig. 9.3. Simplified channel map for rape in Lusaka. See Fig. 9.2 for explanation of abbreviations.

produce alone; more than 100 fresh produce brokers and wholesalers operate in the market. In 2007 and 2008, over 50,000 mt per year of tomato, rape and onion, valued at over US$13m, moved through the wholesale area of the market. Nevertheless, nearly all wholesaling takes place in an uncovered dirt field at one end of the market complex, with no dedicated entry and exit points, very limited storage capacity and no cold storage. The Urban Markets Development Programme, funded by the EU, has made substantial investments in several retail markets in the city, including Soweto, but has now ended without making any improvements in this wholesaling area.

Retail channels for fresh produce include open-air markets, the 'ka sector' (small vendors outside organized marketplaces that pursue sales by locating along busy pedestrian walkways and in residential neighbourhoods; see Fig. 9.2), modern supermarkets and private households producing in or near the city and selling to other households. The main supermarket chain is Shoprite Checkers, which invested in 17 stores across Zambia (five in Lusaka) in 1997; over the past 3–4 years, Spar (a Dutch-owned firm) has opened two outlets, while the local chain, Melissa, now has three outlets. Shoprite Checkers procures fresh produce locally through FreshMark, its wholesale partner.

Figures 9.2, 9.3 and 9.4 were initially compiled based on three independent sets of figures:

- The total reported value of in-home consumption by Lusaka households from February 2007 to January 2008, as estimated from the UCS. This survey allowed total consumption values to be broken down by the retail channel used by households (including their own production and that of other households).

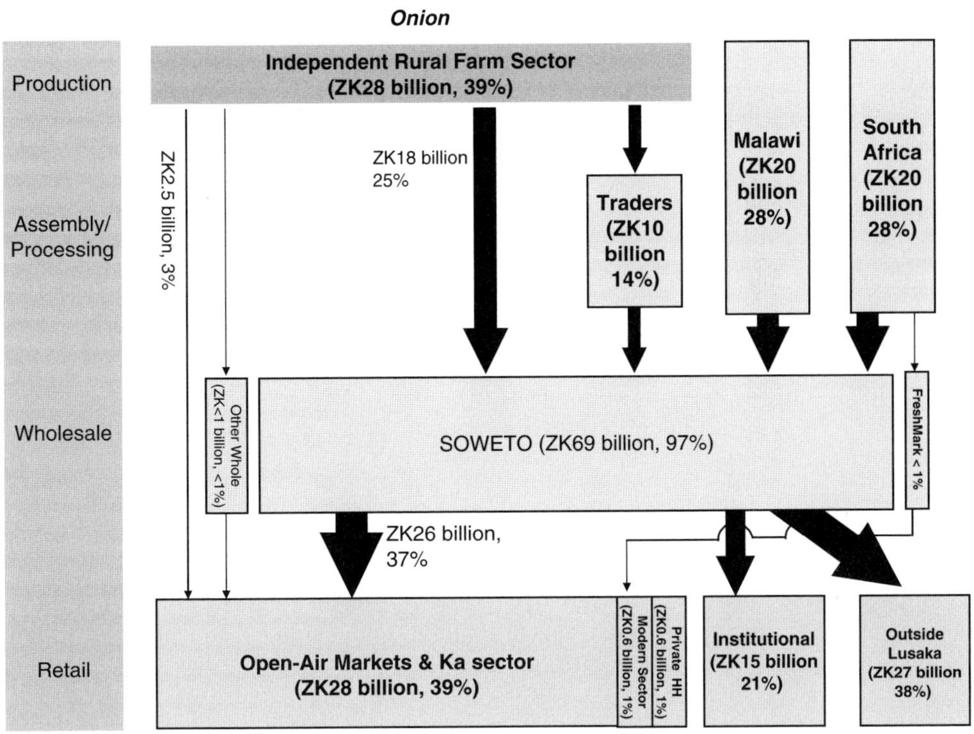

Fig. 9.4. Simplified channel map for onion in Lusaka. See Fig. 9.2 for explanation of abbreviations.

Table 9.2. Key structural characteristics of marketing channels in Lusaka, by crop.

	Crop		
Share[a]	Tomato	Rape	Onion
Marketed directly by farmers to retailers	0.07	0.65	0.03
Flowing first through wholesale markets	0.83	0.35	0.40
sold directly by farmers in wholesale markets	0.48	0.35	0.26
sold first to rural traders	0.35	0.00	0.14
Sold by farmers to 'modern wholesalers' and processors	0.09	0.00	<0.01
Imported	0.00	0.00	0.58
Purchased by consumers	0.52	1.00	0.42
in traditional markets and ka sector	0.48	0.96	0.39
in supermarket chains	0.02	0.01	<0.01
from urban producers	0.02	0.03	0.03
Sold to institutional buyers within Lusaka	0.16	0.00	0.21
Shipped outside Lusaka	0.31	0.00	0.38

[a] Denominator is total value, at weighted average retail prices, of volumes flowing through Lusaka (wholesale and retail markets, modern wholesalers and processors).

- The annualized total volume of product flowing through Soweto from mid-January 2007 to mid-January 2009. This figure is converted to retail value – to be comparable to figures from UCS – using monthly retail prices weighted by monthly Soweto volumes.
- Volumes arriving into Soweto by supply area, and by whether the volumes are first purchased by traders in rural areas or are taken directly to the market by farmers.

Supplies shipped from Soweto out of Lusaka or sold to institutional buyers are based on key informant interviews with brokers and wholesalers in Soweto, while volumes arriving in retail markets from other wholesale markets or directly from farmers are sourced from the market-mapping exercise discussed in the data section. All percentages are based on our estimate of the total value at retail prices of all products flowing through the city; this is the sum of figures in all boxes located in the retail section of each map.

Several patterns emerge. First, marketing channels are short. Rape is the extreme case, with nearly two-thirds being sold directly to retailers by farmers. Tomato shows much less direct farmer-retailer marketing (7%), but nearly half of all its value is taken directly to wholesale markets by farmers. Onion, due to the large market shed, shows the most intermediation, with 72% of all product brought to Lusaka by traders, not farmers (58% imported, plus 14% originating in Zambia). A second pattern is that Soweto serves as a major redistribution market for onions and tomatoes. Over 30% of tomatoes reaching Soweto are exported out of Lusaka to Livingstone in the south, and to the Copperbelt and Democratic Republic of Congo (DRC) to the north. Thirdly, the role of the 'modern' marketing sector is very small. Of the total value of fresh produce purchased by consumers in Lusaka, 92% of tomatoes and onions and 96% of rape are purchased in open-air retail markets or the 'ka sector'.

These exceptionally high shares held by the traditional marketing system prevail more than a decade after Shoprite first invested in Lusaka with 17 stores nationwide, and after two Spar outlets have opened in the city. FreshMark – Shoprite's wholesale division – holds less than a 1% market share in all three products, while processors are important only in tomatoes, with a market share of around 8%; this share is held primarily by Freshpikt, which produces canned tomatoes primarily for the export market.

Fourthly, regional trade dominates the onion system and is important in tomatoes. Over half of the onions reaching Lusaka are imported, and an important share of the 38% that is shipped from Soweto outside Lusaka probably goes to the DRC. Thus, Zambia is probably a net importer of onions, though data for other areas of the country to confirm this could not be found. We find no evidence of imports of tomatoes, but clear indications from brokers and wholesalers in Soweto that significant volumes – probably more than half of the 31% that is shipped out of Lusaka – are exported to the DRC. Zambia is, thus, most likely a net tomato exporter. Only in rape does regional trade play no role.

Finally, evidence of the role of urban horticultural production is mixed. The UCS shows that only 2% of tomatoes and 3% of rape and onions are purchased from other urban households. Independently, the data on volumes entering Soweto show all important supply areas for all three crops to be at least several kilometres outside residential areas of the city. However, 65% of all rape reaching retail markets comes directly from farms (Fig. 9.3), with small-scale retailers typically going to those farms and buying entire plots of rape. These farms are probably close to the city and some may reasonably be considered urban. More information is needed on this aspect of the trade to establish more clearly the importance of urban production of green leafy vegetables. For tomatoes and onions, however, urban agriculture clearly plays only a small role.

The role of brokers

Brokers – agents who arrange sales without taking ownership of the commodity, earning their money on a commission – are a common and frequently controversial presence in the wholesale markets of East and southern Africa.[5] Farmers in Zambia have mixed opinions of brokers. During a rapid appraisal of Zambia's domestic horticultural chain in 2006 (Hichaambwa and Tschirley, 2006), some farmers intimated that they are forced to sell through brokers who threaten to steal their products if they try to sell on their own. However, a group of fresh vegetable farmers, who regularly supply Soweto (and to a lesser extent other wholesale markets), believed that these brokers provide some level of service. Although brokers charged about 10% commission on sales, farmers developed relationships with them over time that provided greater security and sales opportunities for their product, but even this group lodged a common complaint – that brokers sometimes add price markups that they keep for themselves without the farmers' knowledge, in addition to the commission. Those farmers who do not know these agents may be at a higher risk for these problems.

Brokering services can improve market efficiency by economizing on search effort (Gabre-Madhin, 2001). Because brokers do not have to put time and effort into managing the substantial price risk present in fresh produce markets, they can develop great expertise in gathering information on buyers and sellers and bringing them together to effect transactions. In this way, an efficient and competitive set of brokers can match supply with demand at a lower cost than if all sellers and buyers engaged in 'bilateral search' – each conducting their own search. Largely for this reason, all produce moving through South Africa's system of modern wholesale must legally be sold through brokers. The actual performance of a brokering system can be reduced if brokers do not behave competitively or if they are able to impact information – strategically withhold it from sellers and buyers – on supply and demand conditions or on the commissions they are charging. Performance can also be reduced if bilateral search costs are low but buyers and sellers are prevented, either by law or by collusive behaviour among brokers, from conducting their own search and negotiating their own transactions.

Formally testing for the efficiency of a brokering system requires transaction level data on the search costs of buyers and sellers and on the commissions they have paid, and the estimates of the opportunity cost of their time. Higher search costs and opportunity costs of time should increase the use of brokers, and commissions should be in line with the sum of these costs. In the absence of such data, we can nevertheless develop testable hypotheses and may be able to draw insights regarding the efficiency of this process in Soweto.

First, in the absence of legal mechanisms either requiring or precluding brokering, and under the assumption that search costs and opportunity costs of time differ across buyers and sellers, we expect to find a mix of brokered sales and direct sales. Secondly, due to the absence of a cold chain, we expect that the more perishable items are more likely to be sold through brokers, since failing to find a buyer early in the day can result in major financial losses for the seller. This suggests that rape should have the highest rate of brokerage, and onions the lowest. Thirdly, assembly traders selling in the market should be less likely than farmers to use brokers, since traders have more frequent exposure to the market, more opportunity to develop relationships with retail buyers and so should have lower search costs.

Our fourth and final hypothesis is that, if brokerage services are offered competitively, sellers with large quantities are more likely to do so through a broker. We base this hypothesis on several factors. First, large sellers face a higher risk of not selling all of their product and may have higher opportunity costs of time, both of which favour brokering. Secondly, we suspect that such sellers are generally better connected and more knowledgeable of the market and its participants. If so, such sellers will have lower search costs (thus, reducing the likelihood of brokering) and will be less likely to be manoeuvred into a brokered transaction against their will by non-competitive brokers. A finding that large sellers are more likely to sell through brokers may, therefore, be evidence that, at least in their cases, brokers are offering a valued service to sellers. At the same time, such a result could suggest that smaller sellers who may benefit most from access to efficient brokering services (due to high search costs) are not able to gain that access, because brokers focus their efforts on larger sellers.

Other factors that may affect the likelihood of a transaction being brokered, although we cannot form clear hypotheses regarding the direction of effect, are:

- Nearly all brokers in Soweto are male. Might they be more able to foreclose direct sales opportunities among female sellers, increasing the likelihood of brokered sales? On the other hand, females may value their time differently from males, resulting in differing decisions even in the absence of manipulative behaviour by brokers. More conceptual and empirical work is needed to form clear expectations in this regard; here we can only test for the presence and direction of any effect.
- Total volumes arriving on the market will also have an ambiguous effect on brokering. Low volumes will be associated with fewer sellers, suggesting lower search costs and reduced demand for brokering services. However, because fluctuations in daily volumes arriving on the market are unpredictable (see section below, Price Behaviour), the total number of brokers operating in the market on a given day may show little flexibility, or at least little correlation with total volumes. This means that the supply of brokering services *per unit volume arriving on the market* may rise substantially when volumes unexpectedly fall. If this is the case, then the brokers' commission could fall and the likelihood of a brokered transaction could rise.

Tabular analysis confirms our first three expectations (Table 9.3). With the exception of rape, in which essentially all sales are through brokers, we see a mix of brokered and unbrokered transactions; onions, as expected, show the lowest rate of brokered transactions; and, for both tomatoes and onions, farmers are more likely than traders to sell through brokers. The differences between farmers and traders selling tomatoes, however, are not large – 99% of tomato farmers sell through brokers while 89% of tomato traders do so – suggesting that perishability is more important than search costs in driving the seller's decision.

To test our fourth hypothesis, and to explore the relationship between brokering and (i) the gender of the seller; and (ii) the total quantity of product on the market, we conducted a probit analysis using data collected from 29 October 2007 to 16 November 2009.[6] Data are at the level of the individual sales lot, allowing exploration of all the hypotheses and relationships discussed above. We exclude rape because, essentially, all its transactions are conducted through brokers. By including mean daily quantities transacted in the market over the previous month in addition to monthly dummy variables, we controlled for two aspects of potential seasonality: total volumes transacted through the lagged quantity variable, and weather effects (heat and precipitation and their effect on product quality) through the monthly dummies. Because seasonal patterns vary between onions and tomatoes, we ran separate

Table 9.3. Share of transactions using brokers in Lusaka, by crop (%).

	Crop		
	Tomato	Rape	Onion
Total	88.6	100	11.6
Farmer as first seller	98.5	100	86.1
Trader as first seller	85.1	100	2.7

regressions for each, in addition to one pooled regression.[7]

Marginal effects are shown in Table 9.4. Consistent with the bivariate results, farmers in all three regressions are much more likely than rural assembly traders to sell through brokers, and the small number of wholesalers are less likely than rural assembly traders to use brokers. Women in all three regressions are more likely than male sellers to use brokers.

Seasonal patterns (monthly dummies) are not easy to interpret. The use of brokers for selling onions appears to be higher during the hot and wet months of October to March, but this pattern does not hold for tomatoes. These differential results may relate to the fact that tomato volumes in Soweto have shown a much less stable seasonal pattern since 2007 than have onions (see Price Behaviour, below).

Our most important result is that the seller's lot size in all three regressions is positively and significantly associated with the probability of selling through a broker. To probe the robustness of this result, we ran two other regressions, one limited to farmers and the other to traders; in both cases the seller's lot size remained positive and significant. As argued above, this result may suggest that, at least for these larger sellers, brokering services are adding value and involve a free choice by the sellers. The results are consistent with brokers being

Table 9.4. Marginal effects from probit analysis of determinants of selling through a broker.

	Tomato and onion	Tomato	Onion
Number of observations	21,592	13,642	7,211
$P >$ chi^2	0.000	0.000	0.000
Pseudo-R^2	0.797	0.660	0.615
Log likelihood	–2,952	–1,643	–1,058
Variable	Dy/dx	Dy/dx	Dy/dx
Continuous variables			
Log quantity being sold (seller's lot size)	0.082[c]	0.011[c]	0.021[c]
Log total quantity sold in market that day	–0.060[c]	0.008	–0.024[c]
Log mean daily quantity sold in market over past month	–0.042[c]	0.003	–0.023[c]
Trend (week)	–0.000	0.000	–0.000
0/1 variables			
Seller is a farmer (trader excluded)	0.661[c]	0.236[c]	0.756[c]
Seller is a wholesaler (trader excluded)	–0.568[c]	–0.563[c]	–0.020[a]
Seller is female (male excluded)	0.171[c]	0.030[c]	0.108[c]
Product being sold is onion (tomato excluded)	–0.715[c]	–	–
January (relative to June)	–0.002	–0.033[b]	0.036[a]
February (relative to June)	0.068[c]	0.006	0.032
March (relative to June)	0.090[c]	–0.004	0.111[c]
April (relative to June)	0.027	0.015[b]	–0.038[c]
May (relative to June)	–0.025	0.019[c]	–
July (relative to June)	–0.011	–0.018	0.022
August (relative to June)	0.016	–0.023[a]	0.046[a]
September (relative to June)	0.029	–0.015	0.033
October (relative to June)	0.143[c]	0.002	0.141[c]
November (relative to June)	0.170[c]	0.010	0.168[c]
December (relative to June)	0.102[c]	0.011	0.108[c]

Dependent variables: 1, sale made through broker; 0, sale made directly to retailer.
Significant at 10% level[a], 5% level[b], 1% level[c].

more interested in working with larger sellers and perhaps not making these services available to the smaller sellers who may be most in need of them; nevertheless, these are problems that small players face in many markets. On balance the results, although suggestive rather than definitive, paint a less negative picture than is typically held of brokering activities. Key concerns do remain regarding the lack of transparency in commissions, and more understanding is needed regarding the details of the brokering relationships and the level of free choice by sellers who transact through brokers.

Price Behaviour

Seasonality

We estimated seasonal price indices for each crop by running an ordinary least squares regression of daily average prices against monthly dummy variables, saving the predicted value and standardizing it by dividing by the mean price over the entire period of analysis. The seasonal index for a given month is the mean of these daily ratios for that month over the period of analysis. To assess the stability of the seasonal price pattern we plotted three lines: (i) the seasonal index; (ii) the index plus the median negative prediction error; and (iii) the index plus the median positive prediction error.[8] The resulting bound is a 50% confidence interval on the index; 25% of the daily price observations fell below the lower line and 25% above the upper line.

Based on weather patterns and interviews with brokers in Soweto, tomatoes were expected to show two seasonal supply peaks (low price periods), immediately after the wet season during April and May and again from August to October, with low supply and high prices in the rainy season from November to March and during the height of the dry season in June and July. For rape, we expected one long season of high supply and low prices during the dry months of May to October. Onions were expected to be similar to tomatoes, with low supplies and high prices from April to July and another short drop in supplies during December.

Actual price seasonality follows these expectations most closely for rape, followed by onions, with tomatoes diverging widely from expectations (Fig. 9.5). The 50% confidence intervals on the tomato seasonal index were the widest, especially during the anticipated high-price season of May to July. In fact, during only one month (August) did daily tomato prices have a better than 75% probability of lying above the mean price over the period of analysis when the mean index lay above it, or of lying below the mean price when the mean index lay below it.[9] Rape prices reached this threshold during five months (February and December above the mean; June, August and September below it), while onion prices reached the threshold during seven months (April to June above the mean; October to February below it). This instability in tomatoes' seasonal price pattern is linked to known problems with groundwater availability in key producing areas, and potentially to the difficulty in managing disease outbreaks. Rape's seasonal pattern is more stable because most of it is produced in low-lying areas near river banks during the dry season, minimizing the risk of water shortages and (by producing less during the rainy season) avoiding disease problems.

The relatively stable seasonal pattern for onions derives primarily from the ability of traders to draw from a wider geographic range to supply Lusaka. Note, however, that onions show the greatest (though stable) total seasonal variation among the three crops, with seasonal highs of 1.5 times the mean price and lows of 0.5; equivalent figures were 1.20 and 0.62 for tomatoes, and 1.65 and 0.70 for rape.

Price formation: daily and weekly price flexibility

Perishable commodities depend on some combination of cold storage, strong information flows between buyers and sellers, and

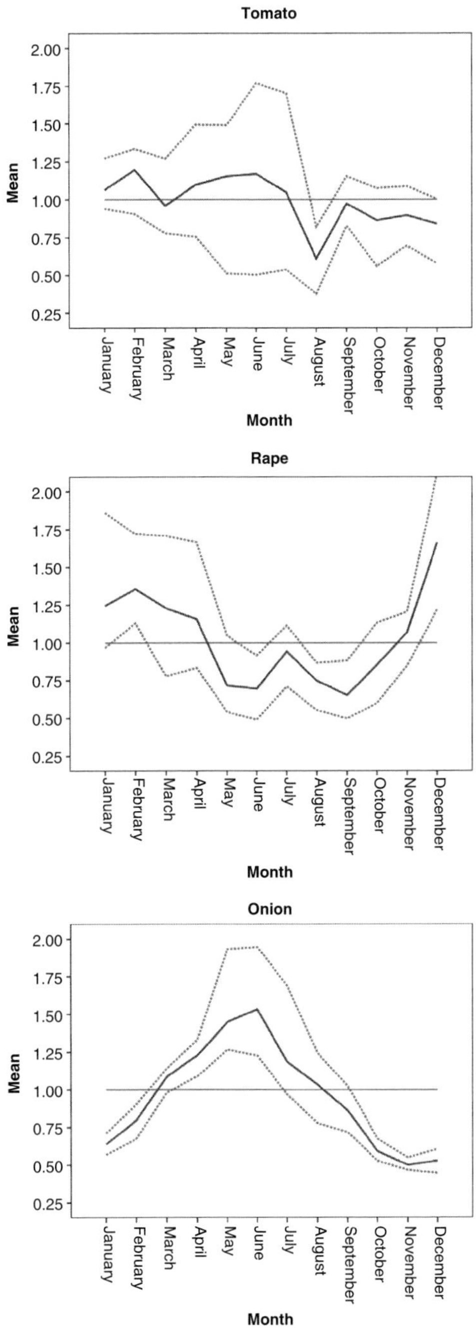

Fig. 9.5. Seasonal price indices for tomatoes, rape and onions in Lusaka.

effective control of production environments (through irrigation and pest management) to regulate the flow of product to the market and avoid dramatic price swings. South Africa has all three: very large producer-shippers who dominate that market have cold transport and access to the inputs and the knowledge they need to avoid the most dramatic fluctuations in output; and brokers who operate in the country's wholesale markets and communicate constantly with producers and buyers to match supply to demand. In Zambia, traditional fresh produce supply chains handle over 90% of all marketed fresh produce.

Production in these chains is dominated by small- and medium-scale farmers, many of whom have poor access to inputs and extension advice and face high variability in yields. Most sales go into an atomistic retail sector that makes it difficult for brokers to anticipate demand, even if farmers were able to respond to attempts at active coordination by the brokers. Finally, these chains have no cold storage, meaning that wholesale prices for products such as tomatoes and (especially) rape must adjust daily to clear the market. Under these conditions, price variability can be extreme, with negative implications for farmers and consumers.[10]

This section first quantifies variability in daily prices and quantities at wholesale in Lusaka. It then estimates price flexibility coefficients for each crop and reaches tentative conclusions regarding market behaviour on this basis. Table 9.5 presents selected measures of variability in the daily average prices and market quantities in Soweto. The coefficient of variation captures both day-to-day fluctuations and longer seasonal variation; the other two measures eliminate seasonal variation by focusing on day-to-day variability. Five points stand out. First, daily quantities arriving in the market fluctuate dramatically for each crop: mean day-to-day changes[11] (absolute value) in quantities arriving in the market are 29% for tomatoes, 32% for rape and 56% for onions. For all three crops, these changes exceed 20% in absolute value more than half of the time. Secondly, this variability in quantities drives great variability in prices for rape

Table 9.5. Selected measures of variability in prices and quantities for tomato, rape and onion (Soweto, 15 January 2007 to 16 November 2009).

	Tomato		Rape		Onion	
Variability measure	Price	Quantity	Price	Quantity	Price	Quantity
Coefficient of variation	0.51	0.31	0.48	0.41	0.41	0.47
Mean day-to-day absolute change (%)	20	29	30	32	7	56
Proportion of day-to-day changes (> 20%)	39	51	52	56	5	64

and tomatoes, which see mean day-to-day absolute price changes of 30% and 20%, respectively. Thirdly, in all cases, prices vary less from day to day than do quantities. This differential is dramatic for onions (which have the highest variability in quantities) but is explained by the storability of this crop, which allows market supply to differ from quantities arriving at the market. For rape and tomatoes, neither of which can be stored to any significant degree, the pattern suggests a perhaps surprising elasticity of demand, at least in the short term. Finally, tomato prices vary substantially less from day to day than do rape prices, despite comparable variability in quantities. The mechanisms behind these last two points are explored below.

We will now examine price behaviour more formally by estimating price flexibilities for each crop. Note that, given the physical characteristics of the crop and of consumer demand, high price flexibilities are associated with less flexible and less stable systems – systems that are less able to slow down or speed up the flow of product from farmer to consumer in response to changing circumstances and in which, for that reason, prices must bear the brunt of most adjustment. High price flexibilities mean high price instability for a given instability in quantities arriving in a market; when these quantities are themselves highly variable, the resulting price instability can impose high costs on farmers and consumers. Marketing systems, as a result, seek ways to reduce price flexibility.

Our regression takes the form:

$$\ln P_t = a + \beta_1 \ln Q_t + \beta_2 \ln P_{t-1} + \beta_{3i} D_i + \ln t + e_i \qquad (9.1)$$

where
ln = the natural logarithm,
P_t = this period's average price,
Q_t = mean daily quantity arriving in the market during this period,
P_{t-1} = last period's average price,
D_i = 11 monthly dummy variables,
t = a time variable and
e_i = a normally distributed error term.

The estimation is in Stata with Prais–Winsten correction for first-order serial correlation and robust standard errors. β_1 is the price flexibility, indicating the percentage change in price for a 1% change in quantity arriving on the market. Because our three crops are 'staple vegetables' consumed on an almost daily basis by most households, we expect their demand to be price inelastic. In the extreme case of zero storage due to perishability, we would expect fluctuations in quantities to drive proportionally larger changes in price; price flexibilities would lie above unity. In the more realistic case where some short-term storage (2–3 days) is possible at some level in the system, flexibilities in high frequency (e.g. daily) data could lie below unity, rising and eventually exceeding unity as data frequency diminishes (e.g. from daily to weekly to monthly).

Storage can, in principle, occur at four levels in the system: at the farm, at wholesale, among retailers and among consumers. Storage on the farm is not relevant to our analysis, since our quantity data are for volumes arriving at the market. These data show that ending stocks of tomatoes and rape average only 15% and 2%, respectively (compared with 89% for onions), of daily volumes arriving at the market; rape is essentially not stored at wholesale, and tomato storage is low. On the other hand, we know that retailers typically keep tomatoes for 2–3 days and may even keep rape until the next day, if they don't sell it all. UCS data indicate that 45% of Lusaka consumers own a refrigerator, suggesting that some of these households may be able to store 2–3 days' supply of rape and perhaps 7–10 days' supply of tomatoes.

Our channel maps show that retailers purchase a large share of their rape directly from farmers. This provides another avenue for reducing price flexibility, as retailers can arbitrage between purchases from both farmers and brokers at Soweto depending on relative prices. We know that retailers buy entire plots of rape and can then harvest it over a period of time, effectively storing it in the field. The channel maps show a final potential mechanism for reducing price flexibility: shipping outside Lusaka. If these shipments can be made at short notice, in response to large arrivals in Soweto or delayed when little product arrives, they will reduce price flexibility/instability; if instead they are contracted and must be shipped on an agreed schedule, they can increase price instability. Given what we know of how the marketing system functions, we expect the former situation to prevail and, therefore, expect that these shipments will reduce price flexibility/instability. Note that rape is not shipped outside Lusaka in any meaningful volume, suggesting another reason, in addition to its greater perishability, to expect high estimated price flexibilities.

We estimate two models for each crop: one with daily data and a second with weekly data (Table 9.6). Based on the discussion above, our expectation was that price flexibilities would be well below unity for onions in both models, would be highest for rape in both models and would be substantially lower for tomatoes and rape in the daily model than in the weekly model.

The results strongly confirm our expectations. Daily onion prices were unaffected by the same day's quantity, being driven entirely by seasonal fluctuations in supply (captured by the monthly dummies) and by the previous day's price; together, these explain nearly all the variation in the daily onion price. The onion's price flexibility becomes statistically significant, but remains small in the weekly model. Rape has the highest flexibility, at –0.79 in the daily model and near unity in the weekly model. The fact that rape's price flexibility does not exceed unity even in the weekly model – well beyond the time that the product can be stored except in refrigerators – suggests that the retail traders' direct purchases from farmers act to stabilize wholesale market prices of this crop. Tomatoes have quite a low flexibility in the daily model (–0.28), suggesting that storage of 2–3 days at retail and consumer levels is common, and probably that shipments outside Lusaka stabilize the market. This crop's flexibility rises markedly in the weekly model, to –0.89.

All in all, this analysis suggests that Lusaka's traditional marketing system, despite no cold chain prior to the consumer level and difficulty coordinating supplies between a large number of small farmers and small traders, has found ways to reduce the effects on prices of fluctuations in quantities arriving in the market. It is notable that flexibilities for tomatoes and rape remain below unity, even beyond the time period that retailers and most households can reasonably be expected to store these products; the shipment of tomatoes outside Lusaka and the procurement of rape directly from farmers are the two most likely mechanisms driving this result.

Nonetheless, even with these stabilizing mechanisms wholesale prices are highly variable due to even greater variability in daily quantities. Key challenges, then, are to assist farmers to control their production environments better (with irrigation, better access to inputs and greater agronomic

Table 9.6. Regression results for estimation of daily and weekly price flexibilities at wholesale in Lusaka.

	Daily			Weekly		
	Tomato	Rape	Onion	Tomato	Rape	Onion
N	357	353	343	139	140	137
R^2	0.92	0.91	0.99	0.85	0.82	0.98
Dependent var. = log(price)						
Independent variable						
Log(quantity)	–0.28[c]	–0.791[c]	–0.001	–0.891	–0.966[c]	–0.091[c]
Log(lagged price)	0.16[c]	0.128[c]	0.615[c]	0.535	0.235[c]	0.932[c]
Log(time)	0.08[c]	0.048[c]	0.008	0.024	0.047[c]	–0.001
January	0.045	–0.040	–0.026	0.032	–0.031	–0.014
February	0.084	–0.023	–	0.006	–0.090[b]	–
March	0.056	0.025	0.058[c]	–0.010	0.012	–0.004
April	0.003	0.048	0.071[c]	0.006	0.018	–0.001
May	0.052	–0.029	0.090[c]	–0.019	–0.004	–0.001
June	–0.020	–0.070	0.087[c]	–0.047	–0.018	–0.032
July	–0.091	0.020	0.043[a]	–0.087	0.029	–0.062[c]
August	–0.166[b]	–0.022	0.032	–0.100	0.036	–0.040[c]
September	–0.098	–0.042	–0.019	–0.048	0.008	–0.067[c]
October	–0.110	–0.021	–0.074[c]	–0.121	0.004	–0.087[c]
November	–	–	–0.077[c]	–	–	–0.036[b]
December	0.049	–0.028	–0.069[b]	0.065	–0.092	–0.014
Constant	3.56[c]	3.56[c]	1.22[c]	5.71	6.13[c]	0.672[c]
Transformed DW	2.00	1.92	1.69	1.98	1.98	2.06

Statistical significance: 0.1[a], 0.05[b], 0.01[c]. Excluding monthly dummy for each crop based on month with mean price closest to overall mean. DW, Durbin–Watson.

knowledge) and to improve the vertical flow of information, so that demand and supply can be matched without the dramatic price adjustments seen in Table 9.5.

Margins

We computed gross margins using the daily average price data at wholesale and retail. Wholesale prices are those paid by buyers – primarily small-scale retailers – in Soweto; retail prices are for sales to consumers in Chilenje. These margins are, thus, an estimate of the gross margins earned by retail traders in Lusaka.

Mean absolute margins are similar across crops, while mean percentage margins are an increasing function of the perishability of the crop (Table 9.7): 131% for onions, 224% for tomatoes and 356% for rape. These dramatic differences in percentage margins suggest the possibility of substantial quality-related loss – perhaps hidden in the form of sharp price discounts at the end of the day[12] – for tomatoes and especially for rape, compared with onions.

Rape's absolute margins are relatively stable, with a coefficient of variation of only 0.15 and only 11% of all day-to-day changes surpassing 20% in absolute value. Tomato and onion margins are substantially more variable.

Comparing day-to-day change indicators in Tables 9.5 and 9.7 allows us to assess the relative exposure of farmers and retail traders to the unpredictable variation in their gross revenue per unit of sale (the sale price for farmers and the gross margin for

Table 9.7. Gross retail margins for tomato, rape and onion in Lusaka.

Indicator	Tomato	Rape	Onion
Mean level (ZK/kg)	2272	2438	2726
Mean (%)	224	356	131
Coefficient of variation	0.29	0.15	0.45
Mean day-to-day absolute (% change)	21	11	22
Share of day-to-day changes (> 20%)	39	14	37

ZK, Zambian kwacha.

retail traders). For tomatoes, we find that the two face nearly identical variation. Rape farmers selling in Soweto, however, suffer much greater unpredictable price change than do retail traders of rape. Since many of these traders bypass Soweto and buy directly from farmers, this pattern raises the question of whether farmers engaged in those relationships with retail traders enjoy substantially more stable prices than do farmers selling in Soweto. Notably, onion retail traders face far more unpredictability than does the mix of farmers and traders selling onions in Soweto, for whom day-to-day price changes are quite modest.

We conducted a regression analysis of margins to explore specific questions about their behaviour:

- Do retail traders stabilize prices to consumers by absorbing some share of any price change at wholesale?
- Do margins show any seasonal pattern?
- Do margins vary by day of the week?

As in the estimation of price flexibility, the Prais–Winsten method is used to correct for first-order serial correlation in errors, and robust standard errors are reported for all coefficients.

We estimated two models for each crop, of the following general form:

$$M_t = \alpha + \beta_1 PW_t + \beta_2 M_{t-1} + \beta_{3i} DM_i + \beta_{4i} DD_i + e_i \quad (9.2)$$

where

M_t = this period's mean gross margin,

PW_t = mean wholesale market price this period,

M_{t-1} = last period's mean gross margin,

DM_i = 11 monthly dummy variables,

DD_i = 2 dummy variables for day of the week and

e_i = a normally distributed error term.

Models are estimated with M_t as a percentage of the wholesale price, and as the natural logarithm. In the first case, PW_t and M_{t-1} are untransformed (linear), while in the second case they are in natural logs. A negative and significant coefficient on PW_t in the percentage regression, combined with an insignificant coefficient in the log–log regression, indicates pass-through of a fixed margin, i.e. no price stabilization by retail traders. Negative and significant coefficients in both regressions indicate that traders are absorbing some of the variability at wholesale, lowering absolute margins when wholesale prices rise and raising them when those prices fall.

Results suggest that rape and onion traders stabilize prices to consumers, while tomato traders fully pass on price changes at wholesale (Table 9.8). Tomato margins show slight evidence of seasonality, being significantly higher in June and lower in September and January. However, the unstable seasonal pattern in tomato prices at wholesale makes this weak pattern difficult to interpret. Rape margins show no meaningful seasonality. Onion margins show pronounced seasonality, being lower from July or August through to January, when wholesale prices are also low. Coefficients for Monday and Wednesday dummies are nearly always negative, and are negative and significant in 4 out of the 12 cases, suggesting some tendency for margins to be higher on Friday.

Summary and Conclusions

This study highlights five findings of potential relevance beyond Zambia. We briefly discuss these here along with potential implications. First, regional trade is an important part of the fresh produce system.

Table 9.8. Regression results for gross margin behaviour of tomato, rape and onion in Lusaka.

	Tomato		Rape		Onion	
	% Margin	Log margin	% Margin	Log margin	% Margin	Log margin
N	386	386	386	386	383	382
R^2	0.54	0.51	0.63	0.50	0.38	0.73
*Independent variable**						
Wholesale price	–0.002[c]	0.029	–0.004[c]	–0.076[c]	–0.001[c]	–0.149[b]
Lagged margin	–0.115[b]	0.613[c]	–0.062	0.594[c]	–0.140[b]	0.692[c]
January	–0.465[a]	0.009	0.424	0.008	–0.182	–0.170[b]
February	–0.172	0.076	0.323	–0.010	0.020	–0.101
March	–0.111	0.067	–0.055	0.009	–0.104	–0.101
April	–		–	–	–	–
May	–0.009	–0.036	0.332	–0.028	0.258[b]	0.016
June	0.855[c]	0.117[b]	0.439	–0.019	0.399[b]	0.107
July	0.291	–0.045	–0.173	–0.008	–0.068	–0.150[b]
August	0.344	–0.031	0.172	–0.047[a]	–0.526[c]	–0.336[c]
September	–0.464[a]	0.028	0.933[a]	–0.012	–0.407[b]	–0.121[a]
October	–0.239	0.047	0.413	–0.032	–0.418[b]	–0.250[c]
November	–0.221	0.070	–0.372	–0.012	–0.269	–0.215[b]
December	–0.139	0.048	–0.090	0.013	–0.209	–0.205[b]
Monday	–0.174[c]	–0.047	–0.231[c]	0.003	–0.021	–0.050
Wednesday	–0.172[c]	–0.006	0.024	0.010	–0.032	–0.070[b]
Trend	0.000	0.000[b]	0.001	0.000[b]	0.001[a]	0.000
Constant	4.663[c]	3.118[c]	6.494[c]	3.65[c]	3.069[c]	3.679[c]
Transformed DW	2.00	2.00	2.19	1.98	1.99	1.99

* Wholesale price and lagged margins are untransformed in percentage margin regressions, natural logarithm in log regressions. DW, Durbin–Watson. Statistical significance: 0.1[a], 0.05[b], 0.01[c].

Other studies have found the same (Tschirley *et al.*, 2004 for Kenya; USAID, 2005 and Louw *et al.*, 2009 stress the importance of regional markets). Thus, regional transport links, harmonization of trade regulations, avoidance of arbitrary border closings, regional market information sharing – all issues typically addressed with vigour in cereals markets – are also important for improving the performance of fresh produce markets, and for reducing and stabilizing prices to consumers.

Secondly, supply chains for these three staple vegetables are short. Together with other findings on the small scale of operation of many traders (Hichaambwa and Tschirley, 2006 for Zambia; Tschirley *et al.*, 2004 for Kenya), this finding suggests that more programmatic emphasis should be placed on helping existing traders scale-up and gain better access to information to do their job more effectively. Helping farmers bypass these traders and market their produce directly will be appropriate in some circumstances, but most farmers will continue to rely on the existing trading system.

Thirdly, we found an uncertain role for urban agriculture. It is clearly marginal for tomatoes and onions but could be important for rape and other green leafy vegetables. More information is needed on this issue.

Fourthly, we found an extreme high price variability driven by an even wider unpredictable fluctuation in quantities. We showed that Lusaka's traditional markets are surprisingly effective in reducing the price response to these quantity fluctuations, but the movements are so large that price variability probably remains a serious problem for farmers, traders and consumers. This finding points to the need for better production technologies and better access to inputs and agronomic advice for farmers, to reduce variability in farm yields and to improve the vertical information flow through the system to match supply more closely to demand in the short term.

Finally, nominal margins appear to be very high, but the lack of reliable benchmarks in other countries at varying levels of development makes it difficult to draw conclusions at this point.

Tschirley *et al.* (2009) highlight the need for 'improved hard infrastructure linked to better management models and improved coordination' in Africa's fresh produce supply chains. They stress that improved physical infrastructure at wholesale is crucial, but that this will do little good if not conceived, built and managed under more innovative public–private collaborative models. They note that legislative changes will often be needed to allow the application of such models but that, if this can be done, new wholesale markets can become focal points for the kind of information flow for improved coordination that we emphasize above.

Further research

This chapter has identified several important research gaps. First, given the widespread presence of brokers and the controversy that they arouse (see Tschirley *et al.*, 2004, for Kenya), rigorous assessment is needed in several regards:

- What economic function do brokers play?
- Do brokers add value and, if so, how and for whom?
- Are abuses such as threatening theft of product or lying about commissions widespread and persistent, or isolated?
- What regulatory framework is needed to assure honest business practices that benefit farmers and consumers?

These are key public policy issues in fresh produce marketing systems, and will only become more important over time.

Secondly, comparative data are needed on gross margins across several countries. At this point in time, it is impossible to judge whether a mean gross margin above 200% for tomatoes (or above 300% for rape) reflects poor performance, and what levels these countries could reasonably expect to reach. Systems for collecting these data must be designed carefully, based on specific knowledge of the market chain being studied, to ensure comparability across countries.

Thirdly, the value chain for rape and other green leafy vegetables is poorly understood, especially the nature of retail trader links to farmers and the quantitative importance of urban agriculture in these chains. Given the importance of green leafy vegetables in consumer diets, the rapid growth in urban populations and concerns about the environmental and health implications of urban agriculture, these chains need to be better understood.

Fourthly, emerging experience with new wholesale markets needs to be documented and analysed. A promising new wholesale market may be emerging outside Nairobi, where private sector stakeholders have collaborated with municipal and national authorities to plan and finance a new wholesale market to be based on shared ownership. Will this become a centre of enhanced services to farmers, as we discussed above? In Maputo, farmers and traders were actively consulted as the municipality planned and built a basic, but well-designed, wholesale market just outside the city. This market, which opened in 2007, has dramatically reduced congestion at the old market within the city and reduced operating costs for traders. Nevertheless, as in so much of Africa, the municipality, which owns and manages the market, has found it increasingly difficult to control the encroachment of petty wholesaling and retailing, which is now disrupting traffic flow and making market maintenance much more challenging. Both these initiatives deserve close scrutiny to draw lessons on how to conceive and create new marketplaces with enhanced services and management that will maintain their value over time.

Finally, it is important to know to what extent trade barriers interfere with regional trade in fresh produce. Such barriers are a persistent problem in cereal markets in East and southern Africa, but less is known about their effect on the fresh produce trade.

Acknowledgements

Many thanks to Kennedy Malambo and Esnart Musukwa for tireless data collection and related survey work in Soweto market. Also thanks to participants in the conference on 'Socio-economic research in vegetable production and marketing in Africa', held in Nairobi, Kenya on 5–6 March 2009, for valuable comments on earlier versions of this research.

Endnotes

[1]For early expositions of this theme on Latin America, see Reardon and Berdegué (2002) for a summary and Alvarado and Charmel (2002), Faigeuenbaum *et al.* (2002), Farina (2002), Ghezán *et al.* (2002) and Schwentesius and Gomez (2002) for country studies; see also Reardon *et al.* (2004). For Asia, see Reardon *et al.* (2003a,b), Hu *et al.* (2004) and Coe and Hess (2005). For Africa, see Weatherspoon and Reardon (2003), Neven and Reardon (2004) and Neven *et al.* (2005).

[2]Soweto's demography is tilted more towards low income, although its size means that it attracts some shoppers from across the income spectrum. Supermarkets serve primarily high-income households.

[3]We limited the period to 24 full months to control for seasonality in supply, necessary in this case to calculate accurate mean yearly figures. All other analysis is based on data from mid-January 2007 to mid-November 2009, i.e. the full data set available at the time of writing.

[4]For rape and tomatoes, these figures are based on straight-line (air) distances from district capital towns to central Lusaka. For imported onions, we used the distance from Johannesburg and mean distance from Blantyre and Lilongwe to Lusaka. Driving distances will be higher.

[5]Gabre-Madhin (2001) analysed brokers' contribution to grain market efficiency in Ethiopia.

[6]Data on whether a seller was selling through a broker began to be collected only at the end of October 2007.

[7]The regression assumes that values of the dependent variable are serially uncorrelated. This may not be the case if some sellers have binding contracts with brokers over given periods of time. While more research is needed, we believe that this problem is small in this particular case because (i) any contracts that do exist are informal, suggesting a more fluid relationship (and less serial correlation) than might exist in the presence of formal written agreements; (ii) few sellers will have such contracts; and (iii) a large number of sellers come to the market, so that the share of transactions showing this lack of independence will be small.

[8]The prediction error is the difference between the observed price and the price predicted from the ordinary least squares (OLS) regression. We use the median positive and negative errors so that the resulting confidence interval includes 50% of all observations.

[9]These months are identified by the top line lying below the mean when the index is below it, or by the bottom line lying above the mean when the index is above it.

[10]See Mwiinga (2009, unpublished MSc thesis) for a detailed assessment of production and price risk for tomatoes in Zambia.

[11]Note that all these calculations are based on Monday/Wednesday/Friday data collection, meaning that two-thirds of the computed 'day-to-day' changes occur over two days, and one-third over three days.

[12]Our retail prices are taken mid- to late morning, to capture the prices at which we believe most products are sold.

References

Alvarado, I. and Charmel, K. (2002) The rapid rise of supermarkets in Costa Rica: Impact on horticultural markets. *Development Policy Review* 20(4), 373–385.

Coe, N.M. and Hess, M. (2005) The internationalization of retailing: Implications for supply network restructuring in East Asia and Eastern Europe. *Journal of Economic Geography* 5(4), 449–473.

Faiguenbaum, S., Berdegué, J.A. and Reardon, T. (2002) The rapid rise of supermarkets in Chile: Effects on dairy, vegetable, and beef chains. *Development Policy Review* 20(4), 317–334.

Farina, E. (2002) Consolidation, multinationalisation, and competition in Brazil: Impacts on horticulture and dairy product systems. *Development Policy Review* 20(4), 441–457.

Gabre-Madhin, E.Z. (2001) The role of intermediaries in enhancing market efficiency in the Ethiopian grain market. *Agricultural Economics* 25, 311–320.

Ghezán, G., Mateos, M. and Laura, L. (2002) Impact of supermarkets and fast-food chains on horticulture supply chains in Argentina. *Development Policy Review* 20(4), 409–427.

Hichaambwa, M. and Tschirley, D. (2006) Zambia horticultural rapid appraisal: Understanding the domestic value chains of fresh fruits and vegetables. Working Paper 17. Food Security Research Project, Lusaka, Zambia.

Hu, D., Reardon, T., Rozelle, S., Timmer, P. and Wang, H. (2004) The emergence of supermarkets with Chinese characteristics: Challenges and opportunities for China's agricultural development. *Development Policy Review* 22(5), 557–586.

Humphrey, J. (2006) The supermarket revolution in developing countries: Tidal wave or tough competitive struggle? *Journal of Economic Geography* 7(4), 449–473.

Louw, A., Chikazungaii, D., Haankukuiii, C. and Ndangaiv, L. (2009) Dynamics of the restructuring fresh produce food markets in the southern African region. Contributed paper prepared for presentation at the *International Association of Agricultural Economists Conference*, 16–22 August. Beijing, China.

Maddison, A. (2009) *Historical Statistics of the World Economy: 1–2006 AD*. IMF, Washington, DC.

Minten, B. (2008) The food retail revolution in poor countries: Is it coming or is it over? *Economic Development and Cultural Change* 56, 767–789.

Neven, D. and Reardon, T. (2004) The rise of Kenyan supermarkets and the evolution of their horticulture product procurement systems. *Development Policy Review* 22(6), 669–699.

Neven, D., Reardon, T., Chege, J. and Wang, H. (2005) Supermarkets and consumers in Africa: The case of Nairobi, Kenya. Staff Paper 2005-04, Department of Agricultural Economics, Michigan State University, East Lansing, Michigan.

Reardon, T. and Berdegué, J.A. (2002) The rapid rise of supermarkets in Latin America: Challenges and opportunities for development. *Development Policy Review* 20(4), 371–388.

Reardon, T., Timmer, C.P., Barrett, C.B. and Berdegué, J.A. (2003a) The rise of supermarkets in Africa, Asia, and Latin America. *American Journal of Agricultural Economics* 85(5), 1140–1146.

Reardon, T., Timmer, C.P. and Berdegué, J.A. (2003b) The rise of supermarkets in Latin America and Asia: Implications for international markets for fruits and vegetables. In: Regmi, A. and Gehlhar, M. (eds) *Global Markets for High Value Food Products*. Agriculture Information Bulletin, USDA-ERS, Washington, DC.

Reardon, T., Timmer, C.P. and Berdegué, J.A. (2004) The rapid rise of supermarkets in developing countries: Induced organizational, institutional, and technological change in agrifood systems. *Journal of Agricultural and Development Economics* 1(2), 168–183 (www.fao.org/es/esa/eJADE).

Schwentesius, R. and Gomez, M.A. (2002) Supermarkets in Mexico: Impacts on horticultural systems. *Developmnt Policy Review* 20(4), 487–502.

Traill, B. (2006) The rapid rise of supermarkets? *Development Policy Review* 24(2), 163-174.

Tschirley, D., Muendo, K.M. and Weber, M.T. (2004) Improving Kenya's domestic horticultural production and marketing system: Current competitiveness, forces of change, and challenges for the future. Vol. II: Horticultural marketing. Tegemeo Institute of Agricultural Policy and Development, Working Paper 8B, Egerton University, Kenya.

Tschirley, D., Ayieko, M., Hichaambwa, M., Goeb, J. and Loescher, W. (2009) Modernizing Africa's fresh produce supply chains without rapid supermarket takeover: Towards a definition of research and investment priorities. Paper prepared for the conference *Towards Priority Actions for Market Development for African Farmers*, sponsored by the International Livestock Research Institute, 15–17 May, Nairobi, Kenya.

United Nations (2007) World Urbanization Prospects: The 2007 Revision Population Database. Department of Economic and Social Affairs (http://esa.un.org/unup/).

USAID (2005) *Global Horticulture Assessment*. University of California, Davis, California.

Weatherspoon, D. and Reardon, T. (2003) The rise of supermarkets in Africa: Implications for agrifood systems and the rural poor. *Development Policy Review* 21(3), 333–355.

10 Value Chains and Regional Trade in East Africa: the Case of Vegetables in Kenya and Tanzania

Thomas König, Jantje Blatt, Kristian Brakel, Kristina Kloss, Thorsten Nilges and Franziska Woellert

Introduction

In recent years horticulture's relevance to the East African economies has increased, and many export-oriented medium- and large-scale enterprises have benefited from this development. Beyond that, in most East African domestic markets horticultural demand has increased as a result of urbanization and the growing middle classes (Weinberger and Lumpkin, 2007). The importance of market coordination and participation has been highlighted, and their absence has been described as one of the most important constraints responsible for the poor performance of the African agricultural sector (Dorward *et al.*, 2005). According to estimations by Kelley and Byerlee (2004), some 60% of the African rural population live in areas of good agricultural potential, but with poor market access. Only 22% live in areas of both good agricultural potential and good market access, while 18% suffer poor market access and poor agricultural potential. Effective markets are required in order to feed the growing number of urban Africans.

In Kenya, currently ten farm households have to feed 2.5 non-farm households. This ratio is expected to change due to the demographic shift; within the next 25 years, ten farm households will have to feed five non-farm households (Tschirley and Ayieko, 2008). A significantly growing urban population, combined with a relatively high income elasticity of demand for vegetables (HDC, 2004; Muendo and Tschirley, 2004), is perceived as a major reason for this development. This link can be expected to have backwards linkages to the rural economy. In Kenya (KE), growth multipliers from the agricultural sector to the overall economy are two to three times higher than those from non-agricultural sectors (Kimenyi, 2002). Since the horticultural sector is labour intensive in production, positive impacts on poverty alleviation can be expected (Weinberger and Lumpkin, 2007). In Tanzania (TZ), an accelerating growth of agricultural sub-sectors, which are not concentrating on large-scale farms producing staple food, is expected to contribute significantly to poverty reduction (Pauw and Thurlow, 2010).

Horticulture in this study refers to the entire value chain, from production to consumption of vegetables. Tomatoes and onions are chosen as the case study crops, because both are important cash crops in Kenya and Tanzania, playing an important role in the domestic and regional fresh vegetable market trade in East Africa. Including the production of tomatoes and onions, the development of local markets and supply chains to markets at the national and regional levels is in line with a poverty reduction agenda (Achterbosch *et al.*, 2005; World Bank, 2005). Specifically, this study addressed the following objectives:

- to investigate the potential for poverty reduction by market-driven development of the respective value chains;
- to analyse the cross-border trade of onions between Tanzania and Kenya; and
- to compare the value chains between the two countries.

The value chain approach considers the value added of a product as well as the actors involved in the chain, and their roles and linkages.

Methodology, Survey Design and Analysis

Analytical framework

The origin of value chain analysis was the business management tool developed by Porter (1985), which has since been adapted for research and development at non-industrial and community levels. The study concept is influenced by general value chain analyses (Humphrey, 2005; M4P, 2005; USAID, 2006; GTZ, 2007) and by the concept of action and decision-oriented research (Fiege and Nagel, 2007, unpublished). The value chain analyses a product's development from input through production and processing levels, transport, assembly and retail, to consumption. This approach reviews the chain from two different perspectives (GTZ, 2007):

- The functional perspective: a value chain is a sequence of related business activities (functions) from the provision of specific inputs for a particular product to primary production, transformation, retail and up to the final consumption of the particular product.
- Institutional perspective: a value chain is the set of enterprises or operators performing these functions. Value chain operators are linked by a series of business transactions in which the product is passed from primary producers to end consumers. According to the sequence of functions and operators, value chains consist of a series of chain links (or stages).

To summarize, the value chain (VC) approach considers both the value added of a product and an insight into the actors' roles and linkages. The study emphasizes the role and linkages of the operators, the bottlenecks they face and the opportunities within the frame of a market-driven development. The perspective of non-value chain actors who might be interested in gaining access to the value chains has not been analysed in this study.

To distinguish the different actors involved in the value chains, the following definitions have been chosen, according to Deutsche Gesellschaft für Technische Zusammenarbeit (GTZ, 2007):

VC actors include all individuals, enterprises and public agencies related to a value chain, in particular the VC operators, providers of operational services and the providers of support services. In a wider sense, certain government agencies at the macro level can be seen as VC actors if they perform crucial functions in the business environment of the value chain in question.

VC operators are all enterprises and persons performing the core functions of a value chain. Typical operators include farmers, small and medium enterprises, industrial companies, exporters, wholesalers and retailers. All of these have in common that each VC operator owns the (raw, semi-processed or finished) product at one stage in the chain. Thus, there is a difference between operators and 'operational service providers', the latter being subcontracted by VC operators.

VC supporters provide VC support services (also called support service provider, like the agricultural extension service) and represent the common interests of the VC actors. They belong to the meso level of the value chain.

Survey design

The collection of empirical data for the analysis of tomato and onion value chains in Kenya and Tanzania took place in selected production areas and markets, as shown in Table 10.1. The study sites were selected for the level of production of both study crops, as well as for the importance of the role of the selected markets within the value chain.

The study team identified different actors at every value chain step in order to conduct interviews with representatives of all important actor groups. There was a wide range of actors with widely differing socio-economic backgrounds who needed to be included, but the sample size at each node (or vertical step) of the VC was small, and therefore the selection was critical. At the production level, respondents were selected and contacted after a transect walk and qualitative interviews with the extension service officer or the village chief had been conducted. For other VC steps the snowball principle or random selection was applied, which meant that the study team either directly asked actors, who named other relevant persons they knew, or asked officials, e.g. at the markets, to make the first contact.

In total, 101 value chain actors/operators (51 in TZ; 50 in KE) completed both the value chain and the livelihood questionnaire. Eleven workers (TZ: six, KE: five) in the production segment and seven workers (TZ: three, KE: four) involved in transportation were asked the livelihood questions only. Finally, the research team carried out the consumer questionnaire by interviewing 16 consumers (TZ: eight, KE: eight). Most of the quantitative questionnaire respondents were selected randomly. Seventy qualitative interviews (TZ: 31, KE: 39) were conducted to complement the individual quantitative interviews. In addition, the data collection was conducted from both ends of the chain at the same time to allow triangulation of findings (e.g. production areas and wholesale markets).

Research results were presented to horticulture experts in the region in a workshop at the end of the fieldwork period, which allowed for the validation of findings.

Data collection

Action and decision-oriented research provided the basis for the development of the research questions and was transferred to the study design, which consisted of the literature review, participatory rural appraisal methods, qualitative expert and focus group interviews, and quantitative VC actors livelihood questionnaires. Additional feedback

Table 10.1. Study locations.

	Kenya		Tanzania	
	Production area	Markets	Production area	Markets
Tomatoes	Kirinyaga District: Mwea Division	Wakulima and Gikomba (both Nairobi) markets: wholesale and retail markets	Arumeru District: Ngari Nanyuki and Nduruma areas	Kilombero market (Arusha): wholesale and retail market
Onions	Nyeri North District: West Division	Thika, Karatina and Kutus markets: smaller local retail and wholesale markets	Karatu District: Mang'ola and Baray area	Central market (Arusha): retail market

and recommendations from experts in value chain analysis, rural development and the East African horticulture sector adjusted the approach of this study.

The qualitative methods applied within this research consist of key resource person interviews and focus group discussions. The latter were supplemented by participatory tools, such as transect walks. In the present context, qualitative interviews are characterized as semi-standardized verbal questioning of the knowledge gatekeepers, who are the actors crucial to or possessing knowledge relevant to the understanding of the tomato and onion value chains in Kenya and Tanzania. The guidelines used in these interviews aimed mostly at identifying the functional and institutional perspectives of the value chains. The focus was on linkages and relations between the research subjects (i.e. actors) on power and profit distribution, and on the constraints and potentials within the VCs.

Key resource personnel from partner institutions (World Vegetable Center (AVRDC), International Centre of Insect Physiology and Ecology (*icipe*)) and other important research institutes, such as the Tegemeo Institute and the Kenya Agriculture Research Institute, ministries and NGOs (e.g. Farm Concern and FAIDA MALI) responded to the qualitative interviews. Their experience at macro and/or meso level was helpful in understanding the chains' performance. Also interviewed were experts who act mainly at the micro level, such as representatives of market authorities, regional, district and village administration, as well as village chiefs, public and private extension services and informal or formal actor groups. The analysis of the qualitative data was an iterative process.

The focus group discussions were targeted at bringing together the different actors of one VC step in order to discuss the study's objectives relevant to those actors. This aimed to achieve a more complex understanding of the chosen subjects by eliciting the different opinions of the actors at the same time. The group discussions were moderated and documented by members of the research team. Some of the focus group discussions were planned and prepared in advance, while others were held spontaneously. In most of the planned group discussions different participatory tools were implemented, in particular Venn-diagram and SWOT (Strength, Weaknesses, Opportunities and Threats) analysis.[1] The objective of this particular part was to elicit more information about the advantages and disadvantages for actors organized in a group or cooperative and to discover the importance of actors working in cooperation.

As a first activity when visiting a village or a marketplace, the team members carried out transect walks. These helped the team members to become acquainted with the location and the local population in order to acquire, through informal talks, some preliminary information about the situation such as characteristics, problems and the potential of that specific area. During the walk, the surroundings, such as infrastructure, people's activities and dwelling conditions, were closely observed. The observations were subsequently discussed and clarified with the local guides.

The quantitative data were collected by standardized questionnaires for all VC actor groups. The most complex one is the value chain questionnaire addressed to actors of every value chain step, such as production, transport and postharvest, assembly and retail. This contained sections related to the functions of the particular value chain step, i.e. information on crop production was asked only of farmers. The questionnaire further included sections that were the same for all value chain steps, in order to compare aspects such as profit distribution and price negotiations.

By combining the value chain analysis with the livelihood approach, this study identified poor actors in the chain, in addition to calculating their income. This enlargement of the value chain analysis has been proved a suitable way to consider the potential for poverty reduction in a broader sense. In this study, poverty is taken as a multidimensional concept, and was assessed by using the livelihood approach. Five assets are normally included for a livelihoods analysis: natural (climate, including water supply), physical (infrastructure, dwelling conditions), financial (access to credit, savings), human (education) and social (social network, membership of organizations) (de Haan, 2000; Dorward *et al.*, 2001).

The livelihood questionnaire of this study covered the physical, financial, social and human assets of the value chain actors. The natural asset was excluded, assuming that the natural capital is quite similar for all relevant actors. The questionnaire was identical for all actors interviewed along the chain, in order to identify the poorest and weakest actors in a comparable and consistent way.

Due to the shortage of time and the broad research approach, the quantitative study sample concerning each value chain step was relatively small and did not aim to fulfil statistical requirements. The quantitative data are related to the qualitative information with examples and are intended to enable a better understanding of the value chains, particularly regarding costs, prices and profits.

Data analysis

Two different programmes were used for the data analysis. One, the atlas-ti programme, connected the qualitative interviews with research questions by keywords or codes. Using these codes, aspects mentioned in different interviews could be compared. In the second, matrices in Microsoft Excel were developed to deal with the quantitative data, which took seasonality into account. The livelihood situation of all interviewees was compared by weighting the responses.

Results: Mapping the Value Chains

This section is organized as follows. First, the institutional perspectives are described; in particular, this section describes the typical steps and activities and provides an overview of the main actors. Secondly, the most typical links within and between the actor groups are analysed. Thirdly, legal regulations and their relevance, particularly with respect to standards, are assessed. Next, the cross-border trade between Kenya and Tanzania is analysed and, lastly, value chains are considered in relation to poverty reduction.

Steps, and activities and actors

Figure 10.1 summarizes the different activities in a vegetable value chain and the actors involved. As a starting point this chart provides a general overview, not reflecting the fact that some actors might operate at multiple levels of the chain.

Inputs

Large companies supplying seeds, fertilizer and pesticides take the part of VC supporters. Input suppliers advertising an intensive use of agrochemicals have promoted the application of chemicals. Both farmers and extensionists in all areas researched report an increasing awareness among farmers regarding the potential misuse of chemicals.

Public and private extension services are VC supporters and provide information and advice. This study found that the extension service available to horticultural farmers is ineffective, primarily due to a lack of personnel capacity within the Ministries of Agriculture of both countries. Most farmers state that they have relied on their neighbours, friends and relatives over the past years for information. Governmental and private credit institutes belong to the VC supporters at this step, but their impact is low, as farmers refrain from taking loans because they fear crop failure and being unable to repay their debts. Another reason that became apparent from qualitative interviews is that the farmers seem to have little information on loan conditions.

Production

At the production level, farmers, farm workers and farmer groups play an important

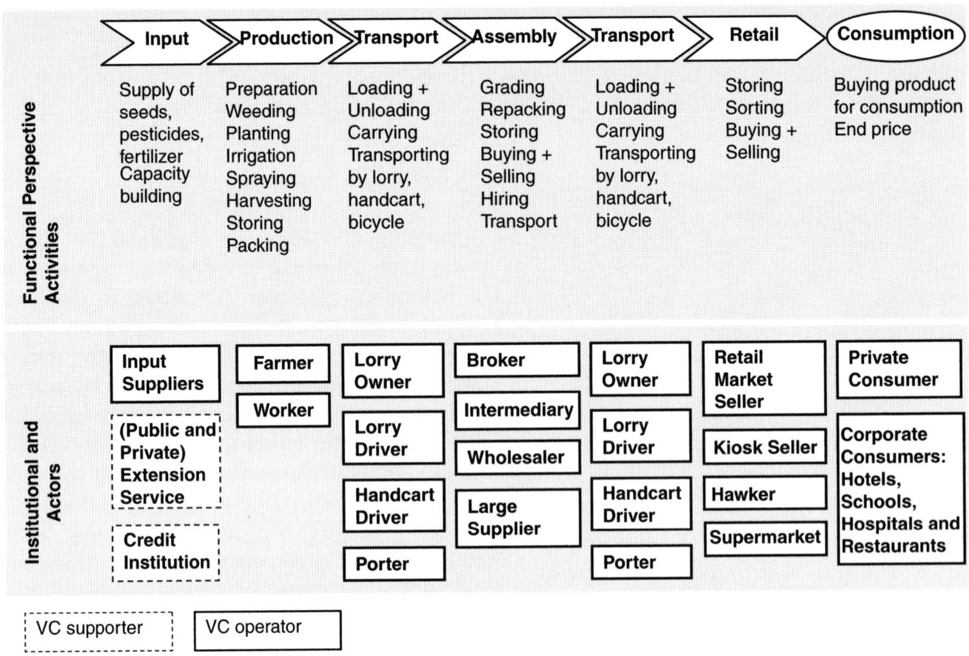

Fig. 10.1. Steps, activities and actors of tomato and onion value chains in Kenya and Tanzania (from König *et al.*, 2008).

role as operators. This study includes only farmers who grow either tomatoes or onions as a main cash crop. An assumption in this study was that, since the cultivation of these crops requires a certain resource endowment and the ability to cover the necessary input expenses, the farmers interviewed did not belong to the poorest households in the research regions. The calculation carried out within the livelihoods questionnaire confirms this assumption, as the farmers rank mid-range compared with all other operators and are distinguished from the poorer ones by higher financial and physical assets. This is reasonable, considering the expenses and costs that tomato and onion farmers incur to run their business. Nevertheless, most of the farmers are small- to medium-scale producers and manage their plot within their family or by employing casual workers. The latter normally belong to other farms but are from poorer households, according to the respondents of the focal group discussions.

Assembly

This level incorporates different kinds of actors for the assembly of the produce. Brokers, being VC operators, do not own the product at any point in the chain and their role is to connect buyers and sellers and, in some cases, to negotiate prices in the name of either the buyer or the seller. They are paid a commission by either party (HDC, 2004; GTZ, 2007; König *et al.*, 2008). It is possible to distinguish between brokers at the farm level, bringing together producers and middlemen and brokers at the markets who, for example, link an intermediary with a wholesaler. In contrast to brokers, intermediaries take ownership of the product at the time they receive it; they neither produce the product nor sell it to the final consumer (GTZ, 2007; König *et al.*, 2008). An intermediary buys directly from farmers, brokers or from other intermediaries at local and regional open markets and he sells to other intermediaries, large suppliers, institutional consumers and entrepreneurs or persons at the retail level. Often an intermediary is responsible for the grading. In this study, a wholesaler is defined as an intermediary operating only at the wholesale market; he buys from intermediaries and often enlists the assistance of a market broker.

Transport

Transport is a cross-cutting issue in the VCs for all actors.[2] The most important actors in transport are entrepreneurs owning trucks, those employing lorry drivers and porters, all of whom are VC operators. The transport entrepreneurs and the drivers are mainly responsible for the transport from the production areas to the markets. Farmers or intermediaries hire the means of transportation plus the drivers and often accompany their produce to the market. Lorry drivers are often employed seasonally.

The porter group comprises handcart drivers lugging the commodities from the wholesale to the retail markets or kiosks. They either own the carts themselves or rent them. The group of porters further includes carriers who load, unload and carry the goods in the production areas and in the markets. In some cases, they do the grading and repacking in the market. Porters are not normally employed permanently. They neither specialize in a certain product nor work on contract relations. They get paid for each trip they do, the price depending on the type and amount of commodity they transport. Given the fact that they do not possess any legal working permits, they are sometimes double-crossed by customers who fail to pay their fees. They occasionally suffer from harassment by officials or the police.

Farmers are involved in transportation only from the field to the farm (if necessary) and, in some cases, they arrange the transport from the farm to the markets, especially to the local markets.

Retail

At the retail level the operators are differentiated according to their location or the volume they sell, or both. There are sellers in open markets, roadside sellers with small wooden kiosks and hawkers. The hawkers sell at the roadside without a booth, walking around and approaching potential customers. While the first two operate in the formal sector, the hawkers work in the informal one. Supermarkets are part of the formal retail sector, but they do not play an important role for the supply of fresh fruit and vegetables (FFV). The smaller supermarkets do not sell onions or tomatoes, but there are kiosks located next to them that do. Most of the vendors who are active in retail markets specialize in some products and change their range only in the off-seasons. The range they offer and the amount they sell depend on the kind of stall they own; kiosk retailers normally sell only FFV and try to have a large selection of products to meet their customers' needs. As hawkers depend on their mobility, in general they have only a few commodities to offer. Large supermarkets, in contrast, often supply 'exotic' FFV, such as apples or strawberries, in addition to the local or seasonal produce.

Consumption

The consumption side focuses more on individual consumers rather than on corporate consumers such as hotels, restaurants, hospitals and schools. Most individual consumers buy FFV at retail markets as well as wholesale markets, at kiosks or from hawkers, while only a few individual consumers purchase them in supermarkets, which confirms the results of Ayieko *et al.* (2005). Decisive criteria when choosing the seller are cheap prices, the quality of the product, the proximity of the vendor and trust. The consumers are unaware of different varieties. The consumers' definition of 'high quality' for tomatoes and onions is medium size, good colour, faultless skin, shape and taste, and they look for storable goods. Few consumers consider organic production or food safety, such as potential pesticide residues. Although most consumers stated that quality is more important than the price, they still respond to price signals and purchase less during the low-supply season when the prices are higher.

Market Linkages

From the institutional perspective, market linkages are characterized by a low degree of cooperation between all actors of the VCs analysed. The market linkages are shown in Fig. 10.2.

Actors prefer to act individually and not to rely on self-help groups or contractual business relations. At the production level, there are few farmer groups in the research regions. Furthermore, for activities such as travelling to the cities to buy input supplies or to sell products, the farmers rarely work together. The few existing linkages are based on family or friendship ties, which is reflected in the low level of cooperation between farmers and other actors of the VC. Contract farming was not practised in any of the analysed VCs. In some cases, farmers depend on the input supply of brokers or intermediaries and then commit themselves to sell their yield to this particular person. The price then is set by the input supplier and often ranges below the market price. Generally, farmers grow produce based on their own financial resources, and so are able to sell flexibly to whoever comes first. In most VCs the farmers depend on the intermediaries for transport to the markets, and the selling price is negotiated between farmers and traders; quite often the traders are more powerful in setting the price.

At the assembly level, information about prices, buyers and markets is essential. But most traders (brokers, wholesalers and intermediaries) do not cooperate with

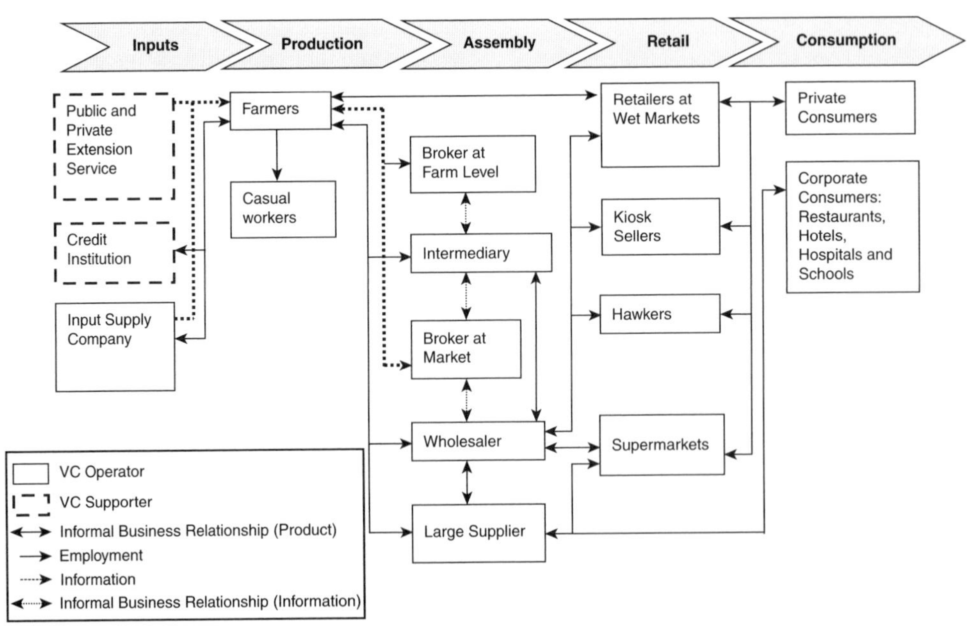

Fig. 10.2. Market linkages in all four value chains (from König *et al.*, 2008).

each other on a regular basis – they prefer to operate individually based on their own knowledge and occasionally receive relevant information from fellow traders. Thus, spot-market relations characterize the link between wholesalers and retailers, as well as between retailers and consumers. Power distributions between retailers and wholesalers are more biased towards the wholesalers, who generally decide the price.

At the retail level, little cooperation exists within or outside the retail markets. Again, information is exchanged only between friends and family members. In contrast to the purchasing price, most retailers are in a strong bargaining position for setting the selling price.

Transport is characterized by little cooperation among transporters, including lorry owners, handcart drivers or carriers. Lorry owners are mostly in a better bargaining position, while handcart drivers and carriers often depend on the goodwill of their clients.

Seasonality and prices

Since this study deals with perishable horticultural crops that require reliable watering, seasonality is considered, and high- and low-supply seasons are distinguished. Seasonality is influenced mainly by weather conditions. The cultivation cycle of tomatoes is only three months long and not all tomatoes mature at the same time. Consequently, the farmers' starting point for the production cycles may vary slightly depending on their access to irrigation, and the same applies to the harvesting period. In contrast, the cultivation cycle of onions is longer, so all onion farmers in a region start more or less at the same time with cultivation and with the harvest. Thus, the availability of storage facilities is important for onion producers. Those farmers who have access to irrigation are less dependent on the seasons, regardless of whether they grow onions or tomatoes. The impact of seasonality on the value chains is summarized in Fig. 10.3.

Due to seasonality, the market prices for both crops fluctuate depending on both the quantity and quality of the products at the market. Especially in the wholesale and retail markets, prices may fluctuate even during one day. Frequently, the constraint on storage space is the reason that traders and retailers try to sell all their commodities by the end of one day, even if they achieve only a low price. Another reason is that most of them suffer from regular cash constraints and therefore they need their money quickly.

Table 10.2 illustrates the fact that farmers are at greatest risk in the value chain since, depending on the country, season and product, they can sell at either a loss or a relatively high profit. The retailers in general appear to have comparatively low risks and good profit levels; however, this picture alters if the volumes are also considered. In this situation the traders claim the highest share of profit for themselves, while in contrast the retailers' share of profit is almost zero. There are many more retailers in the markets than intermediaries or wholesalers, and so they have to share the total volume of goods, and thus the profits. This corresponds with the results of the livelihood analysis.

If one analyses the Kenyan tomato value chain, it demonstrates the difference among value chain actors once volume is taken into consideration. Both the volumes traded and, consequently, the share of profits per season differ remarkably. Wholesalers trade large quantities, usually 10–15 times more per season than farmers and retailers. This means that the profit per season for a wholesaler is higher than that for farmers and retailers. In the high season, when wholesalers' profit per kilogram is already higher than that for the other actors, this unequal profit distribution is further accentuated; wholesalers gain 99% of their profit in the high season, in comparison with retailers (1%). On the median, farmers make no profit in this particular season.

In the low season, it is again wholesalers who achieve the highest profit share (67%). During this period, farmers can increase their profit per kilogram of vegetables traded,

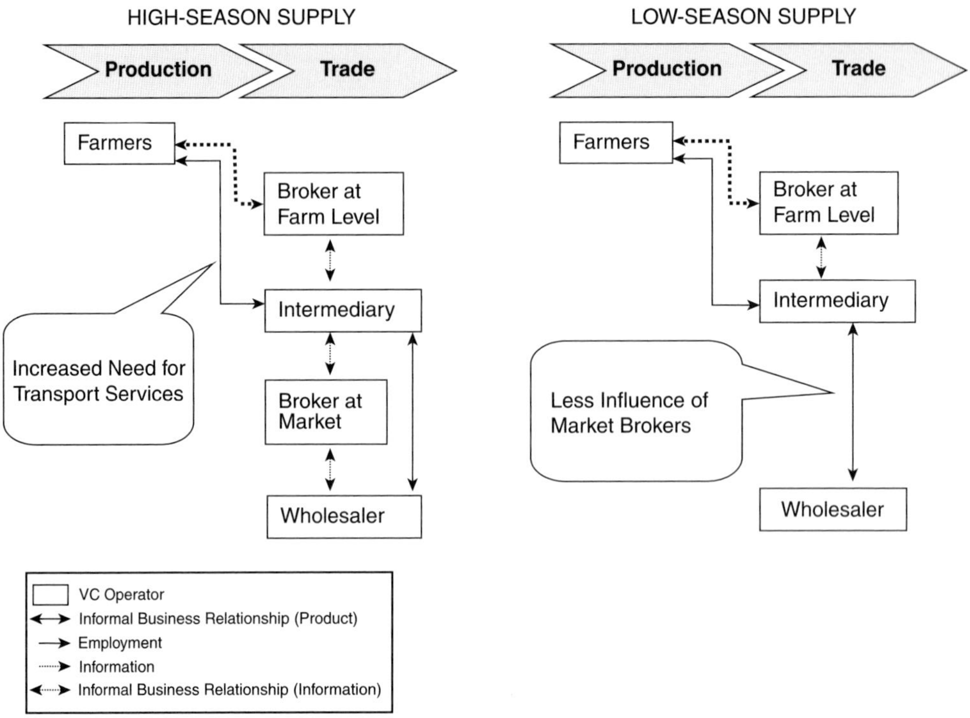

Fig. 10.3. Comparing high (rainy) and low (dry) seasons for tomato and onion (from König *et al.*, 2008).

Table 10.2. Median profit[a] per kg (in US$) for key actors in each value chain, by high- (HS) and low-supply seasons (LS).

	Tomatoes				Onions			
	Tanzania		Kenya		Tanzania		Kenya	
Market actor	LS	HS	LS	HS	LS	HS	LS	HS
Farmers	0.40	–0.01	0.46	0.00	0.25	0.00	0.21	–0.04
Wholesalers/ Intermediaries	0.01	0.06	0.09	0.04	0.05	0.02	0.05	0.03
Retailers	0.11	0.05	0.12	0.02	0.19	0.22	0.08	0.09

[a]Sales revenue minus (total) costs.

which leads to a share of 26% of the total profit to be distributed among all actors. Retailers gain 7% in the low season, although their profit per kilogram is higher compared with the wholesalers' profit, which is explained by the lower quantities traded per person. There are many more retailers in the markets than wholesalers, and so they have to share the total volume of goods and, therefore, the profit.

Standards

Legal regulations, including standards, are applicable in both countries; however, they are not strictly enforced and actors along the value chains are unaware of them. Standards and regulations in Kenya and Tanzania fall into four categories: (i) packaging and measures for selling the produce; (ii) health and sanitary issues, except for chemical residues; (iii) use of chemicals in pre- and postharvest and their residues; and (iv) seeds and chemicals used in the manufacturing industry.

While measuring and packaging standards are widely unregulated (or rules are not consistently enforced), in some instances the market has established its own unwritten rules. For tomatoes, this seems to work better than for onions. Tomatoes are measured in crates, while bags are used for onions. Actors in Kenya have the advantage that they make use of scales more often when trading their products, while this is seldom the case in Tanzania. In Tanzania the situation differs from Kenya: buyers purchase the onions per bag; since the bag size is not standardized, the buyers tend to force the farmers to use larger bags but keep the wholesale price stable. Consequently, the use of these oversized bags causes income losses for the farmers.

As far as the production for the domestic market is concerned, neither farmers nor extension service staff had information about maximum pesticide residue levels. In Kenya, for instance, printed or electronic copies of the standards were not available even at the Ministry of Agriculture, but were available for sale at the Kenyan Bureau of Standards. Consequently, none of the actors in the chain (especially farmers, workers and retailers), as well as the Ministry of Agriculture field personnel (i.e. extensionists), were aware of the existence of any standards for their commodities. Farmers, traders, retailers and consumers were unaware of the issue of chemical residues. The market for organic produce is in its infancy and seems to be an issue mostly for foreigners. Domestic consumers were unaware of food safety issues, and thus there is little demand to enforce such regulations at present. At the production level the use of protective clothing, especially for the sprayers in the fields, is rare. In the informal interviews experts expressed the assumption that, with time, export standards would spill over to the domestic markets – an assumption that could not be proved at the time of the study. Export markets seem to be more profitable; on the other hand, farmers who realize profits in production for the domestic market do not have to pay back the investments necessary for export producer households to comply with food safety standards. Furthermore, the lack of legal requirements to implement food safety standards contributes to the lack of incentive. Hygienic conditions in the markets were rather poor.

Finally, although seed cultivation and the producing/selling of agrochemicals are regulated in the two countries, there are differences. Kenyan national regulations outlaw seed multiplication by farmers for commercial purposes and they have to purchase seeds each season, incurring high input costs.[3] In Tanzania seed regulation differs: local seed production is officially supported in order to reduce input costs at the farm level and to provide locally adapted varieties. As a consequence, seed prices in the two countries differ considerably. For example, the average tomato seed costs US$28.41/ha in Tanzania and US$49.40/ha in Kenya.

Cross-border Trade

Regional trade between Kenya and Tanzania takes place within the framework of the Customs Union (CU), recently established by the East African Community (EAC). The CU is a major part of the establishment of a common market between Uganda, Tanzania and Kenya.[4] The CU has been in place since January 2005 as part of the transition period to develop a common market, until January 2012, when the transition period will be over.

Cross-border trade in this study was analysed for the trade of onions from Tanzania to Kenya. This trade is driven by Kenyan consumers' demand for Tanzanian onions due to their high quality, so a remarkable quantity of onions crosses the border in one direction. In Kenya and Tanzania, cross-border trade is seen as a sub-sector of domestic markets and the same regulations are applied. However, more actors are involved in the governance of this particular value chain, a few of whom are solely involved in cross-border trade (CBT) (clearing agents and the authorities at the border). Other actors, such as wholesalers, who are involved in the domestic onion chain as well as in CBT, achieve higher profits by participating in both chains. Some of the actors who are involved at the later stages of the value chain, such as porters and retailers, are not impacted by the particularities of CBT; they deal with the commodities as with those for domestic value chains. Consumers finally have some influence on the market. Figure 10.4 illustrates the cross-border value chains for Tanzanian onions.

Tanzanian onions produced in Karatu District are traded to Kenya along two different channels, either passing through Kilombero wholesale market in Arusha or going directly from the production site to Kenyan markets. Farmers in Karatu District and traders from Nairobi's Wakulima market describe the Kilombero traders' association as a very powerful organization in CBT; even some VC operators mentioned the negative aspects of the association's monopoly position. However, in general market linkages between Kenyan and Kilombero traders in Arusha are very good, which means that Kenyan traders do not need to travel to Tanzania for every transaction to get the best quality during the high season.

The majority of actors involved gain from CBT. However, the increasing volume of onions produced in Tanzania but consumed in Kenya leads to decreased onion

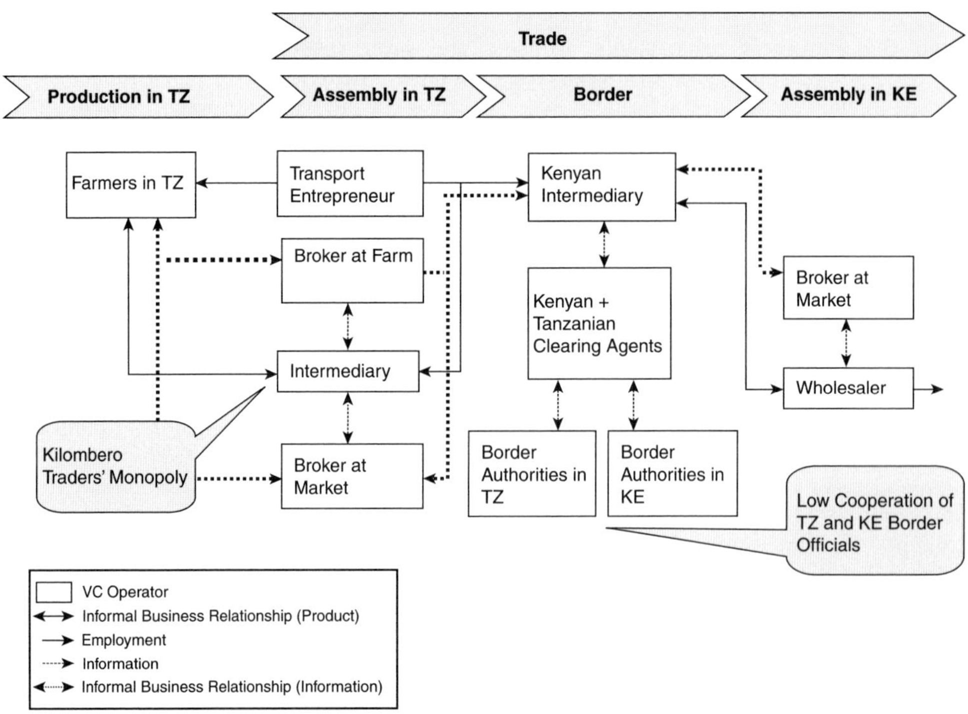

Fig. 10.4. Cross-border trade: value chain operators and market linkages (from König *et al.*, 2008). KE, Kenya; TZ, Tanzania.

prices in Kenya, affecting the Kenyan onion farmers negatively. Tanzanian onion farmers, intermediaries and clearing agents involved in CBT have increased their income through this new opportunity. The strength of the Kilombero traders' association has created a bottleneck for the farmers in Karatu and the traders from Wakulima market. In addition, the Tanzanian farmers face the difficulty that they are urged to use oversized onion bags by the traders. Increased production leads to more work for field workers, porters and other participants in the value chain at that end.

The study team identified the lack of binding measuring standards as one important bottleneck and recommended that both governments implement harmonized standards, which would empower the small actors significantly.

service and are accepted by the market to some extent.

In both countries, the role and the influence of the brokers depend greatly on the season. In times of plentiful supply the brokers' service connecting sellers and buyers is important. Farmers in remote areas are even more dependent on this service, which influences their bargaining power. In both countries the better-graded tomatoes are sold to the important wholesale markets (Arusha, Nairobi), while the poorer grades are sold at the local markets. However, farmers in Tanzania seem to be more involved in assembly and transport than their colleagues in Kenya. That is the reason why they have better and more frequent relationships with their customers (market information) and possess a better overview of prices and the market situation.

Comparing the Value Chains between Kenya and Tanzania

In general, results show many similarities between the two countries, in particular because the level of standards applied is quite low and almost all transactions are of a spot-market nature. Another similarity is the low level of cooperation among the VC actors.

However, the markets in Kenya seem to be much more complex, often including many more actors than in Tanzania, where some of the markets are strictly controlled by cartels. This applies more to the onion chains in both countries, where more brokers and intermediaries are involved.

The input market is a specialized aspect of the chains. Since cultivation and reproduction of seed material is much more liberalized in Tanzania than in Kenya, those inputs are, on average, cheaper in Tanzania. In addition, some farmers have cultivated the seeds themselves to reduce costs. The number of tomato and onion varieties grown is quite limited; however, the tomato varieties Tengeru 97 and Tanya promoted by the World Vegetable Center are produced in Tanzania, supported by the local extension

Comparing tomato value chains

The similarities found in the chains of Tanzania and Kenya are that farmers dominantly plant the varieties Col J and Onyx. There seems to be widespread use of chemicals on farms in both countries, with few producers implementing ecological/organic production methods. In both countries the majority of tomato farmers do not store their produce, chiefly because facilities are not available. The level of cooperation among the actors is low, as they do not consider it necessary to build up long-term relationships. The packaging used for tomatoes is not standardized by law, but has established itself in the market in both countries with a more or less standard weight, even though the produce is not generally weighed. In neither country do brokers play a dominant role, except in the rare times of severe shortage of supply or oversupply. While brokers and other actors elongate the chain in Kenya, tomato farmers in Tanzania are less dependent on them.

At the production level the Mwea production region in Kenya mostly employs rain-fed irrigation, but Tanzanian farmers have access to furrow irrigation systems.

Record keeping was weak for farmers in both countries, but skills seemed to be more widespread in Kenya. At the assembly level, the team found that tomatoes were graded into six grades in Kenya, while only two or three grades were used in Tanzania. Packaging involves different crates in the two countries (TZ: 30–40 kg; KE: 32 kg and 64 kg). The Kenyan farmers prefer to sell most of their commodities directly at the farm gate, while Tanzanian producers sell the majority of their goods at the markets. When brokers buy directly at the farm gate they normally act as wholesalers in the markets of Tanzania, while in Kenya this is often performed by a third party. Tanzanian brokers offer inputs or credits while Kenyan brokers provide important services as information agents, such as providing news about market demand, diseases, etc.

Comparing onion value chains

The differences in the chains are highest at the input level, as the seed quality in Tanzania is better and more adapted to the climate and soil than seeds in Kenya. Farmers' total costs for pesticides are lower in Tanzania than in Kenya, as farmers require less due to the good seed material. However, Tanzanian farmers tend to use more fertilizer, even though the costs are higher. Packaging of the produce in Kenya is standardized in 14 kg bags, but their content generally tends to be more than 14 kg. This is an advantage to the farmers compared with the situation in Tanzania, where farmers are forced to use the large 120 kg bags.

The biggest difference, however, is in the overall quality of the product, which is much higher in Tanzania than in Kenya. Tanzanian farmers have specialized in regional trade and produce directly for it and, as a consequence, the seasons for the farmers are reversed. High season in Kenya is when the supply from Tanzania is low. In the low season in Kenya, the farmers could grow more onions but the market is heavily reduced because of the competition from Tanzanian onions. In the low season in Kenya, most farmers cultivate onions only to maintain their links to the market. They plant fewer hectares, and some farmers do not plant any. Tanzanian farmers profit from a number of advantages, some of which have been stated above. Others include a drier climate, which helps in the drying of onions in the field; better soil; lower labour costs; cheaper land; and better exchange rates.

Value Chains and Poverty Reduction

The tomato and onion value chains in Tanzania and Kenya have the potential to contribute to poverty reduction, but because of their complexity one has to differentiate between possibilities at each step and for each actor group. Vegetable producers in the respective chains are normally not among the poorest, but farm workers, hawkers and, in particular, porters are among the poor actors according to the livelihood analysis of this study.

As an example, Fig. 10.5 illustrates the livelihood analysis: the operators interviewed ranked between 37% and 58% of the maximum points necessary for the four livelihood assets. Their distribution is quite similar, except for the tomato VC actors in Tanzania. The operators with the poorest livelihood situation were casual workers employed on a daily basis according to demand, i.e. carriers at wholesale and retail markets, farm workers and handcart drivers in the markets. Farmers ranked in mid-table compared with all other operators, and were distinguished from the poorer ones by higher financial and physical assets. Actors in the assembly and retail levels were notable for their superior livelihood due to higher financial and physical assets. It should be noted that it was difficult to distinguish between brokers and intermediaries in the urban markets because their functions and activities overlap. It can therefore be concluded that brokers and intermediaries are the least poor actors in the tomato and onion value chains of Kenya and Tanzania, and are characterized

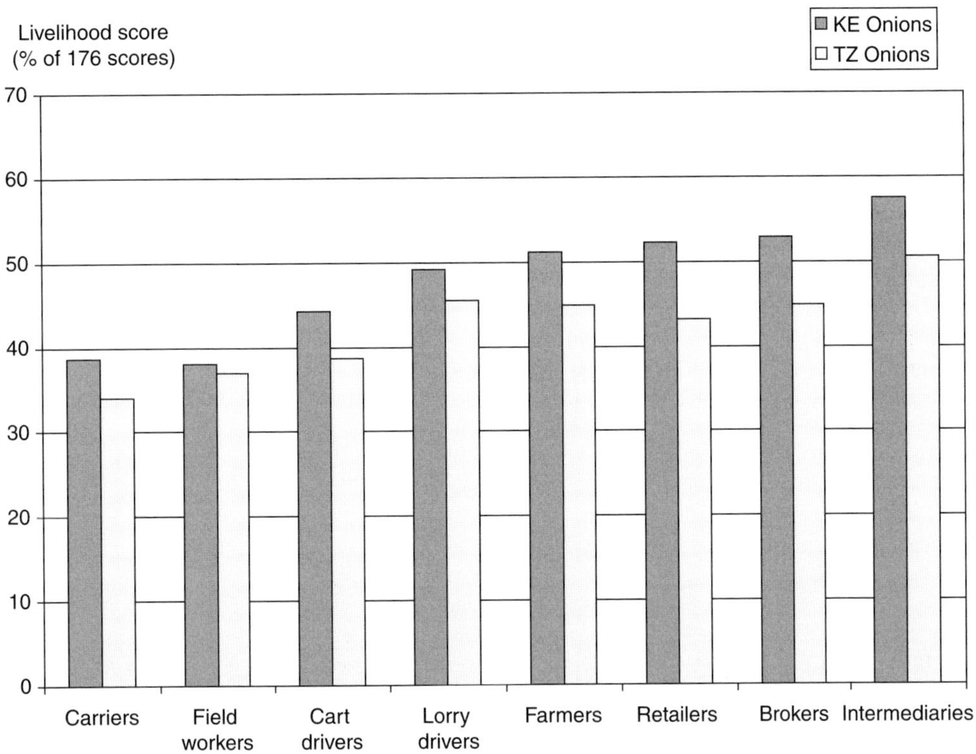

Fig. 10.5. Comparison of the livelihood level of key actors in the onion value chains of Kenya (KE) and Tanzania (TZ) (from König *et al.*, 2008).

by having the highest human, financial and physical assets.

Livelihood analysis of the actors of the tomato VC in Tanzania differs slightly from the other three value chains, the most notable actors here being the market retailers and the carriers. The latter achieve a surprisingly high livelihood level in this VC but, taking into account that only two carriers were interviewed, the authors recommend further validation of these results.

In contrast, for market retailers both qualitative and quantitative data indicate a relatively low livelihood level. This is explained by the fact that the indicator for choosing a retailer to be interviewed was that he or she sells mainly tomatoes or onions (over 50% of the products they offer). It emerged that those retailers who specialize in only one or a few crops belong to the poorest actors on the markets. Consequently, the livelihood scores for the retailers are very low. The less poor retailers in the retail markets in Arusha are those with large stalls and a large selection of different fresh fruit and vegetables; tomatoes are only one among other important crops for their business and, for this reason, they have not been considered in the present research. The same trend, but less pronounced, can be detected in the onion VC of Tanzania. Tomato farmers in both countries reach the highest score among all actors. Tanzanian tomato farmers gained particularly high profits in the spring of 2007, which improved their livelihood situation significantly. In general, tomato farmers in both countries have access to irrigation systems that can be perceived as one indicator of relatively high human physical assets. It has to be noted that brokers in Tanzania are not considered here, because they play a minor role in this chain. Figure 10.6 compares the tomato VC of the two countries.

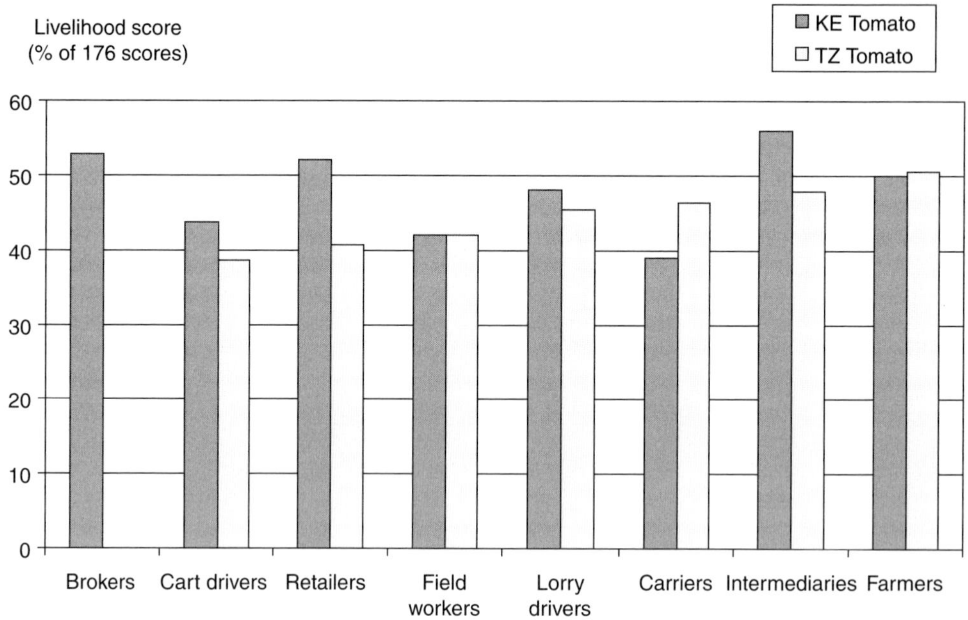

Fig. 10.6. Comparison of the livelihood level of key actors in the tomato value chains of Kenya (KE) and Tanzania (TZ) (from König *et al.*, 2008).

Value chains can contribute directly to poverty reduction in two specific ways. First, they provide employment opportunities for many people, including the less skilled, to work on and off the farm in rural and urban areas. Secondly, because of their informal nature, both value chains have relatively low levels of entry for new actors (except for the transport level). However, existing cartels (especially in the Tanzanian onion sector) do constrain the entry of new actors. Another constraint is the high dependence on middlemen, who often take advantage of their position by either dealing with large volumes of a certain produce or having a monopoly on information. An improved market transparency (chain governance) could help disadvantaged actor groups such as farmers or porters. A number of VC supporter organizations, such as NGOs (e.g. KIT/FAIDAMali/IIRR, 2006) or some authorities, are much in favour of grouping as a means of empowerment of these weak actor groups.

Summary and Conclusions

This study applied the value chain approach combined with a livelihood analysis for assessing tomato and onion value chains in Kenya and Tanzania, including cross-border trade of Tanzanian onions to Kenya. The combination of the two analytical approaches proved to be a valuable method in identifying poor actors along the chains.

In both countries the domestic market is characterized by informal entrepreneurs and on-the-spot market linkages among the actors, particularly in Tanzania. Little external support appears to be available for the small-scale market actors. The study indicates that actors who depend on casual work on farms and markets (e.g. porters) and small-scale retailers are among the poorest, while brokers and transporters profit the most. Farmers were shown to have the highest risks, as their profits vary the most of all the actors, which in some situations includes making a loss rather than a profit.

Although a number of standards and regulations exist at the national level, actors at the value chains' micro and meso levels are unaware of them, except for a few officials. Quality aspects relevant for the trade of tomatoes and onions implemented to date refer only to the outward appearance of the two vegetables, i.e. size and colour. Cross-border trade is considered part of domestic trade because of the common market, and therefore no export regulations are applied. Overall, the analysis of cross-border trade indicates a huge volume in trade of Tanzanian onions to marketplaces throughout Kenya. The major beneficiaries of cross-border trade are the traders and agents with the knowledge of the rules and fees involved at the border, followed by the Tanzanian onion farmers.

In summary, the study team identified several recommendations. At the farm level, cooperation among farmers could improve their negotiating power by marketing larger amounts of produce. Interventions targeting risk management for producers, such as through improved storage capacities during the high-supply season, could contribute to the strengthening of their position in this value chain. Careful application methods for pesticides will improve farmers' and field workers' health. At the produce assembly level, improved market infrastructure could help reduce postharvest losses. More market transparency in terms of prices could help the small retailers to reach a better price, which is also important for farmers. For the poorest actors in the value chains, i.e. porters and handcart drivers, a legalization of their status would improve their situation enormously. They could further benefit from empowerment through the establishment of (in)formal groups, as well as by a better access to market information.

Further research

Based on this study, further research could give more insights into, for example, the economic and legal implications of a higher level of formalization of business interactions within an informal sector. The challenges and implications of regional cross-border trade should be reviewed in more detail. More general research topics could include whether the market for organic products holds economic potential for smallholder farmers; or the assessment of whether contract and/or group farming would be more lucrative than individual farming. Both aspects were widely discussed, but a positive impact was not proved in this study. Finally, recent thinking on conceptual issues in value chain analysis stresses the need for examining the prospects to empower the poor and to serve the poor in exercising their choice as consumers (Prahalad, 2004; Simanis *et al.*, 2008). This issue should be addressed.

Acknowledgements

The authors would like to thank GTZ-BEAF, represented by Michael Bosch, for the commission to carry out the study. Dr Dagmar Mithöfer (*icipe*) and Stefan Pletziger (AVRDC) acted as representatives of our partner institutions in Kenya and Tanzania, and we are very thankful for their support with the logistics and organization, as well as their renowned expertise. Hassan Mndiga (AVRDC) was an important key resource person who provided a great deal of important input for the study. In Kenya we cooperated closely with GTZ-PSDA, represented by Eberhard Krain, Margaret Orina and Heike Höffler. Monica Mueni and Virginia Mwai (both from the Ministry of Agriculture in Kenya) were very supportive during the data collection in Kenya. All the above helped the team to establish contacts for interview partners and provided detailed information on the Kenyan situation.

Our cooperation with the translators was fruitful and effective in carrying out the interviews at all levels, as intended. We would like to thank Gladys Machange, David Solomon, Mayson Ruangisa, Peter Karanka and Francis Imani in Tanzania, and Teresiah Mbugua and Gibson Githaiga in Kenya, for their excellent cooperation.

During our data collection the team split up into subgroups at different production sites and markets to carry out interviews. We appreciate the collaboration with our counterparts in assisting us in contacting all the interview partners and receiving detailed information about the local situation. In Tanzania we cooperated with Simon Loto (Arumeru District), Mr Maganga (Karatu District) and Fatuma Mdibalema (Arusha Municipality Council). In Kenya we were assisted in Kieni West by Peter Ndungu (Farm Concern International) and Peter Makokha (Ministry of Agriculture); in Mwea by Obadiah Mwangi; and in Nairobi by Duncan Murithi (Farm Concern International) and Monica Mueni (Ministry of Agriculture). Regarding the issue of cross-border trade, we offer special thanks to Josphat Mugambi, a Kenyan trader, for his open-minded cooperation.

We talked to more than 100 interview partners in both countries, and this study would not have been possible without their willingness to spend so much time with us.

Endnotes

[1]Further results, which are not presented in this chapter, can be found in König *et al.* (2008).
[2]Therefore, transport is not included in the detailed market linkage description.
[3]In Kenya mostly imported seed is available, but is sometimes contaminated with non-standard seed, which lowers the quality at high-import seed prices.
[4]Rwanda and Burundi recently joined the union and will be integrated into the contractual framework.

References

Achterbosch, T., Allbritton, A., Dang, V.Q., de Jaeger, A., Njue, E., Sonko, R. *et al.* (2005) *EU Assistance to Pro-poor Growth in Horticulture in East Africa and South-East Asia. Summary of Findings*. Wageningen University and Research Centre, Wageningen, the Netherlands.

Ayieko, M., Tschirley, D. and Mathenge, M. (2005) *Fresh Fruit and Vegetable Consumption Patterns and Supply Chain Systems in Urban Kenya: Implications for Policy and Investment Priorities*. Tegemeo Institute of Agricultural Policy and Development, Nairobi, Kenya.

de Haan, L. (2000) Globalization, localization and sustainable livelihood. *Sociologia Ruralis* 40(3), 339–365.

Dorward, A., Anderson, S., Clark, S., Keane, B. and Moguel, J. (2001) Asset functions and livelihood strategies: A framework for pro-poor analysis, policy and practice. Contributed paper to the *EAAE Seminar on Livelihoods and Rural Poverty*. Department of Agricultural Science, Imperial College of Wye, Ashford, UK.

Dorward, A., Kydd, J. and Poulton, C. (2005) Beyond liberalization: 'Developmental co-ordination' policies for African smallholder agriculture. *IDS Bulletin* 36(2), 80–85.

GTZ (2007) ValueLinks Manual. *The Methodology of Value Chain Promotion* (CD) (in particular, ValueLinks Glossary: pdf file on CD). Deutsche Gesellschaft für Technische Zusammenarbeit (GTZ), Eschborn, Germany.

HDC (2004) *Domestic Market for Fruits and Vegetables*. Horticulture Development Centre, Nairobi, Kenya.

Humphrey, J., (2005) Shaping value chains for development: Global value chains in agribusiness. Report prepared for GTZ-Project: Shaping value chains for development. University of Sussex, Institute of Development Studies, Brighton, UK.

Kelley, T. and Byerlee, D. (2004) Surviving on the margin: Agricultural research and development strategies for poverty reduction in marginal areas (mimeo). Agriculture and Rural Development, The World Bank, Washington, DC.

Kimenyi, M.S. (2002) Agriculture, economic growth and poverty reduction. KIPPRA Occasional Paper 3, Nairobi, Kenya.

KIT/FAIDA MaLi/IIRR (2006) *Chain Empowerment Supporting Farmers to Develop Markets*. Amsterdam, Arusha, Tanzania and Nairobi, Kenya.

König, T., Blatt, J., Brakel, K., Kloss, K., Nilges, T. and Woellert, F. (2008) *Market-driven Development and Poverty Reduction: A Value Chain Analysis of Fresh Vegetables in Kenya and Tanzania*. Humboldt University of Berlin, SLE Publication Series, Berlin, Germany.

M4P (2005) Making value chains work better for the poor. *A Toolbook for Practitioners of Value Chain Analyses: Making Markets Work Better for the Poor* (M4P). Hanoi, Vietnam.

Muendo, K.M. and Tschirley, D. (2004) Improving Kenya's domestic horticultural production and marketing system: Current competitiveness, forces of change, and challenges for the future. Vol. I: Horticultural Production Draft for Review, Working Paper 08A/2004, Tegemeo Institute for Agricultural Policy and Development, Nairobi, Kenya.

Pauw, K. and Thurlow, J. (2010) Agricultural growth, poverty, and nutrition in Tanzania. IFPRI Discussion Paper 00947, International Food Policy Research Institute (IFPRI), Washington, DC.

Porter, M. (1985) *Competitive Advantage*. Free Press, New York.

Prahalad, C.K. (2004) *The Fortune at the Bottom of the Pyramid: Eradicating Poverty Through Profit.* Wharton School Publishing, Upper Saddle River, New Jersey.

Simanis, E., Hart, S.L., DeKoszmovszky, J., Donohue, P., Duke, D., Enk, G. *et al.* (2008) *The Base of the Pyramid Protocol: Toward Next Generation BoP Strategy,* 2nd edn. Center for Sustainable Global Enterprise, Cornell University, New York.

Tschirley, D. and Ayieko, M. (2008) *Assessment of Kenya's Domestic Horticultural Production and Marketing Systems and Lessons for the Future*. Tegemeo Institute of Agricultural Policy and Development, Nairobi, Kenya.

USAID (2006) Analysis of the integration of MSEs in value chains. Tanzania Research Protocol, MicroReport 72, Washington, DC.

Weinberger, K. and Lumpkin, T.A. (2007) Diversification into horticulture and poverty reduction: A research agenda. *World Development* 35(8), 1464–1480.

World Bank (2005) *Pro-poor Growth in the 1990s. Lessons and Insights from 14 Countries*. World Bank, Washington, DC.

11 Supply Chains for Indigenous Vegetables in Urban and Peri-urban Areas of Uganda and Kenya: a Gendered Perspective

Katinka Weinberger, Margaret Pasquini, Phyllis Kasambula and Mary Abukutsa-Onyango

Introduction

In this study we focus on the production and marketing of African Indigenous Vegetables (AIV) in urban and peri-urban areas and document how smallholder farmers and other actors along the supply chain gain from being involved in this activity. The contribution of AIV[1] to income and the well-being of small farmers and other actors may be substantial. For instance, in Nairobi it is estimated that 30% of all vegetables sold are AIV produced in the vicinity of the city (Ngugi *et al.*, 2007). Yet little is known about the economic importance of AIV for income generation and livelihoods in urban and rural areas of sub-Saharan Africa (SSA). Writers have highlighted that indigenous vegetables are a valuable food source for the poor, especially in isolated communities (Grivetti and Ogle, 2000), during drought (Humphry *et al.*, 1993) or during social unrest and war (Smith *et al.*, 1995, 1996). However, increasingly there is the realization that (i) indigenous vegetables are a vital component of the rural economy (South Africa: High and Shackleton, 2000); (ii) they contribute to livelihoods through their value for consumption and as income (Tanzania: Weinberger and Msuya, 2004; Weinberger and Swai, 2006); and (iii) there is a large and perhaps growing market for indigenous vegetables (Ngugi *et al.*, 2007; Weinberger, 2007).

Globally, women play an important role in small-scale horticultural production and marketing (Tickner, 1995; Spieldoch, 2007). Women are a strong presence in the sector throughout SSA, especially in marketing and trading (Weinberger and Pichop, 2009). In the urban areas of West Africa, women comprise 60–80% of the workforce in trading (Ruel *et al.*, 1998; Levin *et al.*, 1999). Gockowski *et al.* (2003) estimate that there are at least 4100 female retailers in and around Yaounde, Cameroon, selling African leafy vegetables.

Where women have obligations to provide subsistence needs and men have obligations to provide cash, the production of vegetables is often considered a female domain (Ezumah and Di Domenico, 1995; Dolan, 2001; Howard and Nabanoga, 2007). However, gender norms are complex and can change in response to shifting economic, political and cultural forces, thereby creating new opportunities for both women and men (Quisumbing and Pandolfelli, 2010). Thus, with the increasing importance of horticultural crops in contributing to cash income, changing norms and beliefs about men's and women's involvement in the production process have been reported. Ezumah and Di Domenico (1995) found that, among the Igbo of Nigeria, single men became heavily involved in the production of horticultural crops as they provided additional sources of income. On the Jos Plateau, a significant commercially oriented horticultural production area that supplies markets all over Nigeria and beyond, men dominate the system at both the production and marketing level (Pasquini, 2002, unpublished PhD thesis; Porter *et al.*, 2003). In Cameroon, women have been found to participate less than men in intensive commercial horticulture production (Gockowski and Ndoumbe, 2004). An example of how the increasing importance of horticultural production for export has resulted in power struggles for land and labour is provided by Dolan (2001, 2002) for Kenya, where horticultural production for export has greatly increased over recent decades. Traditionally, horticulture was considered a female domain, with women holding usufruct rights over horticultural lands and controlling the income derived from horticultural produce. However, with the increasing level of horticultural production for export, the pressures on land and the labour force have resulted in a disproportionately increased female involvement in horticultural labour and a loss of control over horticultural production.

Nevertheless, others have shown that increasing horticultural production and the resultant pressure on resources do not necessarily need to be disadvantageous for women, and can create opportunities. Raynolds (2002) describes how, in a predominantly patriarchal environment in the Dominican Republic, women have made use of the fact that labour is the base of horticultural contract farming to strengthen their bargaining power. Because the increase in contract farming has led to an increased demand for female labour, women have been able to claim payments for work in household-based contract farming, which has in cases been extended to compensation for non-contracted crops.

Our study contributes to the literature by considering the gender dimension across the various actors involved in supply chains. Our focus is on supply chains for indigenous vegetables in urban and peri-urban areas of Kenya and Uganda. A supply chain can be defined as 'the network of organizations that are involved, through upstream and downstream linkages, in the different processes and activities that produce value in the form of products and services in the eyes of the ultimate consumer' (Christopher, 1998). We briefly describe the organization of supply chains and the actors involved in both countries, before we turn to an analysis of how the benefits of participation in these supply chains are distributed between men and women. In particular, we assess the degree of involvement of men and women along the chain. We analyse the levels of income, and how these differ for different actors and for men and women across different sites. Finally, we discuss a range of policy options that would support the various actors of supply chains for indigenous vegetables.

Data, Materials and Methods

The results presented here are based on data collected for a research project on indigenous vegetables in urban and peri-urban areas of SSA that covered seven countries, including the west, the east and the south. In this chapter we present results from Kenya and Uganda. In each country, the capital city (representing a 'big' city) and a smaller city were chosen. Interviews were conducted in October and November 2006. The survey team identified vegetable-growing communities in the vicinities of cities, and classified locations as either urban or peri-urban based on factors such as physical, functional and administrative criteria to distinguish them from rural areas. This classification considered roads, housing density, transportation and administrative boundaries. In a second step, a list of vegetable-growing farmers was prepared and producers were chosen at random from this list. A randomized snowballing approach was applied, moving downstream from producers to retailers. Farmers, as the first actors in the supply chain, were interviewed directly on their farms. Each actor of the chain was then requested to provide up to five buyer contacts, and from these contacts a random sample of actors in the next step of the chain was drawn, down to the retail level. Table 11.1 details the sample size. Three standardized questionnaires were used to interview producers, intermediaries and retailers. Since many respondents produce and market several indigenous vegetables, detailed questions about yield, volume, prices and revenue were asked only with regard to those three crops that contributed the most to total turnover for farmers, intermediaries and retailers. For farmers, detailed information on production practices was collected for the most recent complete production season; for intermediaries and retailers, information was collected on a monthly basis. In addition, for all respondents, the total gross income generated from vegetables on a monthly basis was collected for the previous year. Complete information was available for 730 respondents.

We used the following estimation framework to analyse the determinants of the annual income of actors along the supply chain for indigenous vegetables. For all respondents, we aggregated the monthly gross income for the year and converted it into US$. While recognizing that annual gross income is an insufficient measure for the profitability of an enterprise, given the challenges in obtaining reliable cost data we consider this measure adequate to provide us with information about the size of transactions in which an enterprise is involved.

A semi-logarithmic income equation was estimated by ordinary least squares:

$$Y_i = \beta_0 + \beta_1 X_i + \beta_2 Z_i + u_i \quad (11.1)$$

where Y_i represents the annual income of individual *I*, X_i is a vector of observed individual characteristics, Z_i is a vector of observed characteristics of the supply chain and u_i is an error term.

The dependent variable in this analysis is the natural logarithm of annual income of the enterprise, converted to US$. Specifically, we wanted to discover whether female gender negatively affects income. Controlling for other variables and a dummy for gender is, therefore, included in the equation. The vector of observed individual characteristics includes age, education and household size. The vector for characteristics of the supply chain includes a dummy for the position of the respondent along the supply chain (producer, intermediary, retailer) and dummies for the site. It includes a variable indicating whether the respondent operates with a contract, as contract farming has been found to raise household income for farmers in a number of studies (Warning and Key, 2002; Bolwig *et al.*, 2009; Miyata *et al.*, 2009). It is therefore worth asking whether an income effect can also be observed for other actors along the supply chain.

Table 11.2 summarizes separately the mean values of variables for men and women used in our model. In our sample,

Table 11.1. Sample size by country, site, actor and gender.

		Farmer		Intermediate		Retailer		Total		
Country	City	M	F	M	F	M	F	M	F	Sample size
Kenya	Kisumu	33	41	1	60	3	72	37	173	418
	Nairobi	22	34	19	53	5	75	46	162	
Uganda	Kampala	50	20	23	13	2	54	75	87	312
	Mbale	41	26	12		4	67	45	105	
Total		146	121	43	138	14	268	203	527	730

M, male; F, female.

Table 11.2. Mean values of variables, by gender.

	Variable	Male (*n* = 203)		Female (*n* = 527)		Total (*n* = 730)	
		Mean	SD	Mean	SD	Mean	SD
Income (US$)		2296.0	4105.4	1830.6	3477.9	1960.0	3666.2
Household size		7.1	4.3	6.4	3.3	6.6	3.6
Education	Primary and below (%)	59.6		74.8		70.5	
	Above primary (%)	40.4		25.2		29.5	
Age	≤ 30 (%)	18.2		19.5		19.2	
	30–40 (%)	39.9		38.0		38.5	
	40–50 (%)	26.1		24.9		25.2	
	≥ 50 (%)	15.8		17.6		17.1	
Site	Kampala (%)	36.9		16.5		22.2	
	Mbale (%)	22.2		19.9		20.5	
	Kisumu (%)	18.2		32.8		28.8	
	Nairobi (%)	22.7		30.7		28.5	
Respondent type	Farmer (%)	71.9		23.0		36.6	
	Small intermediary (%)	12.3		22.4		19.6	
	Large intermediary (%)	8.9		3.8		5.2	
	Small retailer (%)	3.4		33.6		25.2	
	Medium retailer (%)	2.5		13.7		10.5	
	Large retailer (%)	1.0		3.6		2.9	
Contract (%)		16.3		31.9		27.5	

women on average earned lower incomes, had smaller families and had lower education levels than men. In our sample, approximately 75% of the women had completed only primary education or less, whereas 40% of the men had completed secondary education or above. The age distribution was similar across the two genders. Regarding the position in the supply chains, the representation by gender in our sample is quite different. In our male sub-sample the highest percentage of respondents (71.9%) were farmers, whereas in our female sub-sample the highest percentage of respondents were small retailers (33.6%). Finally, more women than men operated with contracts (31.9% versus 16.3%).[2]

Results

Production of African Indigenous Vegetables

For respondents in our sample, farming was generally the main – and often the only – income source. Ninety-four per cent of farmer respondents in Kenya and 83% of farmer respondents in Uganda gave this as their only income source. However, within farming, vegetable production usually constitutes only one activity among many that include staples (rice, maize, sweet potato), cattle, pigs and poultry and fruit. The average number of different farm activities was 4.5 in Kenya and 3.8 in Uganda. Nonetheless, for nine farmers in Kenya (7% of the sample) and 30 in Uganda (22% of the sample) vegetable production was the only farming activity mentioned, indicating a rather high degree of specialization for these farmers. Farmers in both countries cultivate a range of different AIV. The most important, cultivated by more than half of the farmers, are shown in Table 11.3. Most are of the leafy type except for the African eggplant, which is the most important crop in Uganda but of no importance in Kenya. The total number of different crops cultivated was 14 in Uganda and 12 in Kenya.

Table 11.4 shows a selection of variables that characterized farmers' operations in Kenya and Uganda, by site. In both countries the average area under AIV was

Table 11.3. Most important crops and their share (%) of farmers' total produce.

Kenya		Uganda	
Vegetable cowpea (*Vigna unguiculata*)	80	Amaranth (*Amaranthus* spp.)	84
African nightshade (*Solanum scabrum*/*S. villosum*)	76	African eggplant (*Solanum aethiopicum*/*S. macrocarpon*)	83
Spider plant (*Cleome gynandra*)	72	Ethiopian mustard (*Brassica carinata*)	59
Amaranth (*Amaranthus* spp.)	66	Spider plant (*Cleome gynandra*)	58
Pumpkin (*Cucurbita* spp.)	50	Vegetable cowpea (*Vigna unguiculata*)	50
Total crops grown (*n*)	12	Total crops grown (*n*)	14

Table 11.4. Selected farmer characteristics.

	Uganda		Kenya	
	Mbale	Kampala	Kisumu	Nairobi
Total area under top 3 AIV (m^2)	3514	3790	3373	1853
Share in total vegetable area (%)	98	99	72	84
Production over one season (kg)	3726	1279	625	1662
Revenue over one season (US$)	361	166	274	425
n	67	70	74	56

between 1853 and 3790 m^2. In both countries the area allocated to AIV was larger in the smaller cities (Mbale and Kisumu) as compared with the larger cities (Kampala and Nairobi). Not surprisingly, the area under cultivation was smaller in larger cities because of the higher pressure on land. In both countries AIV made up a large share of the total vegetable area among respondents: between 70% and 85% in Kenya and nearly 100% in Uganda, highlighting the important role that AIV played in overall vegetable production in our sample. Overall average production was much higher in Kampala and Nairobi, as compared with Mbale and Kisumu. Since land areas in our sample were smaller in larger cities, this indicates a higher productivity in larger cities, which might reflect higher prices for land and/or wages and, thus, higher pressure for productivity, or better access to farming inputs. The average revenue from one completed production season was US$361 in Mbale, US$166 in Kampala, US$274 in Kisumu and US$425 in Nairobi.

Organization of the supply chain

In this section we will focus on a brief description of the supply chain for AIV in Kenya and Uganda, and the various actors involved along the chain.[3]

Production of AIV takes place at the farm level, as we did not consider AIV that were being harvested in forests or bush land. Farmers sell directly to either consumers or retailers, or through the intermediary sector. The intermediary sector includes collectors, middlemen or brokers and wholesalers. Collectors aggregate produce into larger quantities by purchasing produce from farms and transporting vegetables to collection points, for instance at wholesale markets. In contrast, middlemen sell to various points along the supply chain, usually with the aim of identifying opportunities for arbitrage along the chain. Wholesalers supply urban and peri-urban retailers who come to purchase produce at wholesale markets.

The retail sector in our sample was dominated by wet market vendors in both countries. The produce from most farmers we interviewed and chose as the base for the supply chains that we tracked had their produce sold in local markets, called wet markets. Only a few retailers in our sample were street vendors or owners of grocery stores. Retailers mostly sold to end consumers, with a small share of wet market vendors also selling to restaurants.

The most frequent supply chain was a relationship where farmers sold to collectors (Kenya) or wholesalers (Uganda) who, in turn, sold to retailers in wet markets, but a certain share of farmers, higher in Kenya than in Uganda, sold directly to households (Fig. 11.1). In Kenya many farmers (approximately one-third) sold directly to households. Other trading relationships involved collectors (21%), wholesalers (17%) and wet market vendors (14%). Both collectors and wholesalers sold mainly to wet market vendors (61% and 62%, respectively) but, for a substantial proportion, households were their main trading partners (26% and 20%, respectively).

In Uganda a small majority of farmers sold to wholesalers (51%), while a smaller proportion sold directly to households (19%) and wet market vendors (18%). Wholesalers sold either to other wholesalers (33%) or to wet market vendors (34%), from where produce was sold to private households (76%).

Characterization of supply chain actors

Incomes in the intermediary sector are generally higher and more variable than those in the retail sector. Intermediaries are made up of a heterogeneous group of individuals, and total income can vary widely and is characterized by high standard deviations, as shown in Table 11.5. Incomes in Nairobi are higher than in Kisumu, and these

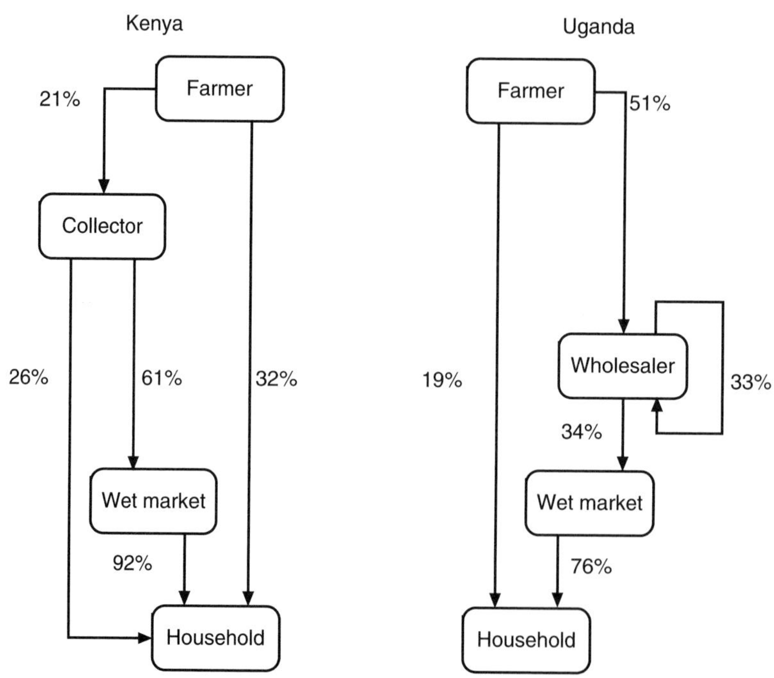

Fig. 11.1. Supply chain channels for indigenous vegetables in Kenya and Uganda. The values do not add up to 100 because this graph summarizes only the most important relationships; other trading relationships are described in the main text.

differences are significant ($p = 0.01$) for both intermediaries and retailers. However, in Uganda, there is no significant difference between the incomes of intermediaries and, for retailers, incomes are significantly higher ($p = 0.01$) in Mbale as compared with Kampala. Overall, incomes in Kenya can be much higher than in Uganda, with the highest reported being US$36,026 and US$8063, respectively.

Due to the number of differences among actors, we clustered intermediaries and retailers separately, based on the total volume of produce sold and the number of supplier relationships. Building typologies through cluster analysis offers a formal way of developing actor profiles (Acosta-Michlik and Espaldon, 2008). For intermediaries the clustering resulted in two categories (small and large), and for retailers three categories (small, medium and large). According to the results of the clustering, in both countries approximately 80% of respondents fell into the category of small intermediaries, and around two-thirds of all respondents (somewhat higher in Uganda at 71%) were small retailers (Table 11.6). In Kenya small and medium retailers sold 80% of all AIV, while in Uganda small retailers sold 67% of all AIV in the market.

Our data show that there is a considerable difference in the overall monthly quantity traded by all retailers between Kenya and Uganda. Overall, the volumes traded vary little throughout the year.[4] The average quantity sold by a retailer in Kenya is approximately seven times that sold by a retailer in Uganda.

The data in Table 11.7 show the participation of women in the various segments of the supply chain in Kenya and Uganda. Overall, the number of women in the entire sector was high in both countries (72.2% average; 80.1% in Kenya versus 61.5% in Uganda). The proportion of women was significantly lower in the farming sector in all locations, as compared with their overall proportion in the sample. In contrast, the proportion of women in retailing was significantly higher than their overall proportion in the sample for all locations. For intermediaries, women were more predominant in Kisumu and Mbale, the two smaller cities in our sample, than in Kampala and Nairobi. In Nairobi, the differences in the proportion of male and female participants among intermediaries were not significantly different from the sample mean.

Overall, our data show that women dominated the supply chain in terms of numbers at the retail level and, to some extent, depending on the location, at the intermediate level, but that men were more prominent in the farming of AIV.

Building on average volumes sold by actors in the supply chain and with regard to the total estimated production of AIV, we attempted to estimate the total number of

Table 11.5. Annual income (US$) of intermediaries and retailers, by country and city.

Type	Country	Site	Mean	SD	Median	Max.	*n*
Intermediary	Kenya[a]	Kisumu	1,943	2,386	772	9881	59
		Nairobi	7,125	8,541	3,597	36,026	72
		Total	4,791	7,006	2,196	36,026	131
	Uganda	Mbale	792	996	314	3,359	11
		Kampala	1,257	1,507	705	8,063	36
		Total	1,148	1,408	552	8,063	47
Retailer	Kenya[a]	Kisumu	1,370	1,435	812	6,862	76
		Nairobi	2,909	3,076	1,436	13,330	80
		Total	2,159	2,533	1,118	13,330	156
	Uganda[a]	Mbale	697	572	476	2,647	73
		Kampala	363	280	238	1,204	57
		Total	550	495	395	2,647	130

[a] $p < 0.01$.

Table 11.6 Respondent clusters and share in sampled market size.

	Kenya			Uganda		
	%	Volume (kg)	Share of total volume (%)	%	Volume (kg)	Share of total volume (%)
Small intermediary	79	22,813	53	79	4,233	35
Large intermediary	21	74,319	47	21	29,682	65
Small retailer	61	16,272	36	71	3,893	67
Medium retailer[a]	29	40,195	44	25	4,702	28
Large retailer[a]	10	56,970	20	4	4,679	5

[a] Volumes sold are similar for medium and large retailers in Uganda; however, these groups differ markedly in the incomes earned, and therefore we decided to place them in different groups.

Table 11.7. Proportion of female actors in supply chains in Uganda and Kenya.

	Farmer		Intermediary		Retailer	
	%	*n*	%	*n*	%	*n*
Kampala	28.6[a]	20	36.1[a]	13	96.4[b]	54
Mbale	38.8[a]	26	100.0[c]	12	94.4[b]	67
Uganda	33.6[a]	46	52.1	25	95.3[b]	121
Kisumu	55.4[a]	41	98.4[b]	60	96.0[b]	72
Nairobi	60.7[a]	34	73.6	53	93.8[b]	75
Kenya	57.7[a]	75	85.0[b]	113	94.8[b]	147

Based on a one-sample *t*-test, comparing the average proportion of female participants in the sample; $p = 0.000$. [a]Proportion of women is significantly lower compared to their average representation across the sample; [b]proportion of women is significantly higher compared to their average representation across the sample; [c]*t*-test not applied because SD = 0.

people partly or wholly reliant on the production or trading of AIV for their livelihood. Since countrywide production or area data for AIV were not available, we based our estimation on total vegetable production data available from FAO (FAO, 2007). Based on these, it was estimated that the total market size for indigenous vegetables was at least 165,000 t in Kenya and 55,000 t in Uganda.[5] Based on the structure of the chain and the average volume dealt with by each agent, as classified in Tables 11.4 and 11.6, the sector is therefore estimated to contribute to the livelihoods of at least 159,000 actors in Kenya and 37,000 actors in Uganda,[6] of whom an estimated 114,000 are women[7] (Table 11.8). Including their dependents, in both countries an estimated 1.3 million people would partly or wholly depend on this sector for their livelihood.[8]

Econometric analysis

Having demonstrated the significant share of women in the sector overall, we now assess the determinants of annual gross income along the supply chain for indigenous vegetables, and whether variations in incomes for men and women vary significantly for different actors along the supply chain, controlling for other variables.

The model appears to be a reasonable fit (Table 11.9). More than 44% of the variance in the dependent variable was explained by the model, and the model is highly significant. Among those variables that describe individual characteristics, household size, gender and age group 30–40 were significant. Incomes in larger families were higher, with men earning significantly higher incomes than women. The estimated

Table 11.8. Total number of people employed in supply chains for indigenous vegetables (from IndigenoVeg survey, 2006 and FAO (2007).

	Kenya		Uganda	
Estimated market volume (t)	165,000		55,000	
Estimated supply chain actors	All	Women	All	Women
Farmers (*n*)	152,074	87,747	21,982	7,386
Intermediaries (*n*)	3,121	2,653	4,659	2,427
Retailers (*n*)	3,863	3,662	10,787	10,280
Total (*n*)	159,058	94,062	37,428	20,093

Table 11.9. Regression results.

	Parameter estimate	SE	Marginal mean (US$)[a]
Household members (*n*)	0.020	0.010*	
Gender			
Male	0.317	0.090***	1012
Female			835
Age (years)			
< 30	0.092	0.117	832
30–40	0.231	0.101*	998
41–50	0.101	0.108	820
> 50			788
Education			
Primary education and less	–0.008	0.077	1077
Above primary			810
Type of respondent			
Farmer			695
Small intermediary	0.327	0.104**	1094
Large intermediary	1.780	0.164***	6499
Small retailer	–0.357	0.103***	442
Medium retailer	0.889	0.127***	1688
Large retailer	1.690	0.215***	4283
Operates with contract (oral or written)			
Yes	0.174	0.084*	1289
No			762
City			
Kampala	–0.958	0.105***	593
Mbale	–1.114	0.110***	500
Kisumu	–0.657	0.093***	789
Nairobi			2019
Intercept	6.810	0.142	906
Adjusted R^2		0.444	
Probability value > *F*		0.0000	

$n = 730$. * $p < 0.05$; ** $p < 0.01$; *** $p < 0.001$. [a] Estimated at mean household size 6.6.

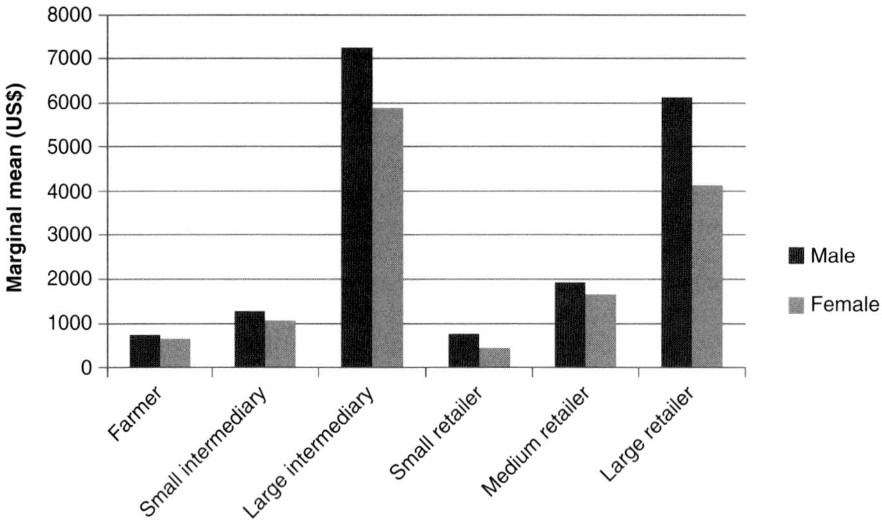

Fig. 11.2. Marginal mean incomes across supply chain actors.

marginal mean income for women and men was US$839 and US$1012, respectively. The income of those in the age group 30–40 was highest, with the lowest being for those 50 years and older. Those supply chain actors who operate with a contract reported significantly higher incomes than those actors operating without contracts. In our sample, incomes were lower in smaller cities (Kisumu and Mbale) as compared with the larger cities of Nairobi and Kampala.

Our results confirm that there were significant differences among the type of respondents, and that small retailers, not farmers, were those with the lowest incomes, with an estimated marginal mean annual income of US$442 versus US$695 for farmers; the highest annual income was for large intermediaries, with an estimated marginal mean of US$6499. The marginal mean income differed between men and women, across all supply chain actors and was always higher for men than for women (Fig. 11.2).

Overall, our results indicate that those with the lowest annual income were female small retailers; however, farmers are only marginally better off.

Summary and Conclusions

Our results confirm that African Indigenous Vegetables are an important means of livelihood for thousands of supply chain actors in Kenya and Uganda, especially for women, who play a central role in the handling and marketing of AIV. We estimate that at least 1.3 million people in Kenya and Uganda partly or wholly depend on this sector for their livelihood.

Various actors are involved along the supply chain and perform a range of different activities. Our analysis has shown that, within these groups of actors, there are differences in incomes that may not be described through their role alone. Large intermediaries have high annual incomes from the indigenous vegetable trade and, in their characteristics, may be more similar to large retailers than to small intermediaries. Along the supply chain, average indigenous vegetable incomes were lowest for small retailers, followed by farmers and small intermediaries.

We assessed gender-based indigenous vegetable income differences among the different groups of supply chain actors by estimating a semi-log income function. Our results showed that women earn the lowest incomes in the supply chain and within each group of supply chain actors. Women, although prominent in the retail sector and in smaller cities in the intermediary sector, on average earn lower annual incomes as compared with male actors in the supply chain, which is true for all segments of the supply chain. Considering the high number of women who are involved in these supply chains and who earn their livelihoods in the production, trading and selling of AIV, policy interventions should specifically consider women's needs, such as access to credit.

Further research

Our findings show that the use of contracts is linked to the higher annual indigenous vegetable incomes of a supply chain actor. However, we did not assess the causality of that linkage. Higher incomes may arise as a result of using contracts in trading arrangements, but that may in itself determine the use of contracts, as higher incomes may serve as an indicator of trust for trading partners. More research is required regarding which factors facilitate the use of contracts, and under what conditions contracts are beneficial for smallholder supply chain actors, as our findings indicate a positive relationship between incomes and the use of contracts in trading arrangements.

Our results highlight that the design of interventions that target supply chains and are aimed at strengthening the role of certain supply chain actors, for instance farmers, should also take into consideration the economic realities of other supply chain actors further downstream. More research is needed to identify how the differences in incomes across gender and actors affect the differences in the relative power of various supply chain actors. One implication of our research is that interventions that aim to strengthen farmers in the supply chain, as opposed to intermediaries and retailers, may be overly simplistic. Small retailers and small intermediaries may need policy and research support to be able to deal with the increasing vertical coordination in horticulture supply chains and its implications for the way trading is being conducted. A question for further research in this respect concerns what institutional arrangements can help small and powerless supply chain actors and how supply chain actors in the intermediary sector can be reorganized.

Acknowledgements

This chapter was prepared when the lead author worked at the World Vegetable Center (AVRDC). The views expressed herein are those of the authors and do not necessarily reflect the views of the United Nations. This study is a product of the contributions of many individuals who have helped directly and indirectly to shape our ideas concerning the organization of supply chains for African Indigenous Vegetables. We acknowledge, in particular, the contribution by Germain Pichop, who organized the supply chain surveys, and Festo Maro and Mei-huey Wu, who assisted in data cleaning and analysis. In the field, this research project could not have been accomplished without the cooperation of the respondents and enumerators. Funding for this research was provided by the European Union under Contract Number 015101, 'Networking to Promote the Sustainable Production and Marketing of Indigenous Vegetables through Urban and Peri-Urban Agriculture in Sub-Saharan Africa', and is gratefully acknowledged.

Endnotes

[1]Indigenous vegetables are either those vegetables that have their natural habitat in sub-Saharan Africa or those that were introduced over a century ago and, due to length of use, have become part of the food culture in the subcontinent (Engle and Faustino, 2007). In the context of this study we considered the following crops: Amaranth (*Amaranthus* spp.); African nightshade (*Solanum scabrum, Solanum villosum, Solanum*

nigrum, Solanum americanum); African eggplant (*Solanum macrocarpon, Solanum aethiopicum, Solanum anguivi*); vegetable cowpea (*Vigna unguiculata*); Ethiopian mustard (*Brassica carinata*); jute mallow (*Corchorus olitorus*); hibiscus (*Hibiscus sabdarifa*); okra (*Abelmoscus esculentus, Abelmoscus caillei*); pumpkin (*Cucurbita maxima, Cucurbita pepo, Cucurbita moschata, Cucurbita mixta*); spider plant (*Cleome gynandra*); crotalaria (*Crotalaria ochroleuca, Crotalaria brevidens*).

[2]These may be either written or oral contracts, with oral contracts being more frequent than written contracts.

[3]For a more detailed analysis please refer to Weinberger and Pichop (2009).

[4]Aggregated values for volumes conceal different varieties of AIV that become available throughout the year.

[5]These values are 10% of the total vegetable production reported by FAO for each country. However, at least in the case of Kenya, this is probably a gross underestimation, since FAO reports detailed data for 13 different vegetable crops for this country. Leafy African vegetables are, therefore, probably aggregated under 'various fresh vegetables', which represents approximately one-third of all vegetable production data. Other sources (see Ngugi *et al.*, 2007) report that AIV represent one-third of all vegetables in markets in Nairobi. The share of AIV in the total volume of vegetables is, thus, probably around 20–30%. However, we decided to base our assumptions on a more conservative estimate.

[6]Based on our results, a certain share of produce is sold directly to households. This share was subtracted from the overall production, and we then estimated the number of intermediaries and retailers based on their overall share in market volumes sold. Since total aggregated volumes sold by the intermediary and retail sectors were very similar, we assumed that all produce sold by intermediaries was sold on to retailers.

[7]Based on the proportion of women for each supply chain category in our sample.

[8]Based on the average household size in our sample.

References

Acosta-Michlik, L. and Espaldon, V. (2008) Assessing vulnerability of selected farming communities in the Philippines based on a behavioural model of agent's adaptation to global environmental change. *Global Environmental Change* 18(4), 554–563.

Bolwig, S., Gibbon, P. and Jones, S. (2009) The economics of smallholder organic contract farming in tropical Africa. *World Development* 37(6), 1094–1104.

Christopher, M. (1998) *Logistics and Supply Chain Management, Strategies for Reducing Cost and Improving Service. Financial Times*, Prentice Hall, London.

Dolan, C.S. (2001) The 'good wife': struggles over resources in the Kenyan horticultural sector. *Journal of Development Studies* 37(3), 39–70.

Dolan, C.S. (2002) Gender and witchcraft in agrarian transition: The case of Kenyan horticulture. *Development and Change* 33(4), 659–681.

Engle, L.M. and Faustino, F. (2007) Conserving the indigenous vegetable germplasm of South-East Asia. *Acta Horticulturae* 752, 55–60.

Ezumah, N.N. and Di Domenico, C.M. (1995) Enhancing the role of women in crop production: A case study of Igbo women in Nigeria. *World Development* 23(10), 1731–1744.

FAO (2007) FAOSTAT data. http://apps.fao.org/page/collections?subset=agriculture (accessed March 2009).

Gockowski, J. and Ndoumbe, M. (2004) The adoption of intensive monocrop horticulture in southern Cameroon. *Agricultural Economics* 30(3), 195–202.

Gockowski, J., Mbazo'o. J., Mbah, G. and Moulende, T. (2003) African traditional leafy vegetables and the urban and peri-urban poor. *Food Policy* 28, 221–235.

Grivetti, L.E. and Ogle, B.M. (2000) Value of traditional foods in meeting macro- and micronutrient needs: The wild plant connection. *Nutrition Research Reviews* 13(1), 31–46.

High, C. and Shackleton, C.M. (2000) The comparative value of wild and domestic plants in home gardens of a south African rural village. *Agroforestry Systems* 48(2), 141–156.

Howard, P.L. and Nabanoga, G. (2007) Are there customary rights to plants? An inquiry among the Baganda (Uganda), with special attention to gender. *World Development* 35(9), 1542–1563.

Humphry, C., Clegg, M.S., Keen, C. and Grivetti, L.E. (1993) Food diversity and drought survival. The Hausa example. *International Journal of Food Sciences and Nutrition* 44(1), 1–16.

Levin, C.E., Maxwell, D.G., Armar-Klemesu, M., Ruel, M.T., Morris, S.S. and Ahiadeke, C. (1999) Working women in an urban setting: Traders, vendors, and food security in ACCRA. FCND Discussion Paper 66, International Food Policy Research Institute, Washington, DC.

Miyata, S., Minot, N. and Hu, D. (2009) Impact of contract farming on income: Linking small farmers, packers, and supermarkets in China. *World Development* 37(11), 1781–1790.

Ngugi, I.K., Gitau, R. and Nyoro, J.K. (2007) *Access to High Value Markets by Smallholder Farmers of African Indigenous Vegetables in Kenya*. Regoverning Markets Innovative Practice series, IIED, London.

Pasquini, M.W. (2002) Soil fertility management strategies in irrigated peri-urban agriculture around Jos, Nigeria: An interdisciplinary approach. Unpublished PhD thesis, University of Durham, UK.

Porter, G., Harris, F., Lyon, F., Dung, J. and Adepetu, A.A. (2003) Markets, ethnicity and environment in a vulnerable landscape: The case of small-scale vegetable production on the Jos Plateau, Nigeria, 1991–2001. *The Geographical Journal* 169(4), 370–381.

Quisumbing, A.R. and Pandolfelli, L. (2010) Promising approaches to address the needs of poor female farmers: Resources, constraints, and interventions. *World Development* 38(4), 581–592.

Raynolds, L.T. (2002) Wages for wives: Renegotiating gender and production relations in contract farming in the Dominican Republic. *World Development* 30(5), 783–798.

Ruel, M.T., Garrett, J.L., Morris, S.S., Maxwell, D., Oshaug, A., Engle, P. *et al.* (1998) Urban challenges to food and nutrition security: A review of food security, health, and caregiving in the cities. FCND Discussion Paper 51, International Food Policy Research Institute, Washington, DC.

Smith, G.C., Clegg, M.S., Keen, C.L. and Grivetti, L.E. (1995) Mineral values of selected plant foods common to southern Burkina Faso and to Niamey, Niger, West Africa. *International Journal of Food Sciences and Nutrition* 47, 41–53.

Smith, G.C., Dueker, S.R., Clifford, A.J. and Grivetti, L.E. (1996) Carotenoid values of selected plant foods common to southern Burkina Faso, West Africa. *Ecology of Food and Nutrition* 35, 43–58.

Spieldoch, A. (2007) *A Row to Hoe. The Gender Impact of Trade Liberalization on Our Food System, Agricultural Markets and Women's Human Rights*. Institute for Agriculture and Trade Policy (IATP), Geneva, Switzerland.

Tickner, V. (1995) Employment and food crop/food marketing. *Food Policy* 20(2), 139–144.

Warning, M. and Key, N. (2002) The social performance and distributional consequences of contract farming: An equilibrium analysis of the arachide de bouche program in Senegal. *World Development* 30(2), 255–263.

Weinberger, K. (2007) Are indigenous vegetables underutilized crops? Some evidence from eastern Africa and South-East Asia. *Acta Horticulturae* 752, 29–34.

Weinberger, K. and Msuya, J. (2004) Indigenous vegetables in Tanzania: Significance and prospects. Technical Bulletin 31, AVRDC, Shanhua, Taiwan.

Weinberger, K. and Pichop, G.N. (2009) Marketing of African indigenous vegetables along urban and peri-urban supply chains in sub-Saharan Africa. In: Shackleton, C.M., Pasquini, M.W. and Drescher, A.W. (eds) *African Indigenous Vegetables in Urban Agriculture*. Earthscan, London, pp. 225–244.

Weinberger, K. and Swai, I. (2006) Consumption of traditional vegetables in central and north-eastern Tanzania. *Ecology of Food and Nutrition* 45(2), 87–103.

12 Private Voluntary Standards, Co-investment and Inclusive Business

Emma Blackmore and James MacGregor

Introduction

The implementation of and compliance with standards and certification are typically associated with increased costs for supply chain actors. These increased costs are incurred through the necessary upgrading of production, logistics and marketing to achieve compliance. Exclusion of certain actors from the supply chain can be one consequence of these increased costs. In particular, the exclusion of small-scale growers as a result of the expansion and proliferation of private 'voluntary' standards (PVS) has been extensively reported (Borot de Battisti *et al.*, 2009).

This chapter gives another perspective on PVS: standards in export horticulture can incentivize a more active role for the private sector in investing in small-scale growers in ways that are mutually beneficial for growers and exporters.

Such co-investment is a feature of trading relationships and business models that are inclusive of small-scale growers – a conclusion that is supported by case studies from other sectors. The chapter concludes with a discussion on wider approaches to linking small-scale producers with formal markets, including domestic markets.

The Proliferation of Private Voluntary Standards and the Exclusion of Small-scale Growers

Private voluntary standards have proliferated in recent years in industrialized countries, as the gap between consumer concerns and public legislation has widened (Homer, 2009). PVS range from collective (e.g. Global Food Safety Initiative) to company-owned (e.g. Tesco Nature's Choice). They can be business-to-business schemes or customer facing, in the sense that they offer visual assurance through the use of labelling and act as a point of difference or unique selling point (e.g. Fairtrade, organic) (Borot de Battisti *et al.*, 2009). PVS can relate to food safety and environmental and social criteria. They affect the entire length of the supply chain and all its participants.

Multiple benefits for retailers and other value chain players can be provided by PVS, including the demonstration of due diligence in compliance with public standards such as the EU food laws (showing the close interrelationship between public and private regulation), the creation of smoother business-to-business transactions and ammunition in the battle between branded goods and retailers' own brands.

Research has long shown that the adoption and implementation of private voluntary standards is costly (e.g. Borot de Battisti *et al.*, 2009). These costs occur at both the macro and micro levels. At the macro level, these costs are in the form of public investments in standards agencies, upgraded skills required for management in agronomy and infrastructure development, and the monitoring of trades. At the micro level, costs are incurred through producers and supply chain actors ensuring and demonstrating compliance in the form of investments in logistics, production and marketing.

In those cases where development benefits can be demonstrated, there are opportunities to leverage donor funding, technical assistance and other support, which can reduce total cost burdens of compliance with standards and certification, but which pose important questions in terms of their long-term sustainability.

Nevertheless, these costs generally present significant barriers to the entry to export markets. Costs of PVS are per certification and the unit is usually the individual farm, regardless of size. In much of the developing world smallholder production dominates, and these small farms face proportionately higher costs per unit area for certification and compliance. Due to these problems, there is often an inherent bias in many standards and certification towards larger farms. However, this is changing, with additions to standards to allow for group certification. There is genuine concern that the proliferation of private standards adds further to the overall costs of, and hence barriers to, trade without adding compensatory value for supply chain participants.

Several researchers have found a falling number of small-scale farmers involved in export horticulture. This decrease has been attributed primarily to the rising costs of production associated with the need to adhere to increasingly stringent standards (Dolan and Humphrey, 2000; Gibbon, 2003; Humphrey *et al.*, 2004). Exporters typically take production in-house to reduce transaction costs and risk, and move away from the more traditional spot-markets that were arguably easier for small-scale growers to access.

Compliance as the driver of co-investment?

Considering all the concern about the exclusionary nature of PVS, perhaps the most surprising feature of horticultural exports in East Africa is the persistence of the market *inclusion* of smallholders. The Fresh Produce Exporters Association of Kenya estimates that there are 6000 small-scale growers involved in the production of fresh vegetables for export markets in Kenya alone. In addition to these 6000, many more are involved in the production of non-fresh, standards-compliant export markets. For example, up to 10,000 small-scale growers are thought to be involved in the production of French beans for canned and frozen exports.

Trading with small-scale growers can have significant advantages for exporters, including offering an effective risk management strategy, for example, and geographical dispersion of growers to reduce environmental risk. Small-scale growers also provide high-quality and cost-effective production for particular produce, such as French beans, which are not well suited to mechanized plantation production (Blackmore, 2007, unpublished MSc thesis), and which can outweigh the economies of scale of larger growers (Reardon *et al.*, 2009).

These incentives explain why the best exporters will co-invest in smallholder compliance with PVS, as part of their cost of doing business.

Evidence from Kenya

Between March 2005 and September 2006, after the introduction of compulsory GlobalGAP[1] certification, a survey of ten exporters was carried out in Kenya (Graffham *et al.*, 2009). These exporters controlled over 50% of the Kenyan export horticulture market. The survey found that there was indeed a drop in the formal participation of small-scale growers in these companies' supplier networks. The survey recorded a reduction of over 40% in their reported small-scale farmer outgrower suppliers to certified markets, from 9342 to 5475.

The authors suggest that the primary reason for this decline in participation in GlobalGAP-compliant export markets is financial rather than technical. The average per-farm initial costs of compliance with GlobalGAP were £1145 (approximately US$2176[2]) and annual recurrent costs were £175 (approximately US$333) (Fig. 12.1). GlobalGAP certification is likely to require far more capital than many small-scale farmers can afford on their own (Graffham *et al.*, 2009). This is where the role of exporters and intermediaries is of significance.

Further analysis of the exporter survey data provides some useful insights into the role of exporters in co-investing in certification, annual monitoring and infrastructure and investment in the necessary technical and managerial support needed for small-scale growers to comply with standards (Figs 12.1 and 12.2). The most successful exporters (in terms of the numbers of small-scale grower suppliers) provide a significant share of the costs of GlobalGAP compliance. Many smaller exporters were highly reliant on donor assistance for adherence to PVS.

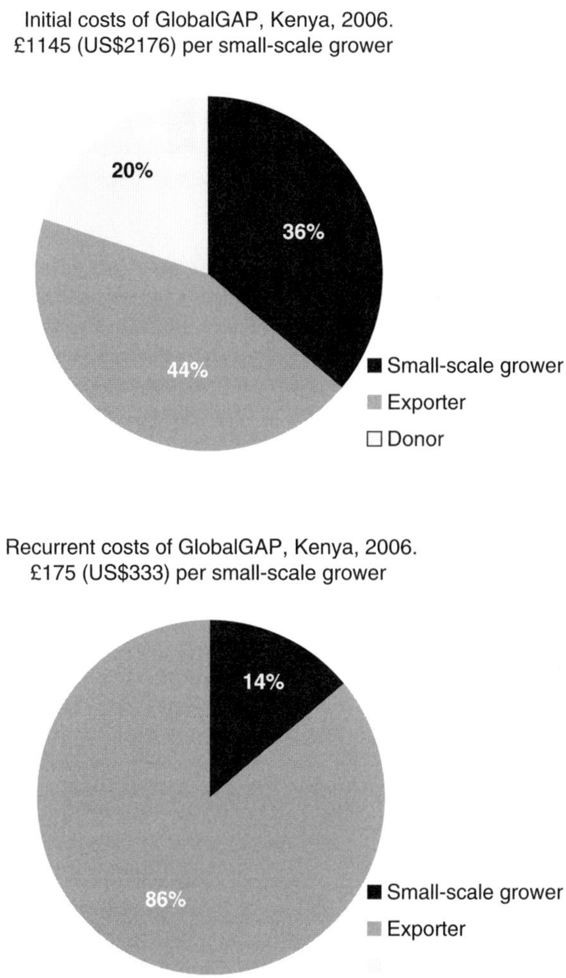

Fig. 12.1. Costs and cost sharing for small-scale grower compliance with GlobalGAP (initial and recurrent costs; from Graffham *et al.*, 2007). NB: the exchange rate in October 2006 was £1 = US$1.9.

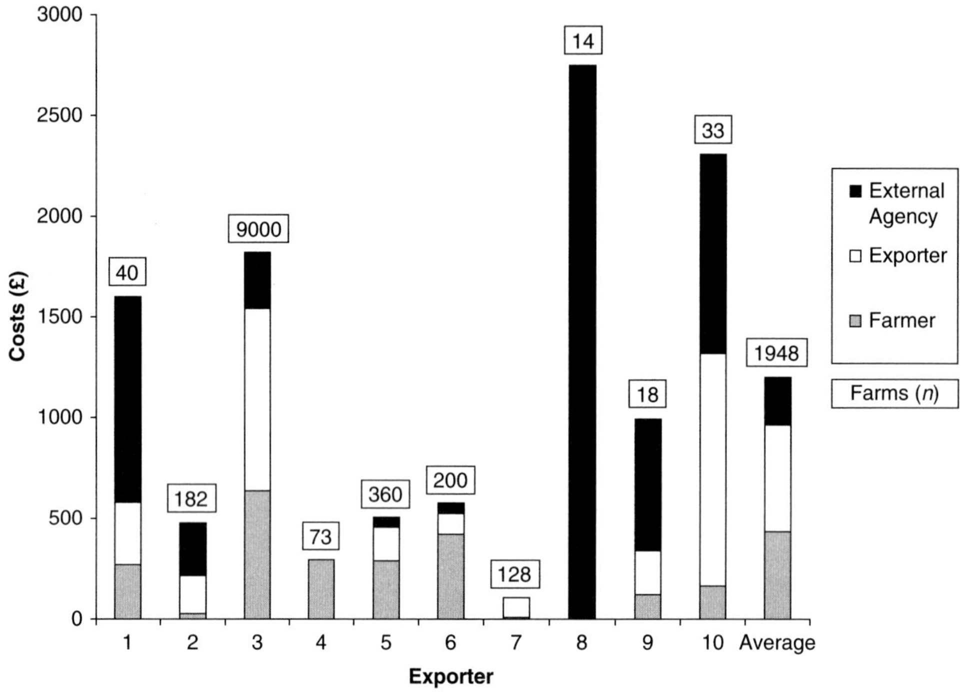

Fig. 12.2. Average initial costs for small-scale growers' GlobalGAP compliance, based on ten exporters and farmers engaged with them in trading relationships, Kenya, 2006 (from Graffham *et al*., 2007). Figures in inset boxes indicate the number of farmers sourced by each exporter. NB: the exchange rate in October 2006 was £1 = US$1.9.

Donor assistance is unlikely to be sustainable in the long term; these companies were frank in saying that they cannot see how the system of smallholder outgrowing can be maintained once donor support is withdrawn (IIED and NRI, 2009).

Those exporters most able to engage successfully with, or procure from, small-scale farmers were those who bore a significant proportion of the initial costs needed to comply with GlobalGAP and relied least on an external agency for assistance. Figure 12.2 shows that exporter No. 3 made the most significant investment in compliance (with some investment from outside agencies and from farmers themselves) and sourced from the largest number of small-scale farmers – a total of 9000 small-scale farms.

Those small-scale farmers who were not well supported by their exporter struggled with GlobalGAP, and this led to farmers either failing to certify or dropping out of the compliance system within one to two years of first being certified (IIED and NRI, 2009).

The survey revealed that over £2.2 million (US$4.18 million) had been invested in getting smallholder farms to a position where they could be audited for GlobalGAP compliance (IIED and NRI, 2009). This may be limited to co-investment in the costs of certification, auditing or monitoring. But it may also involve co-investment in the infrastructure needed to meet standards, such as equipment for postharvest handling, or in technical support in the form of well-staffed and resourced outgrower management teams, annual training programmes, internal auditors and programmes for sampling and laboratory analysis. In this sense, large export companies fulfil the role of a primary marketing organization for the growers, not only providing the necessary managerial, technical and logistical support, but also sometimes representing the growers effectively during the certification audit.

Where small-scale farmers have exited GlobalGAP certification, follow-up research in Kenya has shown that, although some may still be involved with export horticulture, farmers report a much lower level of advice and support from their buyers, a lower price paid per kilogram, less trade and considerably lower income from the sales of export crops (Cooper and Graffham, 2009). Although this could be attributed directly to GlobalGAP, the differences between the farmers still involved in GlobalGAP-certified export and those not involved in GlobalGAP are linked to the differences in the relationship between growers and export companies. Farmers who were GlobalGAP compliant were far more likely to be in a relationship with one of the ten largest export companies in Kenya. These relationships typically bring with them much needed support for farmers (Graffham *et al.*, 2009) and wider development benefits.

In semi-structured interviews carried out in Kenya in October 2006 (IIED and NRI, 2009), all small-scale growers who were GlobalGAP certified and who were interviewed for this project reported that they were generally happy with GlobalGAP. GlobalGAP's greatest benefit for smallholders was the preferential access opportunities that it provides. These include access to markets, credit for inputs, trade credit and quality inputs (high-germination seeds, high-nitrate fertilizer, etc.). Growers perceived considerable non-financial benefits and, to some extent, the use of income or a profit margin as an indicator of success or failure is misplaced. Perceived advantages of GlobalGAP included production of quality produce, improved field hygiene, better knowledge of pesticide use and wider farm management benefits. In truth, many of these so-called non-financial benefits are quantifiable; access to trade credit or higher quality inputs will improve farm efficiency and yields. However, without time-series data, causality is difficult to establish.

Farmers who had attained GlobalGAP certification were clearly benefiting from the adoption of good agricultural practices, record keeping and improved hygiene. Yields were generally higher and input costs reduced as the production process was better managed. Many farmers said that they were using GlobalGAP records to understand their financial viability and run their farms more commercially. Proper handling of pesticides and improved food safety and hygiene had health benefits on the farms, and most farmers said that they had transferred hygiene messages to the homestead with obvious positive implications for family health.

Further benefits accrue through the forms of supply chain relationships that can accompany GlobalGAP certification. For example, contracts enable some smallholders to access trade credit through designated input sellers for seeds, fertilizer or chemicals. The added transaction costs for smallholders are one of the major drivers for cooperative producer management organizations – where one formal contract can be issued by a buyer, and where farmers' names are placed on a list of preferred suppliers who can access particular products at set prices on credit from designated stores.

Even though these models of smallholder outgrowing can offer benefits for small-scale growers, it is important to bear in mind that they are usually exporter-driven and highly paternalistic. Nevertheless, emerging evidence suggests that other, more equitable, trading relationships can exist that facilitate upgrading for small-scale growers as part of sustainable trading relationships for both the export and domestic markets.

Co-investment and access to services through multi-specialized intermediaries

The most significant barriers to the inclusion of small-scale producers and SMEs (small and medium enterprises) in formal markets are typically market linkages. For buyers, market linkages must provide a reliable supply of safe, quality products and services at a competitive price with low transaction costs.

The International Institute of Environment and Development (IIED) has been investigating different models of these business linkages in a number of sectors, in both delivery and contracting models, and is learning important cross-cutting lessons about how they influence the sustainability of outcomes. Market intermediaries are recognized as key to the majority of successful and sustainable trading relationships, as aggregators of supplies, as guarantors of quality and compliance with standards, as capacity builders and as suppliers of finance and technical advice. They may be existing traders or newly specialized intermediaries (Vorley *et al.*, 2008). In many cases the trader is a member of the rural community and has specialized knowledge, information, assets and contacts to facilitate not only commercial ties, but also social support in times of crisis. These intermediaries may play a particularly important role in domestic markets where exporters are absent and where the role of the public sector in service delivery is weak or underdeveloped.

The following case studies provide tangible examples of how multi-specialized intermediaries can facilitate the inclusion of small-scale growers – even in highly demanding export markets.

Cuatro Pinos, Guatemala

Cuatro Pinos is a successful cooperative that has significant experience in the vegetable export business. It has recently succeeded in opening large markets for several fresh vegetable products in the USA, through alliances with a specialized wholesaler and retailers. Demand for produce is significant and exceeds the volumes that the cooperative's current members can provide. As a result, Cuatro Pinos has integrated new producers, organizations and geographies into sustainable trading relationships as a multi-specialized intermediary. Cuatro Pinos works to identify existing farmer groups (including associations, cooperatives and leading farmer networks) and works with them to test production schemes, and contracts those who are able to meet volumes and quality targets. A contract is then signed between the producers and the cooperative that specifies the quantity and quality of the produce that should be delivered, as well as a production schedule. In exchange, farmers are provided with credit (in the form of inputs) and technical assistance, which help them meet the quantity and quality targets. Credit is discounted from the first few product deliveries (Lundy, 2007 in Vorley *et al.*, 2009).

In addition to the credit and technical assistance, Cuatro Pinos provides business and organizational support for its partner organizations to increase their efficiency and to access additional funding from diverse sources for development activities. For example, in 2006, Cuatro Pino partners raised US$1.7 million for investment in irrigation, packing sheds, education and housing. Irrigation, packing sheds and education contribute significantly to the farmers' ability to produce high-quality vegetables, meeting standards laid out by the cooperative in the farmers' contracts.

Over the past 3 years, this model has allowed Cuatro Pinos to achieve a 50% increase in vegetable exports and it has expanded its supplier base from 560 member producers to a network of more than 2000 families. New producers are from regions with poverty levels higher than the national average and with limited access to land (Lundy, 2007 in Vorley *et al.*, 2009).

Normincorp, Mindanao, Philippines

Normincorp is a corporation established by small-scale vegetable farmers in the Philippines, who belong to the Northern Mindanao Vegetable Producers' Associate, NorminVeggies. It acts like a market facilitator linking farmers (who are formed in clusters) directly to buyers. Although it was established as a stock corporation, it functions more like a cooperative or social enterprise. Although it is a business in the strictest sense, it also has full empathy for small farmers.

One of the key aims of NorminVeggies is to:

> Increase and sustain production of safe and healthy, assorted vegetables through sustainable agriculture to meet market demand. Innovative technologies for

> nurseries, tropical rain shelters, the use of mulching and drip irrigation systems are just some of the few adapted technologies used by the farmers of NorminVeggies, to move towards sustainable agriculture.
> (Concepcion *et al.*, 2007, p. 17)

Normincorp has provided the necessary services to try and achieve this aim. Normincorp recognized that changes in the external environment, such as increasing quality and food safety standards, require changes for the suppliers and provided appropriate support, such as putting in significant effort to manage quality: for example, upgrading technology, improving the organizational set-up or increasing capital.

Normincorp serves to ensure that farmer clusters are appropriately managed. It programmes production for farmer clusters (according to marketing plans), helps to ensure quality standards are reached and that postharvest management can be done by each farmer in the cluster. Normincorp is responsible for ensuring that coordination is provided for taking orders, outshipment logistics, billing/charging, collection and remittance to farmers. These services are paid for by a market facilitation fee, which is deducted from the value of sold produce.

The production of high-quality and safe vegetables is incentivized through the Normincorp model. Farmers are given buyer prices and are accountable for the product up to the point of sale – retaining ownership of the product. The final price given to the farmer is the price of all sales after deduction of the market facilitation fee, which is based on the quantity of accepted vegetables. All vegetables rejected by the buyer are charged to the farmer who produced them; this prevents free-riding and encourages all farmers to maximize quality. Products are labelled according to farm/farmer, to provide the necessary traceability (Concepcion *et al.*, 2007).

Mahagrapes, India

Although India is the second largest vegetable producer after China, it is a relatively small exporter. Roy and Thorat (2008) explain that India is a relatively unsuccessful exporter because it is constrained by the inability of its smallholder-based agriculture to meet the food safety and quality requirements of export markets – specifically, its inability to adopt more sophisticated farm practices and make the necessary investments to meet standards.

However, there are success stories – as Roy and Thorat (2008) outline. Mahagrapes is a marketing partner to a farmer cooperative based in the Maharashtra state of India, who produce grapes to GlobalGAP standards. Mahagrapes provides marketing expertise (in both inputs and output markets) for farmers and overcomes the scale diseconomies of small-scale farmers by the bulk-buying or in-house production of inputs, such as fertilizer. Mahagrapes is a for-profit organization. The price each farmer receives is based on the quality of the grape, which can be ascertained due to traceability of the produce. Mahagrapes deducts a facilitation fee per unit of output and passes on the rest to the cooperative, which then keeps a portion of the price for itself and passes the rest to the farmer.

Mahagrapes has had to deal with the rejection of grapes and is, therefore, familiar with produce rejection based on inferior quality. As standards, quality and safety measures change each year, it is vital that Mahagrapes keeps abreast of these changes and amends its own production, processing, storage and testing methods regularly (Roy and Thorat, 2008). In the implementation stage of standards, Mahagrapes provides materials and technical help along with infrastructural support to facilitate the implementation of standards. Indeed, despite the expense of acquiring GlobalGAP certification, Mahagrapes has been able to provide all its associated cooperative societies with certification. Thus, each society becomes certified as GlobalGAP compliant along with member farmers, who now have to pay only US$28 per year – for even the smallest exporter, this represents just 2% of annual export revenues. Having deducted the net costs of compliance, Mahagrapes farmers do earn significantly higher profits compared with those outside the scheme.

The services, support and co-investment offered by Mahagrapes' marketing company are crucial in establishing the necessary economies of scale required to comply with standards and compete with larger farmers. As such, its business model promotes the successful inclusion of small-scale farmers.

These case studies demonstrate the key role that multi-specialized intermediaries can play in linking small-scale growers to exacting export markets. The emerging evidence of similar intermediaries operating in developing countries and serving domestic markets (see Vorley *et al.*, 2009) provides evidence of the ability of the private sector to help fill market failures. Research into the previously informal raw milk market in Kenya (Hooton and Omore, 2007) suggests that partnerships between the private sector, NGOs and governmental agencies can be effective mechanisms for integrating small-scale growers into more formalized markets.

Summary and Conclusions

Private voluntary standards and certification have been viewed as costly and exclusionary. While this is an important part of the story, evidence presented here shows that standards can be utilized to leverage scalable solutions to other barriers faced by small-scale farmers in accessing export markets, and to perceived barriers faced by businesses when seeking to include small-scale farmers in their supply base. Though there are evident costs associated with standards, it is striking that the meaningful inclusion of small-scale farmers is still a possibility.

It can be argued that we need to rethink our idea of the 'costs' of adherence to standards. In most successful cases, small-scale farmers do not have to bear these costs alone, as the survey demonstrates. Rather than labelling exporter investments as unsustainable, it can be argued that this in fact illustrates a healthy and functioning system with two private sector investors – producer and exporter – sharing the costs and benefits as part of a sustainable trading relationship. As Fig. 12.1 demonstrates, the distribution of recurrent costs, in particular, appears more equitable than for initial costs, suggesting that, where small-scale growers can demonstrate compliance, exporters are further inclined to invest in the trading relationship. It is not useful to consider the sharing of costs among participants as a subsidy that has to be covered by farmers in order to prove 'sustainability'.

Standards can, therefore, become the loci of collaboration and co-innovation. They draw business into investing in their supply base, as vertically coordinated chains. They lock producers, exporters/suppliers and retailers into much closer, less opportunistic trading behaviour. Research suggests that these strengthened linkages lead to longer-term, more sustainable trading relationships based on mutual trust and the sharing of risk and cost in compliance, leading in turn to more resilient value chains and trading relationships (MacGregor, 2009). In the absence of public extension services, the private sector can provide technical support for production, compliance and certification, and help improve quality, postharvest handling and storage. Direct extension is limited to producers growing for export, but there are considerable spillovers that benefit small-scale growers supplying the domestic market.

As a response to standards, exporter companies have demonstrated their willingness to form partnerships with small-scale growers by providing high levels of financial, technical and administrative support. Research has shown that in Kenya outgrower production and income from vegetable exports actually increased between 2001 and 2007 (Mwangi, 2009). Indeed, Hoffmann and Vossenaar (2009, p. 140) argue that 'small-scale farmers can be successfully integrated into global supply chains, but as part of well-managed outgrower schemes or producer groups and with close links to exporters'. Minten *et al.* (2009, p. 1729) also found, in their study of small-scale growers in Madagascar, that

> given the right incentives and contract systems, small farmers in developing countries can participate successfully in

emerging supermarket value chains. Thousands of small farmers benefit because of a combination of effects, such as improved access to inputs, credit, extension services, technology adoption, and from productivity spillover effects on other crops and enhanced income stability.

Reardon *et al.* (2009) provide several examples of the successful inclusion of small-scale farmers, through the use of contract farming and the provision of 'resource-providing' contracts that can make small farmers competitive with larger farmers. Indeed, the delivery of these services is particularly beneficial for small-scale producers who often lack the capacity to make similar investments themselves. The authors argue that, in those cases where large companies have sourced from small farmers even when large farmers are available, several distinct advantages for sourcing from small-scale growers emerge – larger farmers may have a broader set of market options, such as exports, and are therefore considered a riskier sourcing option. Smaller farmers may be more able and willing to follow the highly labour-intensive field management practices needed by the company. Small farmers may be able to reduce transaction costs to companies by forming effective marketing cooperatives.

Standards create space for sustainable agricultural learning, replication and scalability (MacGregor, 2009). The examples of small farmers meeting private voluntary standards with support from other players, particularly multi-specialized intermediaries, in the supply chain hint at the significant potential for replication in other supply chains and domestic markets, and for efficient upgraded supply chains providing consistent quality of product from a smallholder supply base (MacGregor, 2009). The sharing of the costs associated with adherence to standards (both direct and indirect) should, in fact, be regarded as a meaningful form of collaboration (through co-investment) and a reflection of a healthy and well-functioning business relationship that implies sustainability and trust rather than the reverse – unsustainability.

High-value international trade is a laboratory for testing how the private sector assesses and works effectively with small-scale growers in developing countries, and provides lessons for wider replication in non-export markets. Standards have driven shifts in the way value chains are governed; they are providing new incentives for the private sector to build trading relationships that extend as far as the small-scale farm, rather than remaining at arm's length.

Innovations in the production and marketing of small-scale producers provide us with fresh insights as to the necessary requirements for successful inclusion of small-scale growers. This is of particular importance for non-export markets in developing countries and in the context of the continued transformation of the agrifood industry in the developing world. As standards become increasingly important, the need for collaboration through co-investment will be essential to ensure the meaningful inclusion of small-scale farmers.

The case studies outlined above detail some of the innovative ways in which small-scale growers can be connected to export markets with exacting requirements. The role of cooperatives and marketing partners in providing services – particularly credit and technical expertise – has been invaluable in overcoming the barriers to working with small-scale growers and filling the gaps typically left by the public sector in the developing world. These studies provide examples of how trading relationships might be organized in export as well as domestic markets in order to facilitate the meaningful inclusion of small-scale growers, particularly as standards also become more prevalent in domestic markets. They can offer a viable alternative to the traditional outgrower model.

Further research

Understanding the benefits to producers of PVS such as GlobalGAP, especially the indirect benefits that derive from improved efficiency and farm management, is key to an informed debate about standards, exclusion and development. Good data can allow policy makers to weigh concerns about PVS as trade barriers against evidence of PVS as one driver of a more efficient agricultural sector. Such data are, however, currently thin on the ground and require further attention.

In terms of reducing the costs and complexity of compliance and levelling the field for small-scale producers, much work remains to be done to (i) find innovative ways to connect small-scale growers with standards-compliant export markets; (ii) adapt GAP standards, audits and certification to the reality of smallholder production (Graffham *et al.*, 2009); and (iii) raise the smallholder voice in standard setting; to date, the process of standard setting has typically been driven by the downstream value chain players, with primary producers as 'standard takers' rather than 'standard makers'.

The costs of certification and standards for small-scale growers could be significantly reduced through the development of common certification protocols that recognize the common elements that exist within the multitude of certification schemes (e.g. the commonalities that exist between GlobalGAP, Utz Certified, Rainforest Alliance, Organic and Fairtrade). This could be achieved without compromising the integrity of these standards, but it has the potential to make a significant difference to the market inclusion of small-scale growers.

Further research is needed to explore the role of donors in supporting small-scale growers' compliance with standards. Typically, where donor support is offered for standards compliance, this compliance lasts only as long as the donor support is provided. How can donor support be designed in a way that promotes enduring compliance?

The role of emerging economies and developing countries in adapting existing standards (e.g. KenyaGAP) and in setting new national standards (ISEAL Alliance, 2010) for domestic markets and south–south trade is likely to have a significant impact on the landscape of PVS, and needs to be considered in future research that considers the relationship between standards and the market inclusion of small-scale growers.

The wider challenge for building inclusive value chains is the development of models and principles for trading relationships in which co-investment in compliance with PVS is just one part of the inclusive business that generates long-term development gains. The type of trader/exporter co-investment described in this chapter usually appears in paternalistic outgrower systems in which producers are locked into an exclusive trading relationship with the exporter, often to the extent that it is the exporter who holds the standards certificate. Further research is needed on forms of co-investment that build, rather than constrain, market choices for small-scale producers. These models can serve as a starting point for the private sector to develop trading relationships that benefit small-scale growers and have wider benefits for development.

Endnotes

[1]Formerly known as EurepGAP.

[2]£1 (GBP) was equal to US$1.9 (USD) in October 2006 (when the original research was carried out). Exchange rate information provided by Oanda (http://www.oanda.com).

References

Borot de Battisti, A., MacGregor, J. and Graffham, A. (eds) (2009) *Standard Bearers: Horticultural Exports and Private Standards in Africa*. International Institute of Environment and Development, London.

Concepcion, S.D., Digal, L. and Uy, C.J. (2007) The Philippines: Keys to inclusion of small-scale farmers in dynamic vegetable markets. Regoverning Markets, Innovative Practice Series (www.regoverningmarkets.org/en/filemanager/active?fid=808).

Cooper, J. and Graffham, A. (2009) GlobalGAP Version 3: Threat or opportunity for small-scale African growers? In: Borot de Battisti, A., MacGregor, J. and Graffham, A. (eds) *Standard Bearers: Horticultural Exports and Private Standards in Africa*. International Institute of Environment and Development, London, pp. 23–25.

Dolan, C. and Humphrey, J. (2000) Governance and trade in fresh vegetables: The impact of UK supermarkets on the African horticulture industry. *Journal of Development Studies* 37(2), 147–176.

Gibbon, P. (2003) Value chain governance: Public regulation and entry barriers in the global fresh fruit and vegetable trade in the EU. *Development Policy Review* 21(5/6), 615–625.

Graffham, A., Karehu, E. and MacGregor, J. (2007) Impact of EurepGAP on small-scale vegetable growers in Kenya. Fresh Insights No. 6 (www.agrifoodstandards.net/en/filemanager/active?fid=83).

Graffham, A., Karehu, E. and MacGregor, J. (2009) Impact of GlobalGAP on small-scale vegetable growers in Kenya. In: Borot de Battisti, A., MacGregor, J. and Graffham, A. (eds) *Standard Bearers: Horticultural Exports and Private Standards in Africa*. International Institute of Environment and Development, London, pp. 53–56.

Hoffmann, U. and Vossenaar, R. (2009) Proactively complying with private voluntary standards: Key findings of country case studies in Ghana, Kenya and Uganda. In: Borot de Battisti, A., MacGregor, J. and Graffham, A. (eds) *Standard Bearers: Horticultural Exports and Private Standards in Africa*. International Institute of Environment and Development, London, pp. 140–143.

Homer, S. (2009) The gap is getting wider: How private standards are filling the void between dynamic public opinion and food safety legislation. In: Borot de Battisti, A., MacGregor, J. and Graffham, A. (eds) *Standard Bearers: Horticultural Exports and Private Standards in Africa*. International Institute of Environment and Development, London, pp. 14–17.

Hooton, N. and Omore, A. (2007) Kenya: Policy innovations in small-scale milk markets in Kenya and East Africa. Regoverning markets: Small-scale producers in modern agrifood markets: innovative policy. Available at www.regoverningmarkets.org.

Humphrey, J., McCulloch, N. and Ota, M. (2004) The impact of European market changes on employment in the Kenyan horticulture sector. *Journal of International Development* 16(1), 63–80.

IIED and NRI (2009) Costs and benefits of GlobalGAP compliance for small-scale farmers: Synthesized findings. In: Borot de Battisti, A., MacGregor, J. and Graffham, A. (eds) *Standard Bearers: Horticultural Exports and Private Standards in Africa*. International Institute of Environment and Development, London, pp. 66–69.

ISEAL Alliance (2010) What will the standards systems landscape look like in 2020? The current sustainability landscape: emerging trends. Presented by S. Courville at *ISEAL Alliance Conference*, 22–24 June 2010, London.

MacGregor, J. (2009) Understanding stakeholder drivers for introducing and complying with private voluntary standards – a fresh produce example. In: Borot de Battisti, A., MacGregor, J. and Graffham, A. (eds) *Standard Bearers: Horticultural Exports and Private Standards in Africa*. International Institute of Environment and Development, London, pp. 10–13.

Minten, B., Randrianarison, L. and Swinnen, J.F.M. (2009) Global retail chains and poor farmers: Evidence from Madagascar. *World Development* 37(11), 1728–1741.

Mwangi, T. (2009) The impact of private agrifood standards on smallholder incomes in Kenya. In: Borot de Battisti, A., MacGregor, J. and Graffham, A. (eds) *Standard Bearers: Horticultural Exports and Private Standards in Africa*. International Institute of Environment and Development, London, pp. 78–82.

Reardon, T., Barret, C.B., Berdegué, J.A. and Swinnen, J.F.M. (2009) Agrifood industry transformation and small farmers in developing countries. *World Development* 37(11), 1717–1727.

Roy, D. and Thorat, A. (IFPRI) (2008) Success in high value horticultural export markets for the small farmers: The case of Mahagrapes in India. *World Development* 36(10), 1874–1890.

Vorley, B., Lundy, M. and MacGregor, J. (2008) Business models for small farmers and SMEs. FAO, All ACP Agricultural Commodities Programme (www.euacpcommodities.eu/files/FAO-models%20for%20Small%20Farmers.doc).

Vorley, B., Lundy, M., Seville, D. and Ferris, S. (2009) Linking worlds: New business models for sustainable trading relations between small-scale farmers and formalized markets. Final draft version (http://usmfiles.s3.amazonaws.com/phpnuigdA/New%20Business%20Models%20for%20Sustainable%20Trading%20Relationships.pdf).

13 An Approach to Strengthening Vegetable Value Chains in East Africa: Potential for Spillovers

Jillian M. Lenné and Andrew F. Ward

Introduction

The aims of this chapter are to outline a new approach to strengthening vegetable value chains in East Africa. This approach is based on identifying and describing examples of potential 'spillovers' of knowledge, technologies and methodologies to address some of the major constraints affecting the efficient functioning and economic growth of domestic vegetable value chains in East Africa. Because many of the examples have been generated by one-off projects, NGO initiatives and serendipitous partnerships, they have rarely been documented in the scientific literature. In many cases, our knowledge of these has been facilitated by the direct contact and analysis generated through comprehensive studies of the vegetable sector in several countries. Wherever possible, references and/or websites will be given for specific examples. In particular, we will emphasize the important role of the export vegetable sub-sector as a source of knowledge, technology, methodology and experience for the domestic sub-sector for both addressing constraints and learning more about potential spillovers. Above all, we believe that an approach to strengthening vegetable value chains in East Africa through spillovers of existing knowledge, methodologies and technologies will be more cost effective and more likely to achieve results in the short to medium term as most, if not all, domestic vegetable production and marketing in sub-Saharan African countries face similar constraints. In this way, spillovers are likely to have a measurable impact on poverty and nutrition of the poor at local, regional and national levels.

Vegetable Production in East Africa

In East Africa, especially Kenya and Tanzania, hundreds of thousands of smallholder farmers grow vegetables for home consumption and cash income (Lenné *et al.*, 2005; Mithöfer *et al.*, 2008). The overwhelming majority produce and market vegetables locally or regionally in the dominant domestic sub-sector. For example, in Kenya, it comprises 90% of the volume and is worth seven to eight times the value of the export sub-sector (Mutuku Muendo and Tschirley, 2004a) (Table 13.1). Not only is the domestic vegetable value chain an important source of nutrition and income, but it is also a source of employment for rural and urban households (Lenné *et al.*, 2005; Lenné and Ward, 2010).

Table 13.1. Annual production (metric tons) of common vegetables in Kenya and Tanzania (from Mutuku Muendo and Tschirley, 2004a; Ministry of Agriculture, 2007; König *et al.*, 2008; MMA, 2008).

Vegetable	Kenya	Tanzania
Tomato	250,000–350,000	150,000
Cabbage	250,000–350,000	42,000
Kale	250,000–350,000	ND
Fine beans (export)	60,000–70,000	ND
Onion	60,000	36,000
Indigenous vegetables	70,000	ND

ND, no data available.

Rapid urbanization is a key driver in the growth of the domestic vegetable sector in East Africa. As much as 80% of the vegetables marketed domestically in this region are sold through major urban wholesale markets such as Wakulima in Nairobi and Kariakoo in Dar es Salaam (Mutuku Muendo and Tschirley, 2004a; König *et al.*, 2008). A strong vegetable sector is an engine for economic growth, significantly contributing to reducing poverty and enhancing the livelihoods of poor smallholders in this region (Lenné and Ward, 2010).

However, in spite of the size of the domestic vegetable sub-sector, the average annual growth in area, production and yield of vegetables has declined in sub-Saharan Africa in sharp contrast to the rest of the world. Table 13.2 gives examples for East and southern Africa. This decline is based on inadequate production technologies limiting yield growth, highly inefficient marketing systems, high wastage, as well as high transaction and marketing costs. Furthermore, the per capita consumption of vegetables is well below the FAO guidelines (discussed below; see also www.who.int and www.fao.org) and the growth in consumption may have declined in line with the growth in supply.

Major Constraints Affecting the Efficiency of the Vegetable Sector in East Africa

Major production and marketing constraints affect the efficiency of the domestic vegetable sub-sector in East Africa. Production constraints include lack of adapted vegetable varieties; lack of access to inputs; poor seed quality; lack of sustainable seed delivery systems; serious pests and diseases; excessive use of pesticides; and lack of organized technical assistance (Global Horticultural Assessment, 2005; Lenné *et al.*, 2005). Furthermore, marketing constraints include disorganized farmer support; market cartels; lack of market information; poor market infrastructure; poor market access; an inefficient regional trade system; high postharvest losses; lack of formal standards; lack of understanding of food safety issues; underdeveloped processing potential; poor transport and infrastructure; lack of credit; high transport and transaction costs; lack of equity and transparency in the chain, resulting in lack of trust; and policy weakness (Lenné and Ward, 2010).

Lenné *et al.* (2005) and Lenné and Ward (2010) have emphasized the critical importance of using the highly efficient and successful export vegetable sub-sector as a role model for tackling these constraints in the domestic sub-sector. From its humble beginnings in Kenya in the 1960s and 1970s, the export sector has successfully addressed *all of the above constraints* to become a vibrant, dynamic, globally competitive and innovative model that other African countries are already beginning to imitate (Minot and Ngigi, 2004; Lenné *et al.*, 2005). A recent article by Mithöfer *et al.* (2008) not only supports this recommendation, but also highlights the important role for the export sub-sector as a model for production standards and

Table 13.2. Average annual growth rate (%) in total vegetable production in selected African countries, 1961–2005 and 1996–2006 (from FAOSTAT, 2008).

Country(ies)	1961–2005	1996–2006
Sub-Saharan Africa		
Area (ha)	2.3	1.6
Production (t)	2.8	2.1
Yield (t/ha)	0.5	0.5
Kenya		
Area (ha)	3.3	1.1
Production (t)	5.3	2.1
Yield (t/ha)	1.9	1.0
Tanzania		
Area (ha)	0.7	0.5
Production (t)	1.4	0.5
Yield (t/ha)	0.7	0.1
Zambia		
Area (ha)	2.2	0.5
Production (t)	2.1	0.3
Yield (t/ha)	−0.05	−0.3
Malawi		
Area (ha)	1.5	0.1
Production (t)	2.1	1.9
Yield (t/ha)	0.6	1.8

certification to the domestic sub-sector. This will be discussed in more detail later in this chapter.

Support for Domestic Horticulture

Despite its proven importance for nutrition, income generation and economic growth, domestic horticulture has been neglected by research investors for many years. It is only during the past 10 years that donors and governments have begun to address the complex and interrelated constraints affecting domestic vegetable value chains in sub-Saharan Africa. During the past 5 years, there has been a plethora of independent studies and projects on vegetable value chains, especially in East Africa – including studies on the same chains in the same countries by different donors, e.g. separate German- and Dutch-funded projects on tomatoes in Tanzania in 2008 (König *et al.*, 2008; MMA, 2008).

Some of these independent projects targeting specific groups of vegetable chain actors, especially smallholder farmers, have identified successful technologies and methodologies that have benefited small groups of farmers and other chain actors, largely those directly involved in these time-bound projects. However, successes have been one-off, relatively isolated and probably transient in some cases. There has been limited opportunity to share experiences and learn lessons. As a result, the application of successful outcomes to similar problems throughout domestic vegetable chains in the same and in other African countries has rarely happened. There has been a failure of governments and donors to identify and use potential spillovers for sector-wide benefits to farmers and consumers. Recent studies on vegetable value chains have placed emphasis on identifying exciting opportunities in East Africa for wider benefits (Lenné *et al.*, 2005; Mithöfer *et al.*, 2008; Lenné and Ward, 2010).

Spillovers Defined

Research and development spillovers of technology, knowledge and methodology among geopolitical entities arise when research conducted by one country confers benefits on other countries that are able to adopt the results (Alston, 2002). Although the past focus has been on spillovers between countries, it is feasible for spillovers from one community to another to happen within one country. Spillovers from agricultural research and development investments have been shown to account for more than half of the agricultural productivity growth globally (Alston, 2002; Alston and Pardey, 2008). In general, the importance of spillovers to agricultural development and increased food production has been greatly under-appreciated.

Most studies of spillovers have been *ex post*, i.e. retrospective analyses of their value and impact nationally and internationally, often many years after the knowledge, technology or methodology was developed. The greatest focus has been on the contribution of improved staple crop technologies, especially for rice and wheat (Pardey and Beintema, 2002). As far as we are aware, there have not been any studies conducted to identify and analyse potential spillovers in the vegetable sector in East Africa.

Characteristics of the vegetable sector that make it attractive for spillovers

Of all agricultural sectors in East Africa, the vegetable sector has many characteristics that make it attractive to facilitate spillovers of knowledge, technology and methodology. First, vegetables are cash crops, and therefore they are important sources of income for many thousands of smallholder farmers (Lenné *et al.*, 2005). Secondly, domestic vegetable value chains and markets, including regional markets, function in all countries, albeit inefficiently, especially due to transaction time and cost (Mutuku Muendo and Tschirley, 2004a,b). Thirdly, farmers growing vegetables tend to be relatively knowledge- and resource-endowed compared with those growing staple food crops (McCulloch and Ota, 2002). In addition, some have previous experience of growing for the highly demanding export sub-sector (Graffham *et al.*, 2007; Mithöfer *et al.*, 2008). Fourthly, smallholders growing vegetables tend to be geographically concentrated in areas most suited to production and, in some cases, with ease of transport to large urban centres, such as in peri-urban areas (König *et al.*, 2008). Geographical concentration will allow new spillovers to reach more smallholders rapidly. Fifthly, private sector traders, transporters and processors are already actively involved in the sector (König *et al.*, 2008). Finally, the facilitation of spillovers in the vegetable sector directly addresses nutrition and health issues, which should be highly attractive to international donors. We feel that the domestic vegetable sub-sector in East Africa possesses unique and positive characteristics to facilitate spillovers compared with most staple crop sectors.

Is it possible to identify potential spillovers?

During the past 10 years, we have been actively involved in reviews and analyses of the vegetable sector in East and southern Africa (Ota and Lenné, 2003; Lenné and Ward, 2004, 2008; Lenné *et al.*, 2005). We have looked at production and marketing issues in vegetable value chains across both the domestic and the export sub-sectors. From these studies, we have identified (i) constraints that limit production and marketing; (ii) opportunities to address these constraints; and (iii) some examples of successful initiatives – mainly time-bound projects targeting small groups – that have benefited stakeholders in these chains: producers, traders and consumers. These initiatives could benefit many more stakeholders through spillovers, and we will use some of them as examples in this chapter.

More importantly, as far as we are aware, there have not been any studies to identify and analyse potential spillovers *ex ante* and to facilitate their uptake, particularly not in vegetable value chains. Facilitating the uptake and capitalizing on potential spillovers are a unique approach that is likely to be the most cost-effective and time-saving means of strengthening vegetable value chains in this region. The approach is likely to be cost effective and time saving, as funding can be targeted at adapting technologies and knowledge and reaching more beneficiaries rather than in developing and testing technologies.

In this chapter, emphasis is given to the postharvest–consumer segment of value chains, including postharvest, value addition, standards, consumption, farmer organization, value chain linkages and policy.

Some Potential Spillovers Targeting Major Constraints Affecting Vegetable Value Chains

Table 13.3 identifies some of the main constraints affecting vegetable value chains in East Africa. It identifies some examples of successful initiatives that have addressed

Table 13.3. Examples of constraints and potential spillovers in vegetable value chains in East Africa.

Topic	Constraints[1]	Potential spillovers[2]
Postharvest	Lack of cold chain facilities, e.g. IVs; lack of drying and storage facilities, e.g. onions; lack of appropriate packing methods for transport	Use of existing cold chain facilities, e.g. Kenya; promotion of storage methods, e.g. Tanzania;[3] promotion of tomato packing cases, e.g. India[4]
Value addition	Lack of use of existing methods, e.g. washing, chopping and packaging; lack of processors and processing facilities, e.g. canning, freezing, drying, sauces and pastes	Promotion of existing methods, especially from export sector, e.g. Kenya;[5] promotion of innovative partnerships between existing processors and sources of raw materials, e.g. Tanzania;[5] public–private sector partnerships to strengthen processing sector
Quality standards	Lack of use of national quality standards, e.g. Tanzania; lack of promotion of organic standards	Promotion of national quality standards, e.g. KenGAP in Kenya;[6] development of national quality standards, e.g. Kenya and Tanzania;[7] support by speciality outlets for promotion of organic standards with assistance from the export sector, e.g. Kenya and Tanzania[3]
Farmer organization	Lack of organization for production, transport and marketing	Promotion of successful methods from export sector, e.g. Kenya;[5] promotion of successful methods from Technoserve, e.g. hub concept;[8] promotion of smallholder x speciality outlet partnerships, e.g. Farm Africa[3]
Chain linkages	Lack of understanding, equity and trust among actors; complex chains	Promotion of participatory dialogues through Participatory Marketing Chain and Participatory Marketing Workshop approaches, e.g. for potatoes (Kenya and Bolivia);[7, 9] models for simple chains from export sector
Consumption	Inadequate consumption of vegetables; lack of studies on demand growth and lack of consumer influence on quality standards, e.g. food safety	Promotion campaigns for vegetable consumption; improved understanding of issues affecting demand; promotion of consumer awareness of quality issues
Policy	Lack of national framework; lack of national or regional policy	Promotion of Kenyan policy and framework as a model;[10] development of regional policy to facilitate and grow regional trade[11]

[1]Selected examples only.
[2]Selected examples – mostly from the region.
[3]Farm Africa (2006).
[4]See www.research4development.info
[5]Lenné and Ward (2008, 2010), MMA (2008).
[6]Mithöfer *et al.* (2008).
[7]Höffler and Maingi (2005).
[8]See www.technoserve.org
[9]Bernet *et al.* (2005).
[10]Ministry of Agriculture (2007).
[11]König *et al.* (2008).

the constraint with one group of chain actors and that have the potential to spillover for the benefit of many more groups and vegetable value chains as a whole (Lenné and Ward, 2010). With further analysis, it is likely that many additional examples for the region, and for the rest of sub-Saharan Africa, will be identified where smallholder vegetable production is important.

Postharvest methodologies and infrastructure

Poor production methods, bad weather (especially at harvest) and deficiencies in postharvest management and marketing often result in significant postharvest losses and wastage, possibly as high as 50% (Global Horticultural Assessment, 2005; König *et al.*, 2008). In particular, the lack of cold chain facilities for perishable vegetables such as leafy greens; the lack of drying and storage methods and facilities, e.g. for onions; and the lack of appropriate packing methods for transporting vegetables, such as tomatoes, are critical constraints causing significant losses. While vertical integration has allowed larger producers and export companies to oversee most of their own postharvest management, small producers usually lack access to critical postharvest knowledge, technology, infrastructure and organization (Mutuku Muendo and Tschirley, 2004b).

Many simple and low-cost technologies have been developed for reducing postharvest losses of vegetables in South-east Asia that have the potential to spillover to East Africa and beyond (Rolle, 2006). Efforts are needed to (i) identify the most appropriate technologies; (ii) build the necessary partnerships; and (iii) secure donor investment to promote them. Within East Africa the export sub-sector has established cold chain facilities, such as cold stores and refrigerated transport, in several countries, and these are particularly well developed in Kenya. First, the possibility of an under-utilized capacity in the cold store networks in vegetable-exporting countries could be actively explored through the Fresh Produce Exporters Association of Kenya (FPEAK). Secondly, governments need to identify financial support to install cold stores at major urban wholesale markets for perishable vegetables. In both cases the methodology is well known and available in-country, but appropriate benefit–cost analysis will be necessary before deciding on the best approach. Thirdly, the export and domestic sub-sectors should discuss how most efficiently to use the existing fleet of refrigerated transport to the benefit of both sub-sectors, and how to expand the facility through joint ventures.

For vegetables that do not require cold chain facilities, such as onions, proven methods of drying and storage need to be more widely promoted. Successful onion drying, storage and marketing methods are widely used in northern and central Tanzania (Lenné and Ward, 2010), which has enabled Tanzanian onion farmers to sell locally and to export profitably to neighbouring countries; for example, 50% of the onions consumed in Kenya are from Tanzania (Mutuku Muendo and Tschirley, 2004b; König *et al.*, 2008). Although these methods are beginning to be used by smallholders in one location in Kenya (Farm Africa, 2009), there is immense potential for wider spillover throughout the region.

Transporting delicate vegetables, such as tomatoes, to market, especially on rural roads, can result in major wastage even before the produce reaches urban wholesale markets. Improved and proven packaging methods could greatly reduce wastage, improving income to smallholders. In Himachal Pradesh, India, a partnership between self-help groups, grass-roots NGOs, farmers, traders, transporters, buyers and a network of box manufacturers was developed to test and promote cardboard transport cartons (see www.research4development.info). As a result, wastage was greatly reduced and all members of the value chain benefited. Such technologies could spillover to tomato-growing communities in East Africa, especially if stakeholders from India could be involved in the process.

Value addition

In East Africa, most vegetables are sold as fresh produce. In Kenya, for example, it has been estimated that less than 5% of marketed vegetables in the domestic sector are processed in any form (Lenné and Ward, 2010). This estimate may not include dried vegetables such as okra, chillies, etc., but we could not find any quantitative data on these products. The problem is not a lack of methodology; proven methodology exists and is used by the export sector in the region and there are even a few examples of its use domestically (MMA, 2008). Hence, the main problem is the lack of promotion and use of existing methods such as washing, chopping and packaging, i.e. value addition prior to marketing. Emphasis should also be given to hygiene (e.g. the quality of water used for washing), and mycotoxins in the case of chillies (Paterson, 2007).

Some domestic vegetables are processed in East Africa by a small number of processors, the most commonly processed being tomato, into sauce and paste. However, the lack of processors and processing facilities for canning, freezing and drying, as well as those for making sauce and paste, significantly limits the level of value addition possible in the region (Lenné and Ward, 2010). Competition from EU-subsidized products such as Italian tomato paste is a disincentive for local processors. Much could be achieved by spilling over appropriate existing value-addition methods from the export sub-sector. Not all of the methods used will be appropriate for the domestic sub-sector. In Kenya, FPEAK could again play a facilitating role.

The range and quantity of imported tomato and other vegetable processed products sold in major supermarkets in Nairobi and Dar es Salaam strongly suggest that there is great potential for expansion of the processing industry in Kenya and Tanzania (Lenné and Ward, 2010). If the vegetable-processing sector is to grow sustainably in East Africa, two key developments are needed. First reliable and equitable partnerships between processors and smallholders are needed to guarantee the consistent supply of high-quality raw product throughout the year (Lenné and Ward, 2008). Currently, tomato processors in Kenya and Tanzania source some, if not all, of their tomato pulp from outside the region. Considering the wastage of tomatoes during the peak season, there is substantial potential for both groups of stakeholders to benefit. The partnership established in Arusha between the Alpha Seed Company for seed and Darsh Industries for pulp is an exciting and mutually beneficial model for spillover to other processors and seed companies in the region.

Secondly, governments need to provide incentives for entrepreneurs to invest in the vegetable-processing sector. Currently, lack of credit and access to the technology, and the physical plant required to establish a processing factory dissuade would-be investors. In addition, governments will need to address the unfair competition provided by subsidized imported products. Innovative public–private sector partnerships are needed to strengthen the processing sector, and many models exist for spillover into East Africa.

Standards: in particular, quality

The lack of national quality standards and food safety protocols and/or the lack of use of existing standards, e.g. from bodies such as the Kenya Bureau of Standards (KEBS), frequently results in poor-quality vegetables being sold in domestic markets (Lenné and Ward, 2008), and there is no regular screening for pesticide residues, which can limit many farmers' access to high-quality markets. With the exception of potatoes in Kenya (Höffler and Maingi, 2005), until recently no formal standards have been implemented in the domestic vegetable sector.

In several countries in the region, such as Tanzania, national organic produce organizations have developed protocols for production of organic produce. However, to date, there have been limited promotion and uptake and there are only a few organic vegetable shops in major cities (Lenné and Ward, 2010). Speciality vegetable outlets, such as organic shops and supermarkets, could play an important role in promoting organic standards in the same way that they have promoted indigenous vegetables (Farm Africa, 2006). Furthermore, the export sub-sector could play a mentoring role domestically based on its comprehensive experience of meeting strict quality standards imposed on export vegetables, e.g. Global Good Agricultural Practice (GlobalGAP). Both FPEAK and Sunripe have indicated a willingness to support such mentoring (Dr Stephen Mbithi, Chief Executive, FPEAK, Kenya; Mr Tiku Shah, Director, Sunripe Ltd, Kenya, 2009, personal communications).

Mithöfer *et al.* (2008) highlighted the important role for the export sector as a model, not only for production standards but also for certification in the domestic sector. A first step was taken in 2008 when KenGAP – a national quality standard for horticulture – was approved based on GlobalGAP. Although still to be operationalized throughout the horticultural production areas in Kenya, these standards provide a model for spillover to other countries in the region.

Consumption

In a recent study, the consumption of fruit and vegetables in ten sub-Saharan African countries was found to range from 27 to 114 kg per person per year, well below the FAO recommendation of 146 kg per person per year (Ruel *et al.*, 2005). However, little research has been done to understand and assess the demand for vegetables in Africa. *Would Africans eat more vegetables if the supply increased, value chains were more efficient, markets had more quality produce on offer and prices were lower?* Recent successful initiatives to promote indigenous vegetables, especially through supermarkets and speciality shops in Kenya and Tanzania (Weinberger and Msuya, 2004; Farm Africa, 2006), suggest that consumption has the potential to increase greatly under the above circumstances, especially if accompanied by attractive promotional campaigns.

For tomatoes, in the peak-supply season when prices to consumers are lowest, postharvest losses are the highest (König *et al.*, 2008; MMA, 2008). This suggests that the market is saturated and/or the market system (transport and market infrastructure) cannot cope with the volume available for sale. In contrast, in the low-supply season, when prices to consumers are highest, it is not known whether supply fails to meet demand. The relative importance of the major constraints to greater consumption of vegetables by the poor and the African population as a whole, such as price, availability, access and dietary concerns, is poorly understood. Clearly, there is a need for a study of demand prior to significant investment in improving vegetable value chains. Until demand is understood, it is premature to expect consumers to have much influence on quality standards.

Table 13.4 shows that the estimated expenditure income elasticity of demand for vegetables in selected countries is lower than 1.0 (0.61–0.91). This indicates that a 10% increase in income is associated with an increase of only 6–9% in the percentage of total household budget allocated to vegetables. The lowest income elasticity of demand is for Kenya and Tanzania. This suggests that vegetables are 'necessities' in the economic sense and that, as income grows, spending will grow, but at a slower pace (Ruel *et al.*, 2005).

Ruel *et al.* (2005) concluded that, as the poorest smallholders allocate a large share of their resources to producing staple food

Table 13.4. Estimated expenditure income elasticity of demand for vegetables and fruit in selected sub-Saharan African countries (from Ruel *et al.*, 2005).

Country	Vegetables	Fruit
Malawi	0.91	0.67
Tanzania	0.65	1.16
Kenya	0.61	1.00
Zambia	0.68	–

crops, it may be unrealistic to advise them to divert household resources to vegetables until they meet their basic energy needs. At the same time, households that have achieved adequate energy intake could be more receptive to promotional efforts to encourage a greater consumption of vegetables (Ruel *et al.*, 2005). Promotional campaigns targeted at providing accurate and useful health information to poorer households could have an acceptable impact on increasing their demand for vegetables, and possibly, on reducing prices. These would not necessarily reduce farmer incomes because, if demand increases, farmers are likely to sell more. There are many models to choose from, as such campaigns have been frequently used in recent years (e.g. the 'Five-a-day' campaign in the UK; see www.5aday.nhs.uk) that could be adapted to suit East African consumers.

Farmer organization

The efficiency and effectiveness of domestic vegetable value chains throughout East Africa are severely hampered by a lack of organization for production, transportation and marketing (Mutuku Muendo and Tschirley, 2004a), resulting in high transaction costs. Large numbers of smallholders selling small quantities of vegetables create major logistical challenges for any buyer (Hichaambwa and Tschirley, 2006). In addition, smallholders lack awareness of farm budgeting and access to market information, are spatially dispersed with poor transport infrastructure and lack knowledge of market and quality requirements (Lenné and Ward, 2010). Efforts are needed to enhance smallholders' ability to act cooperatively to exploit markets better in response to commercial opportunities. Through several NGOs and donor-funded projects some notable successes have been achieved, with organized farmers engaging equitably with buyers and benefiting from collective bargaining power.

For example, through a participatory action approach, the NGO Rural Urban Development Initiatives (RUDI) is a key catalyst for empowering farming communities for improving market linkages and distribution channels for their products throughout Tanzania (see www.ruditz.org). The capacity of farmer groups is built through activities such as organization processes, market linkages, credit linkages and warehouse receipt systems. High-value crops such as vegetables are targeted to raise incomes. Registered producer organizations can link with other producer organizations and, through an apex body, enjoy even greater opportunities. Apex organizations such as the Agricultural Council of Tanzania can then influence policy makers. More information can be found on how farming communities improve their market linkages and distribution channels in Tanzania at www.ruditz.org.

Throughout Africa the NGO Technoserve, which focuses on building enterprises to generate benefits for the rural poor, has developed a 'hub' approach (see www.technoserve.org). In the context of vegetable marketing, this hub is a bulking facility owned by a farmer group. The hub buys the product from the farmers for a fair market-oriented price and then retails the products in larger bulks. The hub has greater negotiating power to command better prices than do individual farmers, and there is the potential for developing hubs for adding value such as washing and packing. Higher prices paid for value-added goods can then be channelled back to the farmers.

There appears to be great potential to spillover these approaches to more farming communities, with increased donor support (Lenné and Ward, 2010). In Tanzania, strengthened linkages between RUDI and Technoserve could further improve farmer organization and links with vegetable markets. Such partnerships have the potential to build platforms to share experiences, learn lessons and to facilitate spillovers. As a result, vegetable value chains should function more effectively to the benefit of more smallholders.

In both Kenya and Tanzania, farmer groups have been organized to sell vegetables to speciality outlets. For example, Family Concern has built the capacity of smallholders to supply Uchumi, Nakumatt and

Shoprite supermarkets, as well as speciality grocers, with high-quality indigenous vegetables (Farm Africa, 2006; Ngugi *et al.*, 2007). Pilot studies have shown that successful farmer groups realize greater profits from these market channels, to the benefit of their families. Similarly, in Tanzania, smallholder high-quality vegetable basket schemes are targeting Shoprite supermarkets, lodges, restaurants and hotels in Dar es Salaam and northern Tanzania (MMA, 2008).

Above all, the organization of smallholders for the collective marketing of vegetables to export companies, especially in Kenya, is a proven role model for the domestic sub-sector (Lenné *et al.*, 2005). Building the capacity and group organization of smallholders in a similar manner to meet the quality standards and procurement conditions of large domestic buyers will not only allow them to sell to supermarkets and other speciality shops, but should also contribute to improved efficiency in domestic marketing systems, as many smallholders grow vegetables for both markets.

These lessons learnt from the effective development of smallholder organizations in East Africa have the potential to be spilled over more widely throughout vegetable value chains in this region.

Value chain linkages and information systems

Domestic vegetable value chains face many constraints between harvest and consumer. Farmer marketing is informal through spot-markets at the farm gate; chains are inefficient and long, being fragmented at both the producer and retailer end; there is a lack of organization resulting in inefficiencies and inequalities throughout the chain; market information systems are poor or non-existent; and, importantly, there is a lack of trust (even among family members) among chain actors (Mutuku Muendo and Tschirley, 2004a; Hichaambwa and Tschirley, 2006; König *et al.*, 2008; MMA, 2008; Lenné and Ward, 2010). In addition, cross-border trade is severely hampered by complex marketing chains with poor linkages, in particular low cooperation between cross-border officials and poor price information systems for urban markets (König *et al.*, 2008). These constraints result in inefficient chains and, as a result, all chain actors' incomes from vegetables are reduced, consumers' access to quality produce is compromised and the value chain is far from reaching its full economic potential.

Vegetable smallholders generally lack access to market information systems, recently highlighted by Evans (2009). Even where market information systems exist, such as KACE in Kenya, smallholders do not appear to be aware of how to access them or the systems do not penetrate into rural areas (Weber *et al.*, 2005; Tollens, 2006). These existing systems can potentially provide market information on vegetables. Priority should be given to (i) incorporation of information on major traded vegetables into these systems; and (ii) improving access for smallholders, whether through radio, newspaper or mobile phones. In South Africa, SAFEX already provides market information on vegetables (see www.safex.co.za) and there are other successful models from South Asia, e.g. Bangladesh (www.electronicgovernment.se/AMIS).

Value chain analyses with actors provide a sound basis for improving the chain linkages. Although examples are rare in the domestic vegetable sub-sector in East Africa, several good initiatives have been developed in the potato sector. In Kenya, the participatory market chain-mapping workshop approach is already improving potato value chains (Höffler and Maingi, 2005). Representatives from all the actor groups participate in developing a joint understanding of the value chain. This reveals the strengths and weaknesses of the chain and serves as a starting point for mutual trust building. Perceptions are then demystified and improvements can be made. Similarly, participatory market chain approaches were developed independently for potato value chains in Latin America through the INNOVA and PAPA-ANDINA programmes (Bernet *et al.*, 2005). The INNOVA approach from Bolivia is being replicated in potato systems in Uganda with great success and is being tested in Bangladesh for rice and pulse value chains (Lenné and Ward, 2010). These examples provide potential spillovers more widely throughout domestic vegetable value chains in East Africa.

Above all, the export vegetable sub-sector, with its simple and efficient value chains, especially in Kenya, is a proven role model for the domestic sub-sector of effective linkages between smallholders and export companies (Lenné *et al.*, 2005; Lenné and Ward, 2010). Building the capacity of chain actors to aspire to this model could have a far-reaching positive impact on the efficiency of domestic vegetable marketing systems.

Coordination and policy

The lack of both national and regional coordination for vegetables is a key constraint to further development of the sector in East Africa. This has been a major contributing factor to the lack of knowledge sharing among the many recent isolated and duplicating studies on vegetable value chains, especially in East Africa. At the national level, governments and other appropriate bodies need to play a more proactive role in coordinating such activities. At the regional level, the Global Horticultural Initiative could facilitate better coordination and information sharing of horticultural activities.

Furthermore, the lack of both national horticultural frameworks and national or regional policy for horticulture has hampered improvements in the vegetable sector. Governments should look at recent national coordination efforts in Kenya as a potential model to improve coordination in other countries. Kenya has a comprehensive national operating structure for horticulture, a national task force and a national horticulture policy (Ministry of Agriculture, 2007). Although the performance of these entities is only beginning to be tested, these are necessary first steps for other countries to take to begin to address the serious deficiencies in domestic vegetable value chains.

Capitalizing on Spillover Opportunities

In this chapter, we have suggested not only the potential to capitalize on spillovers to strengthen vegetable value chains in East Africa, but also the existence of a number of proven successful technologies and methodologies that could be immediately promoted throughout value chains in the region. At the same time, we should acknowledge that, when a new approach to development is being suggested, there is a danger that previous valuable experience is not built into the approach. A value chain approach does not mean that all elements of the value chain need to be strengthened at once; it requires that the actors in the value chain identify entry points together to increase the efficiency of the chain. Emphasis should be given to the weakest links. Through the actors' better understanding of the value chain, the ownership and appropriate enabling environment are more likely to be created for catalysing potential spillovers from previous successes in the vegetable sector. In addition to the direct benefits to the actors, such an approach will build confidence for the actors to continue to work together to strengthen the value chain.

We would like to suggest the following four steps for identifying and promoting the best spillovers for the vegetable sector in East Africa. First, comprehensive sector-wide research and analysis is needed to identify successes, opportunities and potential spillovers. Secondly, consultation is needed with chain actors on successful initiatives to better understand the reasons for their success and, especially, any shortcomings. Information is likely to be more readily available on recent successes, but there are some historical successes that are worth investigating, such as onion storage in Tanzania. Thirdly, a concerted effort will be needed to generate awareness, facilitate ownership by key stakeholders and identify champions of the best potential spillovers. This should be done together with potential donors so that they understand the probable investments. Fourthly – and it is hoped that this will be spontaneous – partnerships, especially between the public sector and private sector entrepreneurs and directly between smallholders and the private sector, should be fostered and built to capitalize on potential spillovers.

Finally, governments have an important role to play in strengthening the policy and regulatory environments, including support for smallholders and consumers' needs and chain linkages, so that the considerable benefits from wider spillovers of knowledge, technologies and methodologies can be enjoyed throughout the vegetable sector.

Export Vegetable Sub-sector: an Important Role Model

A focus on East Africa has one significant advantage: the region, especially Kenya, has a vibrant, highly efficient, innovative and successful export vegetable sub-sector. This provides not only a role model for tackling production and marketing constraints, but also a suite of potential production and marketing spillovers (Lenné *et al.*, 2005; Lenné and Ward, 2010). A previous report showed that many of the lessons of Kenyan horticulture can be applied elsewhere in Africa provided that support and time are given to the development of the marketing institutions that underpin the sector (Minot and Ngigi, 2004). A recent article by Mithöfer *et al.* (2008) highlights the important role for the export sub-sector as a model for production standards and certification to the domestic sub-sector. The export sub-sector's successful diversification from fine beans to many other vegetables is a prime example of the kind of spillovers that may be achieved in the domestic vegetable sub-sector. Not only is the export sector a good model for the aspirations of the domestic sector, but it is also available, accessible and close by. It can play a key role through bodies such as FPEAK in developing national policy. There is the potential to appeal to its philanthropic values. Recent discussions between FPEAK, Sunripe, KARI and the authors confirm overwhelming support for such an approach (Dr Lusike Wasilwa, Head of Horticulture, KARI, Kenya; Dr Stephen Mbithi, Chief Executive, FPEAK, Kenya; Mr Tiku Shah, Director, Sunripe Ltd, Kenya, 2009, personal communications). The domestic value chain does not have to look further for experienced capacity for its efficiency to be improved.

Conclusions and Further Research

In this chapter, we have outlined a new approach to strengthening vegetable value chains in East Africa and have suggested the first steps in identifying the wealth of potential spillovers of knowledge, technology and methodology available in East Africa. We suggest that the greatest chance for immediate success will be to target simple spillovers, e.g. onion storage technology or improved packaging for transporting tomatoes within and between neighbouring countries in East and southern Africa. Farmers already using these technologies could help to mentor adopting farmers. Due to the breadth of constraints faced by domestic vegetable value chains, partnerships between government, private sector and donors will be essential to mobilize the financial resources and knowledge to capitalize on potential spillovers. Although we have suggested the next steps needed in priority to identify and promote spillovers, innovative thinking and further research will be required for selecting the best-bet spillovers and designing investment programmes to further enhance smallholder participation in vegetable value chains, while reducing overall marketing costs and prices to consumers. In particular, local research, development and adaptation are likely to be needed for spillovers from one country, region or continent to another. Strong leadership will be essential, as well as national and regional policy support. A recent comprehensive analysis of successful agricultural research and development initiatives that have fed millions highlighted the critical confluence of *science*, *policy* and *leadership* (Spielman and Pandya-Lorch, 2009). Strengthening vegetable value chains in East Africa through spillovers has great potential to create a measurable impact on the poverty and nutrition of the poor at local, national and regional levels in the short to medium term.

References

Alston, J.M. (2002) Spillovers. *The Australian Journal of Agricultural and Resource Economics* 46, 315–346.

Alston J.M. and Pardey, P.G. (2008) Public funding for research into specialty crops. *HortScience* 43, 1461–1470.

Bernet, T., Devaux, A., Ortiz, O. and Theile, G. (2005) Participatory market chain approach. *BeraterInnen News* 1, 8–13.

Evans, A. (2009) The feeding of the nine billion: Global food security for the 21st century. Chatham House Report, London (www.chathamhouse.org.uk).

FAOSTAT (2008) Food and Agriculture Organization of the United Nations, Rome (http://faostat.fao.org/default.aspx).

Farm Africa (2006) Improved health and livelihoods with African indigenous vegetables. *Appropriate Technology* 33, 10–12 (www.appropriate-technology.org).

Farm Africa (2009) Kenyan farmers reap benefits from better marketing. *Appropriate Technology* 36, 24–26 (www.appropriate-technology.org).

Global Horticultural Assessment (2005) Report, USAID Award #EDH-A-00-04-00006-00, University of California, Davis, California.

Graffham, A., Karehu, E. and MacGregor, J. (2007) Impact of EurepGAP on small scale vegetable producers in Kenya. Fresh Insights 6, Chatham Maritime, Chatham, UK.

Hichaambwa, M. and Tschirley, D. (2006) Understanding Zambia's domestic value chains for fresh fruit and vegetables. Policy Synthesis 17, Michigan State University Food Security Research Project, Lusaka, Zambia (http://www.aec.msu.edu/agecon/fs2/zambia/index.htm).

Höffler, H. and Maingi, G. (2005) Rural urban linkages: Promoting agricultural value chains. *Entwicklung und Ländlicher* 5, 26–28 (www2.gtz.de/snrd/english/topics/awf.html).

König, T., Blatt, J., Brakel, K., Kloss, K., Nilges, T. and Woellert, F. (2008) *Market-driven Development and Poverty Reduction: A Value Chain Analysis of Fresh Vegetables in Kenya and Tanzania*. SLE Publication Series, January 2008, Nairobi, Kenya and Berlin, Germany.

Lenné, J.M. and Ward, A.F. (2004) Lesson learning study from the vegetable cluster with special emphasis on the links with the private sector. A report commissioned by the DFID Crop Protection Programme, Natural Resources International, Aylesford, UK.

Lenné, J.M. and Ward, A.F. (2008) Exotic and indigenous vegetable value chain analysis for selected countries in East and southern Africa. A report commissioned by the Bill and Melinda Gates Foundation, Seattle, Washington.

Lenné, J.M. and Ward, A.F. (2010) Improving the efficiency of domestic vegetable marketing systems in East Africa: Constraints and opportunities. *Outlook on Agriculture* 39, 1–40.

Lenné, J.M., Pink, D.A.C., Spence, N.J., Ward, A.F., Njuki, J. and Ota, M. (2005) The vegetable export system: A role model for local vegetable production in Kenya. *Outlook on Agriculture* 34, 225–232.

McCulloch, N. and Ota, M. (2002) Export horticulture and poverty in Kenya. IDS Working Paper 174, University of Sussex, Brighton, UK.

Ministry of Agriculture (2007) National Policy on Horticultural Industry Development, February, 2007. Republic of Kenya.

Minot, N. and Ngigi, M. (2004) Are horticultural exports a replicable success story? Evidence from Kenya and Côte d'Ivoire. EPTD Discussion Paper 120; MTID Discussion Paper 73, International Food Policy Research Institute, Washington, DC.

Mithöfer, D., Nang'ole, E. and Asfaw, S. (2008) Smallholder access to the export market: The case study of vegetables in Kenya. *Outlook on Agriculture* 38, 203–211.

MMA (2008) Fresh fruit and vegetable sub-sector: Value chain analysis – Tanzania. SCF small and medium-sized enterprise competitiveness facility. Implemented by Match Maker Associates Ltd, Tanzania.

Mutuku Muendo, K. and Tschirley, D. (2004a) Improving Kenya's domestic horticultural production and marketing system: Current competitiveness, forces of change, and challenges for the future. Vol. I: Horticultural Production. Tegemeo Institute of Agricultural Policy and Development, Working Paper 8A, Egerton University, Njoro, Kenya.

Mutuku Muendo, K. and Tschirley, D. (2004b) Improving Kenya's domestic horticultural production and marketing system: Current competitiveness, forces of change, and challenges for the future. Vol. III: Horticultural Research and Input Sector Regulation in Kenya and Tanzania. Tegemeo Institute of Agricultural Policy and Development, Working Paper 8C, Egerton University, Njoro, Kenya.

Ngugi, I.K., Gitau, R. and Nyoro, J.K. (2007) Innovative practice Kenya: Access to high value markets by smallholders of African indigenous vegetables. Regoverning Markets (www.regoverningmarkets.org).

Ota, M. and Lenné, J.M. (2003) Effectiveness of commercial horticulture networks as pathways for poverty alleviation, exploitation and promotion of pest management technologies in Kenya. A report commissioned by the DFID Crop Protection Programme, Natural Resources International, Aylesford, UK.

Pardey, P.G. and Beintema, N.M. (2002) Slow magic: Agricultural R&D a century after Mendel. Working Papers 14364. Centre for International Food and Agricultural Policy, University of Minnesota, Minnesota.

Paterson, R.R.M. (2007) Aflatoxin contamination in chilli samples from Pakistan. *Food Control* 18, 817–820.

Rolle, R.S. (2006) Postharvest management of fruit and vegetables in the Asia-Pacific region. Reports of the APO seminar on *Reduction of Postharvest Losses of Fruit and Vegetables* held in India, 5–11 October 2004 and *Marketing and Food Safety: Challenges in Postharvest Management of Agricultural/KHorticultural Products in the Islamic Republic of Iran*, 23–28 July 2005. FAO, Rome, Italy.

Ruel, M.T., Minot, N. and Smith, L. (2005) Patterns and determinants of fruit and vegetable consumption in sub-Saharan Africa: A multi-country comparison. Background paper for *FAO/WHO Workshop on Fruit and Vegetables for Health*, September 2004, Kobe, Japan. World Health Organization, Geneva, Switzerland.

Spielman, D.J. and Pandya-Lorch, R. (2009) *Millions Fed: Proven Successes in Agricultural Development.* IFPRI, Washington, DC.

Tollens, E. (2006) Market information systems in sub-Saharan Africa: Challenges and opportunities. Contributed paper prepared for presentation at the *International Association of Agricultural Economists Conference*, 12–18 August 2006, Gold Coast, Australia.

Weber, M.T., Donovan, C., Staatz, J.M. and Dembélé, N.M. (2005) Guidelines for building sustainable market information systems in Africa with strong public–private partnerships. Policy Synthesis for Cooperating USAID Offices and Country Missions No. 78 (www.aec.msu.edu/agecon/fs2/psynindx.htm).

Weinberger, K. and Msuya, J. (2004) Indigenous vegetables in Tanzania – significance and prospects. Technical Bulletin 31, The World Vegetable Center, Shanhua, Taiwan, pp. 1–70.

14 Challenges for Economic Impact Assessment of Classical Biological Control in Kenya and Tanzania

Anna Asfaw, Dagmar Mithöfer, Bernhard Löhr and Hermann Waibel

Introduction

In eastern and southern Africa, cabbage (*Brassica oleracea* L. var. *capitata*) is one of the most widely grown vegetables. For small-scale farmers, it is an important source of vitamins, minerals and income. Cabbage production is, however, constrained by pest problems. The diamondback moth (DBM, *Plutella xylostella* (L.)) is cabbage's major pest, and occurs wherever *Brassica* crops are cultivated (Talekar and Shelton, 1993). In East Africa, most of the farmers depend solely on the use of synthetic pesticides to minimize damage and output losses due to DBM (Varela *et al.*, 2003).

Increasing difficulties in DBM control can arise because DBM quickly develops resistance to pesticides (Gelernter and Lomer, 2000). Thus the control of DBM becomes difficult and, without alternatives, farmers tend to overuse chemicals by increasing pesticide quantity and spray frequency, as well as applying pesticide cocktails (Varela *et al.*, 2003). Since chemical control measures are associated with negative external effects such as a build-up of resistance and the negative impact on human and environmental health, biological control is gaining attention (Margni *et al.*, 2002; Hajek, 2004). Classical biological control (BC) is the introduction of an exotic natural enemy into a new environment. It involves releases of an appropriate BC agent, which spreads on its own and is ubiquitous when successfully established. Because the pest control is permanent and requires no investment from individuals, it is extremely appropriate for pests that affect resource-poor farmers (Hajek, 2004).

In the highly intensive vegetable production systems of Asia, problems in the chemical control of DBM were reported for many years. However, although classical BC[1] programmes of DBM were successful in several regions (Talekar and Shelton, 1993), evidence of their economic impact has not been documented to date. Although the intensity of pesticide use in the vegetable production systems of East Africa is generally lower than in Asia, Kibata (1996) found that, in Kenya, pesticide resistance had developed in DBM because some popularly used pesticides did not provide the desired level of control. However, the assessment of resistance is complicated due to heterogeneous pesticide quality caused by inappropriate storage, high temperatures and other factors.

To control DBM in the major cabbage-producing highland areas of Kenya, Tanzania and Uganda, the parasitoid *Diadegma semiclausum* (Hellén) was imported from

Taiwan and introduced by the International Centre of Insect Physiology and Ecology (*icipe*) in 2001 at pilot sites. For Kenya, an economic *ex ante* impact assessment was conducted during the pilot phase of the project. Using a number of assumptions based on the available literature and experimental results on the potential impact of the BC agent, Macharia *et al.* (2005) calculated a benefit–cost ratio of biocontrol investment of 24:1. This study aims to supplement the *ex ante* analysis by verifying the assumptions made previously. It represents a short- to medium-term *ex post* impact assessment for the BC in Kenya and Tanzania. Specific questions addressed in this chapter are (i) does the presence of the BC agent result in reduced pesticide use?; and (ii) how does the presence of the BC agent affect the output of cabbage production?

Study Area and Data

Study areas in Kenya and Tanzania were selected and data collected following a 'with and without' survey design. The survey followed a two-stage sampling procedure. The first stage involved the selection of major cabbage-producing areas with and without the presence of the parasitoid. Because cabbage is usually produced at an altitude between 800 and 2000 m above sea level (Macharia *et al.*, 2005), the highlands of Kenya and Tanzania were of special interest. As depicted in Fig. 14.1 and Table 14.1, selected study sites were the Central Province (Nyandarua, Nyeri and Kiambu districts) in Kenya and the Northern Zone (Arusha and Tanga regions) in Tanzania. The BC agent was present in South Kinangop (Nyandarua District), Kieni East (eastern part of Nyeri District), Kikuyu (Kiambu District), Limuru (Kiambu District) and the Arumeru Region.

In addition to being major cabbage-producing regions within their countries, Central Province and Northern Zone have other similarities. They have a common dry and rainy season, relatively high rainfall of bimodal pattern and fertile soils, which makes them well suited for agricultural production. A large share of the population is employed in the agricultural sector, emphasizing its importance for income generation for the rural population (FAO, 2006a,b). However, these areas are characterized by high population density and competition for limited land resources (Ministry of Agriculture (MoA) Kenya, 2005; United Republic of Tanzania (URT), 2007).

A list of cabbage farmers, compiled by national extension officers, served as a sampling frame to select farmers randomly on the sub-location/ward level at the second sampling stage. The sampling procedure resulted in a total sample of 1291 randomly selected cabbage farmers covering two production seasons, for whom 1250 complete data sets were retained for the analysis. In Kenya the sample consisted of 634 farmers, 496 from areas where the BC agent was present and 138 from areas without the parasitoid. In Tanzania the total number of farmers sampled was 616, with 354 from areas with and 262 from areas without the BC agent. Data were collected in several steps by means of face-to-face interviews. First, a recall survey for a dry season crop during October 2004 to March 2005 was conducted. The second survey wave was based on a season-long monitoring, involving three farm visits for the rainy season crop from March to July 2005. Both surveys were based on the same structured questionnaire, capturing information about household characteristics and the technical and economic parameters of cabbage production.

Pest Control Methods and Crop Productivity

In estimations of the productivity of pest control measures, especially of pesticides, a production function framework with an integrated damage control function is frequently used. To establish an attribution and analyse the effect of the presence of the BC agent, in this study the parasitoid was incorporated in the damage control function.

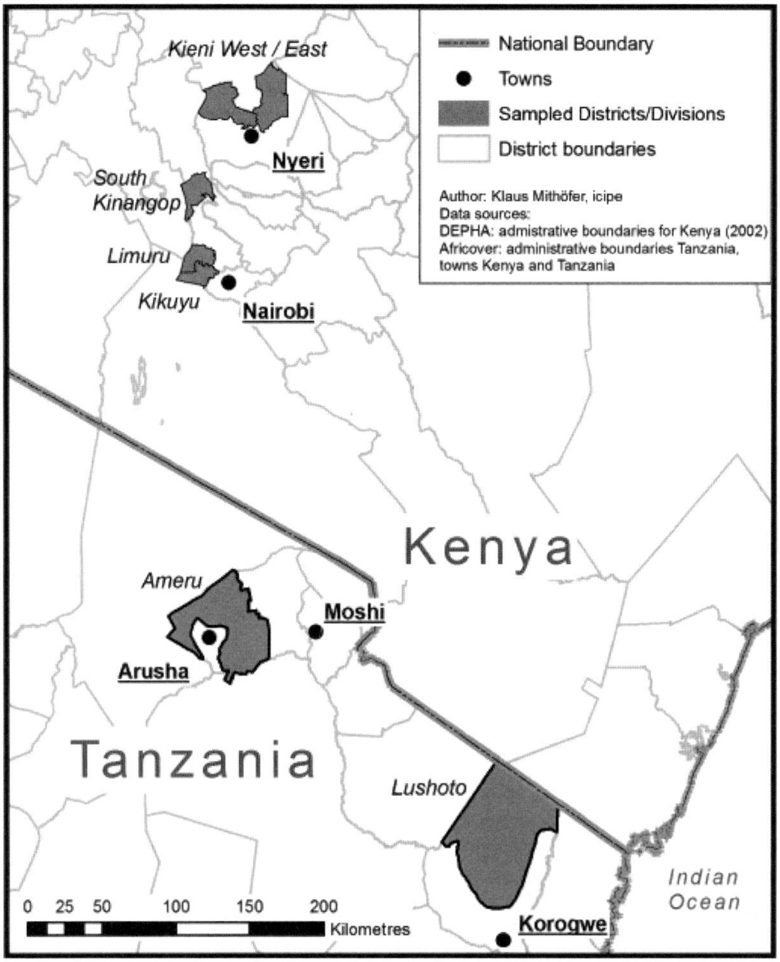

Fig. 14.1. Map of study sites in Kenya and Tanzania.

Table 14.1. Study sites in Kenya and Tanzania.

Country	District/Region[a]	Division/District[b]	Presence
Kenya	Nyandarua	South Kinangop	w
	Nyeri	Kieni East	w
	Nyeri	Kieni West	w/o
	Kiambu	Kikuyu	w
	Kiambu	Limuru	w
Tanzania	Arumeru	Arumeru	w
	Lushoto	Lushoto	w/o

[a] District for Kenya, region for Tanzania.

[b] Division for Kenya, district for Tanzania. w, with presence of BC agent; w/o, without presence of BC agent.

In the analysis of pesticide productivity, the use of a standard Cobb–Douglas function is criticized for treating pesticide as a yield-increasing production factor and for not capturing knowledge about physical and biological processes of pest control agents. Lichtenberg and Zilberman (1986) showed that using a standard Cobb–Douglas function results in overestimation of the productivity of damage control inputs, while underestimating the productivity of other factors. To address this problem they introduced the concept of damage abatement, which has one key feature, namely the distinction between inputs in standard factors of production (e.g. land, labour, capital) and in damage control agents (e.g. pesticides, BC agents). This distinction is important, because control agents do not enhance productivity directly as do standard inputs, but contribute indirectly to the actual output by preventing output losses.

Lichtenberg and Zilberman (1986) proposed different specifications of damage control functions; however, even now no agreement on the most suitable functional form has been reached. They noted that damage abatement functions may not only include damage control agents but also other exogenous state variables, which affect the damage abatement effort but are uncontrollable. Blackwell and Pagoulatos (1992) modified the original approach, explicitly accounting for natural abatement, so that the productivity of damage control agents included in the production function would not be overestimated. Using a process-modelling method, they derived a dynamic Cobb–Douglas production function that is generated by biological and physical processes within the ecosystem and, thus, includes state variables, which have been excluded in earlier damage control functions. Carpentier and Weaver (1997) demonstrated the substantial restrictions of the specifications of the Blackwell and Pagoulatos (1992) model and proposed a more general specification form, which accounts for interactions between direct and damage control inputs. Oude Lansink and Carpentier (2001) also applied this approach. However, statistical evidence was weak since none of the estimated parameters were significant.

The present study explicitly accounts for at least one controllable natural damage abatement agent, since the BC agent, *D. semiclausum* is a natural enemy of DBM and its introduction contributes to natural damage abatement, so that overall the upward bias in the productivity of pesticides will be reduced. Figure 14.2 illustrates the effect of damage control agents on yield and the theoretical effect of the combined use of pesticide and BC. Due to the existing biodiversity in most cases, without pest control measures ($G = 0$), the minimum yield level Y^{min} is greater than zero. Y^{min} depends on several factors, such as the effectiveness of natural pest control. Y^{max} denotes the potential yield, i.e. if pest control is 100% successful or if the pest is absent. The difference between Y^{max} and Y^{min} denotes the yield loss that can be saved by introducing a pesticide as a control agent. The pesticide productivity curve shows the effect of pesticides on damage reduction. In the most favourable case, without losses ($G = 1$), Y^{max} becomes the actual yield and equals the potential yield (Ajayi, 2005; Pemsl, 2006).

The concept illustrated in Fig. 14.2 has been adopted from the literature on Bt-cotton (Pemsl, 2006), where the Bt trait constitutes an alternative to chemical pest control. Similarly, BC is an alternative technology. If, in addition to pesticides, a BC agent is introduced as a damage abatement agent in the production system, three issues have to be considered. First, although the population density of the parasitoid does affect its effectiveness against DBM, for this study, information on population density was not available and thus only the presence or absence of the BC agent is considered. However, it should be noted that the full potential is reached only when the parasitoid is well established and the relationship between parasitism and pest density is negative. Secondly, the impact of the BC agent on damage abatement is fixed because the parasitoid can only control damage caused by DBM. Due to improvements in biodiversity and increased self-regulation through

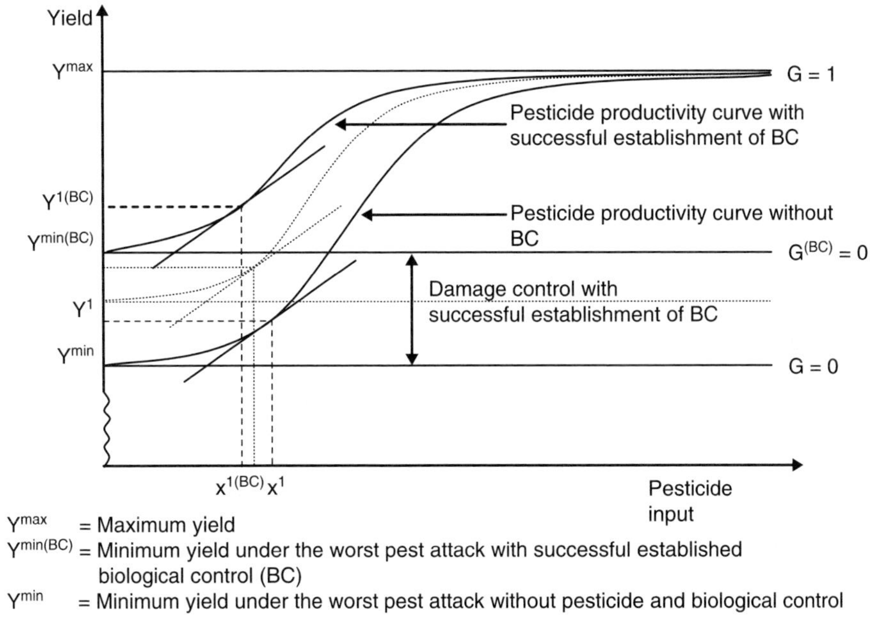

Fig. 14.2. The effect of damage control agents on yield.

the presence of a well-established BC agent, the minimum yield obtainable without the use of pesticide shifts upwards from Y^{min} to $Y^{min(BC)}$. The distance between maximum and minimum yield decreases, indicating that the economic productivity of pesticides will be lower when the parasitoid is present.

If selective pesticides for pests other than DBM are used appropriately, the effect of the pesticide and BC agent is complementary and, with a successfully established parasitoid, a lower amount of pesticide needs to be applied. This is indicated by the decrease in the optimal input level of pesticide, from x^1 to $x^{1(BC)}$. However, thirdly, a potential negative impact of pesticide use on the BC agent must be accounted for. Since broad-spectrum pesticides can affect BC agents, their use can hamper successful establishment and/or maintenance in the production system, which could decrease the level of output losses prevented by the BC agent to lower levels. Therefore, the minimum yield with unsuccessfully established BC agents can vary between $Y^{min(BC)}$ and Y^{min}, as indicated by the dotted pesticide productivity curve line in Fig. 14.2. To enhance the establishment of the parasitoid, several integrated pest management (IPM) methods, such as the use of biological (selective) pesticides and spot spraying, can be applied.

Estimation of Production Function with Damage Control Function

Integrating a damage abatement function, *G* in the Cobb–Douglas production function leads to a cabbage output function, *Y*, that can be written as

$$Y = a\left[\prod_{i=1}^{n}(x_i^D)^{\beta_i}\right] * G(x^P)^{\gamma} \qquad (14.1)$$

where *Y* is the output of cabbage production; *a* the intercept x_i^D; I = 1, 2, ...; *n* are standard inputs (e.g. seed, fertilizer and labour); β_i are the coefficients to be estimated; and x^P is a vector of control agents (pesticides, BC agent) within the damage abatement function. While past studies showed that results can be sensitive to the functional form of the damage control function (Carrasco-Tauber and Moffitt, 1992; Fox and Weersink, 1995; Pemsl, 2006), in this study a widely applied exponential function[2] was used:

$$G(x^P) = 1 - \exp(-\lambda_1 x_1^P - \lambda_2 x_2^P - \lambda_3 x_1^P x_2^P) \quad (14.2)$$

where x_1^P is the pesticide input; x_2^P the presence of the BC agent; and $x_1^P x_2^P$ an interaction term for both damage control agents. The production function is estimated by using its logarithmic form:

$$Log(Y) = Log(a) + \sum_{i=1}^{n} \beta_i Log(x_i^D) + Log(1 - \exp(-\lambda_1 x_1^P - \lambda_2 x_2^P - \lambda_3 x_1^P x_2^P) + \varepsilon \quad (14.3)$$

The error term ε is added to the equation and represents factors other than the explanatory variables that affect the dependent variable but are unobserved (Wooldridge, 2003). Because the damage control function is non-linear, non-linear regression methods are applied.

One major problem in estimating production functions is that all production factor variables are treated as exogenous, although farmers decide their level. While the problem is apparent for all inputs, it is special for pesticide use. Due to the nature of potentially omitted variables and correlations, by not accounting for endogeneity parameter estimates of the impact of pesticide use on the output of cabbage production could be biased. As in Bt-cotton studies (e.g. Huang *et al.*, 2002; Pemsl, 2005), in this study the problem of endogeneity, in common with that of damage abatement models, is solved using a two-stage least squares (2SLS) estimator. First, a pesticide use function is estimated by adopting an instrumental variable approach. Secondly, the predicted values of pesticide use are applied to estimate a production function. As long as the pesticide use equation in this procedure incorporates a set of variables that explains actual pesticide use but that does not have an independent explanatory power on the cabbage output, the instrumental approach allows for a better examination of the effects of damage control agents on output (Huang *et al.*, 2002). Towards that end in this study, the pesticide price is used as the instrumental variable.

In a standard Cobb–Douglas function, the estimated coefficients of input factors are identical to their production elasticities. However, with an interaction term, the partial effect or the elasticity of the dependent variable with respect to an explanatory variable in the interaction term depends on the magnitude of the other explanatory variable included in the interaction term. Therefore, the coefficients cannot be directly interpreted. The partial effect can be computed, *ceteris paribus*, at interesting values and/or for dummy variables plugging in zero or one using the appropriate derivation (Wooldridge, 2003). Using a standard Cobb–Douglas function as an example, the partial effect of pesticide use on cabbage output, which is created with or without the presence of the parasitoid, is computed as $\beta_1 + \beta_3 * x_2^D$, where β_1 is the estimated coefficient of pesticide use, β_3 is the estimated coefficient of the interaction term and x_2^D denotes the presence of the BC agent. The effect of the presence of the BC agent on cabbage output is expressed as $\beta_2 + \beta_3 * x_1^D$, with β_2 as the estimated coefficient of the BC agent, β_3 denoting the estimated coefficient of the interaction term and x_1^D is any interesting value of pesticide use to be plugged in, such as the mean or median value of the sample. Since the dependent variable and continuous explanatory variables used in the model are log-transformed, the value of pesticide use needs to be converted into its logarithmic form.

Furthermore, due to the log-transformed dependent variables, the interpretation of dummy or count variables is not straightforward. Multiplying the estimated coefficients of these variables by 100 and interpreting them directly as the percentage effect on the dependent variable will cause an approximation error. Following Kennedy (1981), to obtain an exact percentage change one can compute

$$g = 100 * \left[\exp\left(\beta_d - \frac{1}{2} V(\beta_d) \right) - 1 \right] \quad (14.4)$$

where g is the percentage change in the level of the dependent variable, β_d the coefficient of the dummy or count variable and $V(\beta_d)$ the estimated variance of β_d.

Descriptive Analysis of Cabbage Production in Kenya and Tanzania

To describe the cabbage production system, an extensive comparison is conducted. First, to cover country specific characteristics (farmers by country), cabbage farmers from Kenya (K) are compared with farmers from Tanzania (T). To capture characteristics regardless of political borders, farmers from areas with BC (w) are compared with farmers from areas without the occurrence of the parasitoid (w/o) (farmers by presence of BC). Additionally, cabbage farmers operating with and without the presence of the BC agent are compared within the countries.

The survey found that Kenyan households had, on average, more members, and that household heads were on average about 5 years older and with a higher education level than Tanzanian households. In financial terms, Kenyan households were better off because, on average, they obtained a higher income from farm and off-farm activities, owned more land, possessed higher values of livestock, lived in better housing conditions and were more likely to have at least one permanent employee who works on the farm. The share of farmers from Kenya who participated in agricultural training was 35% and, thus, about twofold higher than the proportion of Tanzanian farmers who attended training.

As shown in Table 14.2, the major input factors for cabbage production are land, seeds, labour, fertilizer, irrigation, pesticides and the management of pest problems. In Kenya, the surveyed households held on average 2.44 ha of agricultural land, while in Tanzania this was significantly lower with an average of 0.55 ha per household. The average size of a cabbage field was 0.21 ha for the whole sample, which did not differ significantly between areas with and without the presence of the BC agent, or between Kenya and Tanzania. The survey revealed that in both countries the cabbage field occupied significantly higher proportions of total landholdings in areas where the BC agent was present. This observation is in line with the study of Macharia *et al.* (2006), who found that about two-thirds of interviewed farmers from areas where the BC agent had been introduced intended to increase their cabbage production in the future due to reduced production costs. However, because no before-and-after comparison was available, it was not possible to attribute this clearly to the project intervention.

A few cabbage varieties show a higher tolerance to different pests; however, the more tolerant varieties are generally not preferred and not cultivated because of other, poorer agronomic features (Lim, 1992). In the study area the majority of farmers use two varieties, both of which are similar standard varieties and are more susceptible to different pests. Although cabbage can be cultivated under rain-fed conditions, growing cabbage under irrigation conditions is common and 56% of Kenyan farmers and 88% of Tanzanian farmers do so. Within Kenya and Tanzania, as well as overall, a significantly higher share of farmers from areas where the BC agent is not present irrigates the cabbage crop. A further analysis of the irrigation system showed that, in Tanzania, manual irrigation is practised almost exclusively, while in Kenya 32% of farmers irrigate manually and 55% use sprinklers.

Seed costs are, on average, significantly higher in areas where the BC agent is present. For Kenyan farmers these are higher than for those in Tanzania, where expenditure on seed is higher for farmers from areas with the presence of the BC agent than for farmers who operate in areas without. Labour costs are aggregated for different crop management tasks such as labour for field preparation, planting, irrigation and weeding. Labour for pest control measures and associated costs do not differ significantly

Table 14.2. Input use in cabbage production with and without the presence of the biological control (BC) agent in Kenya and Tanzania (*n* = 1250).

	Farmers, by country		Farmers, by presence of BC		Kenya		Tanzania	
Variable	K	T	w	w/o	w	w/o	w	w/o
N	634	616	850	400	496	138	354	262
Size of cabbage field (ha)	0.20	0.21	0.21	0.19	0.19	0.21	0.24	0.18
	(–1.34)		(4.20)***		(2.07)*		(–0.55)	
Tolerant variety (%)	12.0	3.9	8.6	6.8	14.8	2.2	0.0	9.2
	(27.88)***		(33.78)***		(1.27)		(16.24)***	
Irrigation (%)	55.8	87.5	59.7	96.5	45.8	92.0	79.1	98.9
	(153.53)***		(53.74)***		(181.06)***		(93.70)***	
Seed (US$/ha)	124.6	138.9	145.1	102.7	123.7	128.0	174.4	88.9
	(–2.06)**		(7.88)***		(5.74)***		(–0.43)	
Fertilizer (US$/ha)	212.8	190.5	208.5	186.5	218.8	190.7	194.9	184.4
	(1.63)		(0.52)		(1.49)		(1.21)	
Labour (US$/ha)	394.7	138.0	261.9	280.0	358.1	527.3	127.9	151.7
	(12.07)***		(–1.30)		(–0.75)		(–3.68)***	
Pesticide applications (*n*)	2.0	1.6	1.7	2.0	1.9	2.5	1.4	1.7
	(7.9)***		(–4.6)***		(–4.7)***		(–5.4)***	
Pesticide (US$/ha)	35.8	19.2	23.3	36.7	31.3	51.9	12.2	28.6
	(7.07)***		(–5.90)***		(–5.25)***		(–4.63)***	
Pesticide quantity (kg/ or l/ha)	1.9	1.9	1.5	2.7	1.5	3.1	1.4	2.6
	(–0.31)		(–4.61)***		(–6.41)***		(–4.71)***	

K, Kenya; T, Tanzania; w, with presence of BC agent; w/o, without presence of BC agent. Figures in parentheses are *t*-test values for continuous and Chi-square test values for categorical variables and statistically significant at the 0.01 (***), 0.05 (**) and 0.1 (*) levels of probability.

among the groups, and are included in the overall labour cost. Costs of family labour are approximated with its opportunity costs based on average wages, and added to the costs of hired labour. The total labour costs are about threefold higher in Kenya than in Tanzania. This is due to a higher wage rate in Kenya, but also to higher labour input in terms of time. Within Kenya, farmers from areas where the parasitoid is present have significantly lower labour costs.

The application of fertilizer is necessary to supplement the nutrients available in the soil (Ghosh and Madhavi, 1998) and, even under natural conditions, fertile soils can suffer from degradation due to intensive cultivation. The fertilizer input includes farmyard manure and is measured in US$/ha. In cases where the farmyard manure was not purchased but self-produced, the respondent was asked for the replacement value of manure.

Dixon (2007) argues that availability of herbicides for *Brassica* crops is limited and, in fact, farmers in the study area do not use herbicide in cabbage production. Thus, the variable pesticide consists of the aggregated costs of insecticide and fungicide. In this study, insecticide costs make up a large share of pesticide costs. For the whole sample, the average pesticide expenditure is approximately US$28/ha, with Kenyan farmers having higher pesticide costs than those in Tanzania. In both countries, farmers from areas with the presence of the BC agent spend less on pesticides than farmers without the presence of the parasitoid, which is reflected in the quantity of pesticides used. When one includes all farmers from areas where the BC agent is present, the pesticide use is on average 1.5 l or kg/ha, which is approximately half the amount that farmers without the BC agent use. Comparing the two countries the pesticide quantity used is almost the same, but the higher pesticide expenditure of Kenyan farmers shows that in Kenya the average pesticide price is higher than in Tanzania.

In our survey it was found that the level of pesticide use in Kenyan cabbage production for farmers without the BC agent was lower as compared with the estimations in an *ex ante* study (Macharia *et al.*, 2005, 2006), which ranged between US$99 and US$198/ha. However, the results of the present study are based on a larger sample size, and are further supported by another recent study that showed, on average, a low pesticide use of about 2.15 kg or l/ha, with associated costs of approximately US$15/ha, in vegetable production for the Kenyan domestic vegetable market (Asfaw *et al.*, 2009).

Our survey showed that pesticides are applied with a knapsack sprayer, and around 30% of the farmers spray a mixture of several pesticides; in several cases farmers mix pesticides with foliar fertilizer. The average number of pesticide applications differs significantly between Kenya, where on average pesticides are applied about twice per season, and Tanzania, where farmers apply pesticides on average 1.6 times. For Kenya, Macharia *et al.* (2006) found that the average number of sprays was 4.5 before the introduction of the BC agent, and this decreased to 1.3 after the BC agent had been introduced. In both countries farmers from areas with the presence of the parasitoid apply pesticides significantly less frequently than farmers from areas without. Chemical pest control can be complemented or substituted by practices compatible with the requirements of BC, such as biological pesticides and spot spraying. However, only 5% of the sampled farmers apply such measures.

Most pesticides interfere with various biochemical and physiological processes that are common in a wide range of organisms (Pretty and Hine, 2005). They differ in their active ingredients and toxicity level and, because some carry higher dangers for human and environmental health than others. It is helpful to investigate to what level of toxicity the farm household and the BC agent are exposed. The World Health Organization (WHO) classifies pesticides into four categories based on the toxicity of the technical compound and formulation (WHO, 2005). The classes are (Ia) extremely hazardous; (Ib) highly hazardous; (II) moderately hazardous; and (III) slightly hazardous. The WHO provides a list of pesticides that are unlikely to present acute hazards in normal use. For our analysis, these pesticides are grouped into the category 'Toxicity level IV', and, due to the minor use of pesticides belonging to classes Ia and Ib, these classes are merged. Figure 14.3 indicates that the surveyed farmers used low quantities of pesticides belonging to categories Ia, Ib, III and IV. Thus, the share of moderately hazardous pesticides dominates the pesticide use of farmers from areas without the presence of the BC agent, but also of farmers who operate in areas where the parasitoid is present.

Overall, farmers reported the major three insect pest problems as DBM, cabbage aphid and cutworm. In general, plant diseases were reported less frequently than insect pest problems, but black rot,[3] *Alternaria* leaf spot and anthracnose can be considered the main plant disease problems. However, 17% of farmers face pest pressure above the normal level. Pest pressure refers to the average pest pressure for a variety of pests and diseases, as perceived and measured on a three-point scale by the farmers in the sample. The survey revealed that sources of information on the so-called safe use of pesticides are government extension officers, pesticide salesmen, label instructions, other farmers, family members and

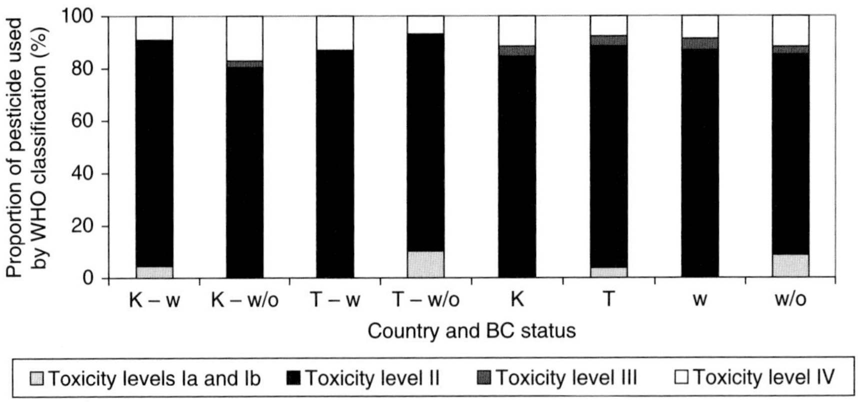

Fig. 14.3. Proportion of pesticide quantity used, by WHO classification. K, Kenya; T, Tanzania; w, with presence of biological control (BC) agent; w/o, without presence of BC agent.

media (TV, radio). In general, extension officers and pesticide salesmen are the most frequently reported information sources. The amount of protective clothing worn during pesticide application shows that farmers make attempts to protect themselves against pesticide exposure and to use masks (cloth to cover the mouth and nose), gloves, long-sleeved coats, long trousers, rubber boots and other items such as hats. The average amount of protective clothing used varies between two and three individual items, with the most common combination being long trousers and rubber boots. It was found that 8% of all farmers in the sample do not use any protective gear during pesticide application.

Table 14.3 displays a comparison of the output of cabbage production across the different survey groups of farm households. Cabbage output is given by the total market value in US$/ha.[4] This includes marketed and non-marketed output valued at current market prices. As shown in Table 14.3, output per hectare did not differ between the two countries. Within Tanzania, farmers from areas without the BC agent had, on average, a significantly higher output than farmers from areas where the parasitoid was present. In general, farmers from areas where the parasitoid was present had a lower cabbage output value than farmers from areas without the BC agent. The variable cabbage yield is measured by the number of cabbage heads per hectare, and shows that in areas where the parasitoid was present the number of heads was on average higher than in areas without the BC agent. This contradiction is explained by the fact that farmers who operate in areas with the presence of BC receive a significantly lower price per cabbage head and, therefore, have a significantly lower value of cabbage output. With respect to individual country differences, this observation differs for Kenya, where farmers from areas with the presence of the BC agent have on average a lower cabbage yield than farmers from areas without the parasitoid. Although Tanzanian farmers from areas with the BC agent have, on average, about 10,000 heads more per ha, their average cabbage price per head is half the price that farmers from areas without the BC agent receive and the value of the cabbage output as well as the net income[5] is significantly lower. In Kenya, farmers from areas without the BC agent have a significantly higher cabbage yield, but also receive a significantly lower cabbage price. Overall Tanzanian farmers obtain a significantly higher net income per hectare than their counterparts in Tanzania.

In Table 14.4, cabbage output and pesticide use are compared for farmers who use different pest management strategies.

Table 14.3. Cabbage production output with and without the presence of the biological control (BC) agent in Kenya and Tanzania (N = 1250).

	Farmers, by country		Farmers, by presence of BC		Kenya		Tanzania	
Variable	K	T	w	w/o	w	w/o	w	w/o
Output (US$/ha)	1499	1479	1316	1649	1528	1396	1262	1790
	(0.23)		(–2.57)**		(0.97)		(–4.19)***	
Net income (US$/ha)	614	978	516	570	653	476	706	1391
	(–4.45)***		(–1.26)		(1.25)		(–6.06)***	
Yield (cabbage heads/ha)	26,490	28,439	28,523	25,083	25,661	29,423	32,490	22,653
	(–1.48)		(2.42)**		(–2.25)**		(4.32)***	
Cabbage price (US$/head)	0.06	0.07	0.06	0.08	0.06	0.05	0.05	0.10
	(–2.42)**		(–7.25)***		(3.00)***		(–11.07)***	

K, Kenya; T, Tanzania; w, with presence of BC agent; w/o, without presence of BC agent. Figures within parentheses are *t*-test values and statistically significant at the 0.01 (***), 0.05 (**) and 0.1 (*) levels of probability.

Table 14.4. Comparison of cabbage output and pesticide use for farmers using different pest management strategies: biological control (BC), pesticides or a combination of the two (n = 1194).

	Pest management strategy			
Variable	BC[a] (n = 64)	Pesticide[b] (n = 365)	Combination[c] (n = 765)	F-test
Output (US$/ha)	1524.44	1668.95[c]**	1407.33[b]**	3.99**
	(1780.22)	(1598.17)	(1356.16)	
Pesticide use (US$/ha)	0.00[bc]***	38.19[ac]***	25.32[ab]***	28.40***
	(0.00)	(47.64)	(39.96)	

Means in rows followed by different letters indicate a statistically significant difference between the type of strategy using the Tukey test. Statistically significant at the 0.01*** and 0.05** levels of probability. Figures in parentheses are standard deviations of the means.

ANOVA analysis is conducted to compare three groups: (i) farmers who rely exclusively on the BC agent; (ii) farmers who use pesticides exclusively; and (iii) farmers who use a combination of both control measures, i.e. who apply pesticide and have the BC agent present. The analysis shows that farmers using only pesticides have higher pesticide expenditure, but also a significantly higher cabbage output as compared with farmers using the combination of both control measures. However, the output does not differ significantly for farmers relying exclusively on BC, which could be attributed to a high standard deviation in the output. However, in absolute terms there is an observable tendency that farmers using the combination have lower cabbage output than farmers using only one type of control measure.

Estimation of Production Function with Damage Abatement Function: Pesticide Use and Production Function Model

At the first stage of the two-stage estimation, a pesticide use function is estimated.

The dependent variable pesticide is the logarithm of expenditures on insecticides and fungicides in US$/ha. The explanatory variables are the presence of the BC agent, pest pressure, pesticide prices and products, tolerant variety, training and good practices, and season and district dummies are included. The district dummies serve to capture the variation in agroclimatic conditions and infrastructural settings that could affect the price and marketing of cabbage. The base district is Nyeri in Kenya, which has divisions with and without the BC agent. Due to resource constraints, the presence of the parasitoid was not measured at each cabbage plot, but was measured at division level in Kenya and district level in Tanzania, using information on release points and findings on the migration of the parasitoid.

At the second stage, the production function is estimated. The dependent variable output represents the total value of cabbage production in US$/ha, converted to its logarithmic form. Output is expected to be directly and indirectly influenced by different input factors such as seed, fertilizer, labour, season, age of farmer, district, pesticide, presence of the BC agent and the interaction between these two damage control agents.

Table 14.5 summarizes quantitative and qualitative variables of the two models. The table can be divided into three parts, presenting variables (i) used only in the pesticide use function; (ii) included only in the production function; and (iii) used in the estimation of both functions. For an overview, the variables are presented with their definitions and measurement units, sample mean and standard deviation.

As shown in Table 14.6, the coefficient of the presence of the BC agent in the first-stage model suggests that, in areas where the parasitoid is present, pesticide expenditure is significantly lower than in areas

Table 14.5. Overview of variables used in two-stage least-squares estimates (n = 1250).

Variable	Description (dummy = 1)	Mean	SD
Pesticide use function			
Pest pressure	Above normal level	0.17	–
Tolerant variety	Variety used is tolerant to different pests	0.08	–
Products	Different pesticides used (*N*)	2.09	1.37
Training	Attended agricultural training	0.26	–
Good practices	Adopted good practices	0.05	–
Pesticide price	Pesticide price (US$/kg/l)	19.50	28.36
Production function			
Output	Market value of cabbage production (US$/ha)	1438.90	1459.40
Seed	Seed costs (US$/ha)	131.66	121.48
Labour	Labour costs (US$/ha)	267.70	396.51
Fertilizer	Fertilizer costs (US$/ha)	201.60	236.45
Age	Farmers' age (years)	43.10	12.63
Both functions			
Pesticide	Pesticide costs (US$/ha)	27.57	42.23
BC	Presence of BC agent	0.68	–
Season	Rainy season	0.47	–
Kiambu (K)	Inhabitant of Kiambu (BC present)	0.09	–
Nyandarua (K)	Inhabitant of Nyandarua (BC present)	0.22	–
Arumeru (T)	Inhabitant of Arumeru (BC present)	0.28	–
Lushoto (T)	Inhabitant of Lushoto	0.21	–

BC, biological control agent; K, Kenya; T, Tanzania.

Table 14.6. Estimated parameters for production function with damage control function (2SLS) (n = 1250).

	Pesticide use function		Production function	
Parameter	Coefficient	SE	Coefficient	SE
Intercept	1.073***	0.183	4.380***	0.360
Pest pressure	0.245***	0.078		
Tolerant variety	–0.016	0.121		
Training	–0.093	0.070		
Good practices	–0.190	0.134		
Products	0.325***	0.024		
Pesticide price	0.563***	0.048		
Seed			0.284***	0.035
Fertilizer			0.128***	0.029
Labour			0.069***	0.028
Age			0.039	0.081
Season	–0.048	0.060	–0.079	0.049
BC	–0.354***	0.123		
Kiambu (K)	–0.118	0.1456	0.3845***	0.101
Nyandarua (K)	–0.280	0.115	0.457***	0.074
Arumeru (T)	–0.465***	0.113	0.111	0.089
Lushoto (T)	–0.431***	0.107	0.687***	0.088
Damage control function				
Pesticide			0.613***	0.188
BC			0.698**	0.341
Interaction term			–0.378*	0.229
R^2	0.446		0.250	
Adjusted R^2	0.439		0.241	

BC, biological control; K, Kenya; T, Tanzania. Statistical significance at the 0.01 (***), 0.05 (**) and 0.1 (*) levels of probability. The functions were tested for collinearity and heteroskedasticity; however, none of the problems was apparent in the estimation.

without its presence.[6] *Ceteris paribus*, the presence of the BC agent reduces pesticide expenditure by around 34%. Furthermore, operating in an environment in which the farmer considers the overall pest pressure to be above normal level increases pesticide cost per hectare by about 23%. A higher pesticide price, as well as the use of a higher number of different pesticide products, increases pesticide expenditure significantly. The district dummies depict that farmers from Nyeri District (Kenya) spend significantly more on pesticides compared with farmers from both districts of Tanzania (Arumeru and Lushoto), but no significant differences were found between the Kenyan districts.

The second-stage estimation results show that the estimated production function has an adjusted R^2 value of 0.24. The significant positive coefficients of the direct inputs such as seed, fertilizer and labour indicate that higher input intensity increases cabbage output. The negative coefficient of the season dummy shows that the cabbage output is lower in the rainy season. Regarding the districts, Nyeri farmers had a significantly lower cabbage output compared with other surveyed districts in Kenya and in Lushoto district in Tanzania. Exclusive reliance on BC shows the highest positive impact on output, as shown by the coefficient of the damage control function, followed by exclusive reliance on pesticides when the BC agent is absent. When both damage control agents are relied upon, their interaction needs to be accounted for.

The interaction term between the two damage control agents is negative and significant at a 10% level of probability. Thus, it is possible that there are some negative effects of the farmers' pesticide use practices on the BC agent. It is known that the frequent use of non-selective pesticides can be toxic for the parasitoid and, thus, hamper its ability to prevent output losses (Varela *et al.*, 2003).

As shown in Table 14.7, the partial productivity effect of pesticides on cabbage output with the presence of the BC agent is lower. This could be an indication of two phenomena: (i) pesticide productivity is lower in ecosystems where the BC agent is present; and (ii) there is a reliance on the parasitoid by farmers with less favourable production conditions and poorer socio-economic status. This conclusion is further supported by the model results showing that the significantly positive effect of the exclusive use of BC turns negative when taking into account the interaction with chemical pesticides and computing it at the average level of pesticide use in the sample. Computing it at the median value of pesticide expenditure, which is about US$15/ha for the whole sample, leads one to the same conclusion.[7]

Contrary to conventional pest control strategies, which primarily rely on the use of chemical pesticides, the effect of the parasitoid on cabbage output depends highly on the level of establishment and maintenance of the BC agent in the cabbage field, which is hampered by commonly used broad-spectrum pesticides. For example, in the Cameron Highlands of Malaysia, BC initially was unsuccessful because pesticides were widely used by farmers, thereby keeping the parasitoid population low. Development of pesticide resistance of DBM and the rejection of cabbage at the major market due to high insecticide residues induced a shift to the use of biological pesticides. As a result of this shift, the level of parasitoid establishment increased, the need for insecticide applications drastically declined and yield increased (Talekar and Shelton, 1993). On the other hand, several studies on the effect of insecticides on key pests and natural enemies in maize production showed an increase of the pest after partial removal of BC agents by applying sublethal dosages of pesticides (Kfir, 2002; Cugala *et al.*, 2006). Sublethal doses, typically applied by farmers without the necessary training, can cause higher mortality among natural enemies that are more susceptible than the pest. The pesticide use function showed a significant reduction in pesticide costs due to the presence of the BC agent, which supports the argument of sublethal doses leading to high output losses. This situation may be aggravated further if DBM builds up resistance to pesticides, and also by the poor quality of pesticides. Unfortunately for this study, no information on these issues is available, and hence the low cabbage output in areas where the parasitoid is present might be explained by either of these factors.

To measure the full impact of the BC agent requires capturing other benefits such as the reduction of negative externalities from pesticides, including human health and environmental hazards. However, to quantify such costs and benefits for a BC programme was not possible in this study because of the non-market nature of such effects (Pretty and Waibel, 2005).

Table 14.7. Computed coefficients of damage control agents incorporating interaction ($n = 1250$).

Parameter	BC agent present	At average level of pesticide use	At median level of pesticide use
Pesticide	0.236	–	–
BC	–	–0.553	–0.319

BC, biological control.

Summary and Conclusions

This study showed that the introduction of a BC agent in Africa can be bound by constraints, even in production systems where pesticide use has not yet reached the high levels of Asia. The estimated pesticide use function showed that the presence of the BC agent leads to a decrease in pesticide expenditure. Moreover, the estimation results of a production function indicated that either using pesticides or the presence of the BC agent has a positive impact on cabbage output. However, using both pest control strategies at the same time results in lower cabbage output. This suggests that the economic benefits of the BC project, as measured by its impact on cabbage outputs in Kenya and Tanzania, are lower than expected based on the *ex ante* impact assessment. It further highlights the complexity of classical BC programmes in intensive cropping systems and the importance of accounting for the interaction between damage control agents. The expected positive interaction between pesticides and the BC agent can be hampered by several factors. First, farmers' cabbage production systems differ, including their use of BC-friendly pest management techniques. Secondly, pesticide use is characterized by the widespread use of broad-spectrum pesticides, which may have negative side effects on BC agents. If no pesticides are being used, the presence of the parasitoid demonstrates its positive impact on cabbage output and, thus, its potential to contribute to poverty alleviation and its relevance for farmers with financial constraints. However, since vegetable production generally relies on the intensive use of agrochemicals, stakeholders need to know of the negative relationship between chemical control and the natural enemy and that using both strategies, especially sublethal doses of broad-spectrum pesticides, may lead to output losses.

Since BC agents take time to become fully established, complete information on the project impact can be obtained only after many years (Van Driesche and Hoddle, 2000). Hence, since this study was conducted 2–3 years after the release of the parasitoids, the results presented here may not yet reflect the long-term impact. However, an effective BC agent can release farmers from dependency on pesticides, reduce their input costs and, via decreased pesticide use, reduce potential negative effects on environmental and human health.

Further research

Compared with conventional pest control strategies, which primarily rely on the use of chemical pesticides, the role of biological control re output depends largely on two major factors: first, the level of establishment of biological control in the cabbage field, and, secondly, the level of DBM's resistance to pesticide. For this study, no information was available to establish the facts on these two issues; however, future research could attempt to capture this information.

It would be interesting to discover whether the introduction of classical biological control reduces negative pesticide effects on human health and, as a result, whether farm households experience fewer health hazards due to pesticides. The relevance of the use of biological control against pest problems increases as concerns about the negative effects of pesticide use gain more attention. Nevertheless, although natural enemies gain importance, farmers might lack knowledge of the role of these beneficial insects. To understand and conserve classical biological control agents, knowledge gaps must be filled. As an approach in educating farmers about biological control farmer field schools (FFS), which use a participatory learning approach, could be conducted. If farmers are to benefit from the impact of this renewable resource, they need to be educated in the science of biological control to allow full advantage to be taken of beneficial insects (Ooi and Kenmore, 2005).

Endnotes

[1]Classical biological control is hereafter referred to as biological control.
[2]For other specifications, see Lichtenberg and Zilberman (1986).
[3]Massomo *et al.* (2004) outline that black rot in Tanzania is especially widespread in the two surveyed districts of Arumeru and Lushoto, and has caused high cabbage yield losses in the past.
[4]Given by the number of cabbage heads produced, sold, consumed, fed to livestock and given away (as a gift or as return for a service), and the market price of cabbage.
[5]Net income is computed as total value of cabbage production minus variable costs, including opportunity costs of family labour.
[6]As an entry point of the 2SLS estimation, two alternative pesticide use functions were estimated by ordinary least squares (dependent variables being pesticide cost (US$/ha) and pesticide quantity (kg or l/ha) converted to their logarithmic forms). Both functions show a highly significant negative effect of the presence of the BC agent on pesticide use, while the coefficients of other variables are similar in the two models.
[7]In order to explore the robustness of the estimated models, rather than district dummies an alternative country dummy was used for the 2SLS estimation. The estimation results confirm previous results; however, the R^2 value of the production function decreases, indicating that the country dummy explains less of the variation in cabbage output.

References

Ajayi, O.C. (2005) Biological capital, user cost and the productivity of insecticides in cotton farming systems in sub-Saharan Africa. *International Journal of Agricultural Sustainability* 3(3), 154–166.

Asfaw, S., Mithöfer, D. and Waibel, H. (2009) Food-safety standards, pesticide use and farm level productivity: The case of high-value crops in Kenya. *Journal of Agricultural Economics* 60(3), 645–667.

Blackwell, M. and Pagoulatos, A. (1992) The econometrics of damage control – Comment. *American Journal of Agricultural Economics* 74(4), 1040–1044.

Carpentier, A. and Weaver, R.D. (1997) Damage control productivity: Why econometrics matters. *American Journal of Agricultural Economics* 79(1), 47–61.

Carrasco-Tauber, C. and Moffitt, L.J. (1992) Damage control econometrics: Functional specification and pesticide productivity. *American Journal of Agricultural Economics* 74(1), 158–162.

Cugala, D., Schulthess, F., Ogol, C. and Omwega, C. (2006) Assessment of the impact of natural enemies on stemborer infestations and yield loss in maize using selected insecticides in Mozambique. *Annales de la Société Entomologique de France* 42(3/4), 503–510.

Dixon, G.R. (2007) *Vegetable Brassicas and Related Crucifers*. CAB International, Wallingford, UK.

FAO (2006a) *Compendium of Food and Agriculture Indicators 2006, Country: Kenya.* www.fao.org/es/ess/compendium_2006/pdf/KEN_ESS_E.pdf (accessed 27 March 2008).

FAO (2006b) Compendium of Food and Agriculture Indicators 2006, Country: Tanzania. www.fao.org/es/ess/compendium_2006/pdf/URT_ESS_E.pdf (accessed 27 March 2008).

Fox, G. and Weersink, A. (1995) Damage control and increasing returns. *American Journal of Agricultural Economics* 77(1), 33–39.

Gelernter, W.D. and Lomer, C.J. (2000) Success in biological control of above-ground insects by pathogens. In: Gurr, G. and Wratten, S. (eds) *Biological Control: Measures of Success*. Kluwer Academic Publishers, Dordrecht, the Netherlands, pp. 297–322.

Ghosh, S.P. and Madhavi, D.L. (1998) Cabbage. In: Salunkhe, D.K. and Kadam. S.S. (eds) *Handbook of Vegetable Science and Technology: Production, Composition, Storage, and Processing*. Marcel Dekker Inc., New York, pp. 299–321.

Hajek, A. (2004) *Natural Enemies: an Introduction to Biological Control*. Cambridge University Press, Cambridge, UK.

Huang, J., Hu, R., Rozelle, S., Qiao, F. and Pray, C.E. (2002) Transgenic varieties and productivity of smallholder cotton farmers in China. *The Australian Journal of Agricultural and Resource Economics* 46(3), 367–387.

Kennedy, P.E. (1981) Estimations with correctly interpreted dummy variables in semilogarithmic equations. *American Economic Review* 71(4), 801.

Kfir, R. (2002) Increase in cereal stemborer populations through partial elimination of natural enemies. *Entomologia Experimentalis et Applicata* 104, 299–306.

Kibata, G.N. (1996) The diamondback moth: A problem pest of brassica crops in Kenya. In: Sivapragasam, A., Kole, W.H., Hassan, A.K. and Lim, G.S. (eds) *The Management of Diamondback Moth and Other Crucifer Pests, Proceedings of the 3rd International Workshop*. The Malaysian Agricultural Research and Development Institute, Kuala Lumpur, Malaysia, pp. 47–53.

Lichtenberg, E. and Zilberman, D. (1986) The econometrics of damage control: Why specification matters. *American Journal of Agricultural Economics* 68(2), 261–273.

Lim, G.S. (1992) Integrated pest management of diamondback moth: Practical realities. In: Talekar, N.S. (ed.) *Diamondback Moth and Other Crucifer Pests, Proceedings of the 2nd International Workshop*. Asian Vegetable Research and Development Center Publication, Shanhua, Taiwan, pp. 565–576.

Macharia, I., Löhr, B. and De Groote, H. (2005) Assessing the potential impact of biological control of *Plutella xylostella* (diamondback moth) in cabbage production in Kenya. *Crop Protection* 24(11), 981–989.

Macharia, I., Mithöfer, D. and Löhr, B. (2006) Update of the ex-ante impact assessment of the biological control of *Plutella xylostella* (Diamondback moth) in Kenya. Paper presented at the *5th International Workshop on Diamondback Moth and Other Crucifer Insect Pests*, 24–27 October. Chinese Academy of Agricultural Sciences, Beijing, China.

Margni, M., Rossier, D., Crettaz, P. and Jolliet, O. (2002) Life cycle impact assessment of pesticides on human health and ecosystem. *Agriculture, Ecosystem and Environment* 93(1), 379–392.

Massomo, S.M.S., Mortensen, C.N., Mabagala, R.B., Newman, M.A. and Hockenhull, J. (2004) Biological control of black rot (*Xanthomonas campestris* pv. *campestris*) of cabbage in Tanzania with bacillus strains. *Journal of Phytopathology* 152(2), 98–105.

Ministry of Agriculture (MoA) Kenya (2005) Horticulture Division Annual Report 2005. Ministry of Agriculture, Nairobi, Kenya.

Ooi, P.A.C. and Kenmore, P.E. (2005) Impact of educating farmers about biological control in farmer field schools. Paper presented at the *Second International Symposium on Biological Control of Arthropods*, 12–16 September, Davos, Switzerland, pp. 277–289.

Oude Lansink, A. and Carpentier, A. (2001) Damage control productivity: An input damage approach. *Journal of Agricultural Economics* 52(3), 11–22.

Pemsl, D.E. (2005) Why do some Bt-cotton farmers in China continue to use high levels of pesticides? *International Journal of Agricultural Sustainability* 3(1), 44–56.

Pemsl, D.E. (2006) Economics of agricultural biotechnology in crop protection in developing countries: The case of Bt-cotton in Shandong Province, China. Pesticide Policy Publication Series Special Issue 11, Pesticide Policy Project, Hannover, Germany.

Pretty, J. and Hine, R. (2005) Pesticide use and the environment. In: Pretty, J. (ed.) *The Pesticide Detox: Towards a More Sustainable Agriculture*. Earthscan, London, pp. 1–22.

Pretty, J. and Waibel, H. (2005) Paying the price: The full cost of pesticides. In: Pretty, J. (ed.) *The Pesticide Detox: Towards a More Sustainable Agriculture*. Earthscan, London, pp. 39–54.

Talekar, N.S. and Shelton, A.M. (1993) Biology, ecology and management of the diamondback moth. *Annual Review of Entomology* 38, 275–301.

United Republic of Tanzania (URT) (2007) National Website. www.tanzania.go.tz/issuesf.html (accessed 26 March 2008).

Van Driesche, R.G. and Hoddle, M.S. (2000) Classical arthropod biological control: Measuring success, step by step. In: Gurr, G. and Wratten, S. (eds) *Biological Control: Measures of Success*. Kluwer Academic Publishers, Dordrecht, the Netherlands, pp. 39–75.

Varela, A.M., Seif, A. and Löhr, B. (2003) *A Guide to IPM in Brassicas Production in Eastern and Southern Africa*. International Centre of Insect Physiology and Ecology (ICIPE), Science Press, Nairobi, Kenya.

Wooldridge, J.M. (2003) *Introductory Econometrics: A Modern Approach*, 2nd edn. South-Western College Publishing, Manson, California.

World Health Organization (WHO) (2005) The WHO recommended classification of pesticides by hazard and guidelines to classification 2004. WHO, Geneva, Switzerland.

15 Indirect and External Costs of Pesticide Use in the Vegetable Sub-sector in Kenya

Ibrahim Macharia, Dagmar Mithöfer and Hermann Waibel

Introduction

The negative effects of pesticide use received increasing attention after the publication of *Silent Spring* by Carson (1962). This book documented the detrimental effects of pesticides on the environment, as well as on humans. Unfortunately, research and documentation of such effects are sparse in developing countries, despite extensive documentation of the unsafe use and handling of pesticides (Yassin *et al.*, 2002; Salameh *et al.*, 2004). The few studies conducted in developed as well as in Asian and Latin American countries have shown substantial external costs associated with the use of pesticides (Waibel *et al.*, 1999; Pretty *et al.*, 2000; Azeem *et al.*, 2003; Tegtmeier and Duff, 2004; Pimentel, 2005).

To date, few studies have been conducted in Africa and, of those, most have concentrated on the effects of pesticides on human health (Ajayi, 2000; Ohayo-Mitoko *et al.*, 2000; Maumbe and Swinton, 2003; Okello, 2005; Williamson, 2005; Jankowski, 2007; Ngowi *et al.*, 2007; Asfaw, 2008). Fewer studies still have been conducted on the negative effects on the African environment, excluding human health. Among those that have been conducted are Ajayi *et al.* (2002) on the cotton sub-sector in Mali, Houndekon *et al.* (2006) on locust control in the Sahel in Niger and Leach *et al.* (2008) on locust control in Senegal. While dealing with the effects on human health, these studies were expanded to include the effects on livestock and livestock product losses, the pesticide-related destruction of natural enemies, the development of pesticide resistance, crop and crop product losses, the effects on birds and fish, and governmental expenditures to reduce the environmental and social costs. Further studies in Kenya have documented pesticide residues in eggs of free-range chickens (Mugambi *et al.*, 1989), fish killed by leaching of pesticides in ponds in the Nyeri district (Waikwa, 1998) and organochlorine pesticide residues found in breastfeeding mothers (Kinyamu *et al.*, 1998). Of these studies, none established the costs associated with the negative effects documented.

Building on those previous studies, the objectives of this study were to (i) account for the external and indirect costs related to pesticide use in the vegetable sub-sector in Kenya; (ii) contribute to the sustainable use of pesticides; and (iii) the design of a policy favouring sustainable pesticide use.

 Vegetable Production and Marketing in Africa: Socio-economic Research (eds *D. Mithöfer* and *H. Waibel*)

Methodology

Conceptual framework

A framework for evaluating the impact of pesticide use in agriculture was established by Antle and Capalbo (1995). Pesticide costs can be grouped into three categories: (i) the direct costs paid by farmers in terms of pesticide purchases, costs of personal protective equipment (PPE) and labour for application; (ii) the indirect costs borne by pesticide users in terms of their own health costs and livestock losses; and (iii) the indirect costs borne by society, i.e. consumers, tax payers, governments and health service providers. The last category is defined as externalities, as these are borne by a third party and not the user. Such externalities include (i) pesticide residues in food and water; (ii) secondary pest outbreaks due to the pesticide-related destruction of natural enemies; (iii) the development of pesticide resistance; (iv) honeybee losses; (v) yield losses resulting from the reduction in pollinating insects and pesticide drifts; (vi) losses of birds, fish, wildlife and biodiversity; and (vii) governmental expenditures to reduce the environmental and social costs.

Indirect, as well as external, costs are more difficult to assess than direct costs since, for some of the effects, no market prices exist. Valuation of indirect and external effects varies according to the type of market to which they are linked and the complex interaction within the ecosystem. Valuation approaches can be grouped into three broad categories: (i) those that follow an 'accounting approach' based on actual market prices, with scientific evidence of the externalities; (ii) those that utilize the 'economic approach', based on hypothetical and surrogate markets in the absence of market prices; and (iii) those that combine both approaches.

The studies that utilize the accounting approach follow three steps: identification, quantification and monetization. Identification involves the collection of all available scientific evidence from different sources to establish the attribution; quantification involves the establishment of the quantities of the physical impacts, while monetization involves putting a value to the quantities estimated and then summing the totals. A common methodology for the accounting of externalities is the simple partial equilibrium approach, whereby a volume change in the externality is multiplied by its marginal value.

Studies based on the 'economic approach' treat pesticide externalities as non-market goods because they cannot be bought or exchanged in a market situation and, thus, use hypothetical or surrogate market prices. Non-market valuations can be either direct or indirect (Hanley *et al.*, 1997). Direct methods, also referred to as stated preference methods, include the contingent valuation (CV) and the choice experiment (CE). CV tries to establish individuals' valuation for non-marketed goods by directly asking respondents their 'willingness to pay' (WTP) for improvements in non-market goods and services or their 'willingness to accept' (WTA) the loss of well-being associated with a change. CE is a relatively new technique that allows respondents to trade off between the attributes of a non-market good and indicate their most preferred alternative in each choice set. Indirect methods, also called revealed preference methods, observe behaviour in related markets and use the data as proxies (a surrogate market). These include hedonic valuation, travel cost approach, aversive behaviour and dose response (Hanley *et al.*, 1997; Farber *et al.*, 2002). Examples of the various study approaches are provided in Table 15.1.

These efforts illustrate the difficulties of developing a uniform, comprehensive and clear framework for valuation of pesticide indirect cost and externalities. The method applied in this study is a modification of the accounting approach and uses market prices for valuation. This study analyses the indirect costs of pesticide use, as well as the external cost of pesticide use according to the definition of Antle and Capalbo (1995), and refers to both categories as 'externalities' for simplification.

Table 15.1. Summary of valuation studies of pesticide indirect cost and externalities.

Approach	Country	Description	Authors
Accounting	Mali	Estimated the costs in cotton-growing area by considering official data and extrapolation; annual indirect and external costs of pesticide use equalled US$10 million	Ajayi *et al.* (2002)
	Niger (Sahel region)	Health costs, defined as medical expenses plus the value of time lost, were found to increase by US$0.46/ farmer/year of pesticide use; livestock losses due to intoxication were valued at US$0.33/ha treated; costs of destroying obsolete pesticides estimated at US$0.06/ ha treated	Houndekon *et al.* (2006)
	Thailand	Estimated external costs of pesticides to be almost equal to their market value or US$228.9 million/year	Jungbluth (1996)
	USA	Costs based on readily available data on a number of cases of human health, costs of hospitalization, biodiversity impacts, veterinarian costs, additional pesticide costs from pesticide resistance and loss of natural enemies, compensation costs and costs from loss of honey production; total costs estimated at US$9.6 billion	Pimentel (2005)
Economic (direct methods: CV)	Nicaragua	Farmers' valuation of health effects of chemical pesticides measured as their WTP for low-toxicity pesticides; results indicated that farmers were willing to spend additional 28% of pesticide expenditure to avoid health risks	Garming and Waibel (2009)
Economic (indirect methods: aversive behaviour)	USA	Study examining public expenditure on apples over 6-year period in the context of the controversy over the use of the growth regulator Alar; percentage changes expenditure estimated for Alar and Alar-free apples; study found that consumers were WTP up to 31% extra to avoid Alar in fresh apples	van Ravensway and Hoehn (1991)
Accounting plus economic	Germany	Estimated external costs at DM252 million/year;[a] figures included the costs of monitoring drinking water, the processing of water to remove unwanted chemicals, poison damage to honeybees and working days lost as a result of pesticide-related ill health; WTP for species conservation considered for loss of wildlife estimations	Waibel *et al.* (1999)
	UK	Study estimated only externalities giving financial costs, i.e. treatment, prevention and administration costs and costs of restoring species and habitats as a proxy of the costs of wildlife and habitat losses; costs for pesticide removal from drinking water estimated at £120 million/ year,[b] for pesticide monitoring in food at surface and groundwater at £5.4 m and at £4.75 m, respectively	Pretty *et al.* (2000)

[a] 1 US$ = 0.541 DM (1999).
[b] 1 US$ = 0.689 GBP (2000).

The present study started with the identification and classification of the relevant externality categories that were expected in the Kenyan vegetable sub-sector and assigned possible valuation methods. The impact categories were then classified according to those that could be identified, quantified and valued and those that could only be quantified or identified (Table 15.2). The study followed three stages: (i) identification of the externalities relevant in the vegetable sub-sector through literature review, group discussions and expert discussions; (ii) quantification and valuation via analysis of the existing information/inventory taking, farmers interviews, pesticide residue analysis and expert interviews; and (iii) validation of the estimates through expert workshops.

Data Collection

Secondary data/inventory taking

Review of official and research documents was the main source of secondary data during the inventory taking. These data were collected from official documents from the various ministries, non-governmental organizations (NGOs), veterinary officials, hospitals, the Kenyan Central Bureau of Statistics and other relevant sources. Research articles from various

Table 15.2. Pesticide indirect cost and externalities and possible methods for valuation in the sub-sector.

Externality	Type	Possible assessment	Method[a]
Valuation	Human health	Cost of illness approach[b]	Farmer interviews
	Residues in vegetables	Value of produce that has to be withdrawn from the market due to exceeding maximum residue levels Avoidance costs: extra costs incurred to obtain pesticide-free vegetables	Residue analysis, experts
	Livestock poisoning	Treatment cost, value of lost livestock, loss of produce, e.g. milk	Farmer and veterinarian interviews
	Residues in water	Mitigation costs (costs for cleaning up drinking water); cost of alternative sources	Residue analysis Consumer interviews
	Pesticide resistance	Additional cost of pest control	Farmer and expert interviews
	Loss of bees	Loss of honey production and productivity loss from reduced pollination	Farmer and expert interviews
	Loss of beneficial organisms	Production value (lost production due to increased pest outbreaks and costs of additional pest control)	Farmer and expert interviews
	Damage prevention costs	Fraction of budgets from research institution and government geared towards pesticide risk reduction	Expert interviews
Quantification	Fishery losses	Farmers' and experts' estimates	Farmer and expert interviews
	Birds and soil biota losses	Farmers' and experts' estimates	Farmer and expert interviews
Identification	Biodiversity loss	Farmers' and experts' experiences	Farmer and expert interviews

[a] In all cases, inventory taking is the first step.
[b] Costs incurred through medical consultations, hospitalizations, opportunity costs of traditional medicine, medications, transport to and from clinics, dietary expenses from illness resulting from, for instance, drinking milk, and workdays lost.

organizations, including NGOs that had been involved in pesticide studies, were reviewed. From government ministries secondary data were collected at the provincial, district and divisional levels in major vegetable production districts.

Farmer survey

In 2008 interviews were conducted in seven major vegetable-producing districts of the central and eastern provinces of Kenya: Kirinyaga, Kiambu, Nyandarua, Nyeri North, Meru Central, Makueni and Muranga. The sample was a random sub-sample of 425 farmers, sampled by probability proportional to size (PPS) from two previous surveys conducted by the International Centre of Insect Physiology and Ecology (*icipe*): the Diamondback Moth Biological Control Impact assessment survey conducted in 2004/2005 (DBM, with 295 farmers), and the Good Agricultural Practices assessment survey (GlobalGAP,[1] with 544 farmers) conducted in 2005/2006.

In both surveys, a multi-stage sampling procedure was employed to select districts, sub-locations and farmers. First, districts were purposely sampled according to the intensity of vegetable production and agro-ecological zones. Lists of farmers compiled by extension workers at the sub-location level served as a sampling frame, from which the farmers were randomly sampled using the PPS procedure.

The present farmer survey covered a wide range of topics, such as cropping systems, demographics, common farming practices, pesticide use and handling practices, the type and quantities of pesticides sprayed and pesticide-related health problems (frequency, related costs of treatment, lost days due to the illness). Information was collected on the poisoning of livestock, the mortality of natural enemies (ladybird beetles, praying mantis, spiders, dragonflies and fire ants), soil biota (earthworms, millipedes and crickets) and loss of bees and birds due to pesticides used in vegetable production. For all the indicators listed, only the effects observed during the spraying operation or within 24 h of spraying were considered.

To test the farmers' recognition of the natural enemies and soil biota, a self-made arthropod zoo (a clear plastic jar containing the most common natural enemies and soil biota) was shown to them. For the identification of farmland birds, farmers were offered a list of the most common species in the local language and asked to name the species they knew to be visiting their vegetable fields. In both cases, they were encouraged to add to the list.

In addition, to confirm and ascertain the farmers' pesticide container disposal methods, a walk through the farm was conducted and the number of empty pesticide containers lying in the vegetable field, near the vegetable field, in ponds or nearby streams was assessed and counted. Respondents were asked to show samples of pesticide containers or labels they had used, to facilitate the accurate recording of the names of pesticide products applied.

Vegetable residue survey

Vegetable samples from plots that were ready to be harvested were randomly sampled from half of the surveyed farmers (208) for pesticide residue analysis. Following the protocol of the Kenya Plant Health Inspectorate Service (KEPHIS), the samples were drawn from the upper, middle and lower segments of harvested produce (mostly from crates for tomatoes and French beans, as well as peas, and bags for cabbage, kale and spinach). The samples were then put in sterile polythene bags, appropriately labelled and placed in large cool boxes filled with freezer-pack ice and transported to the KEPHIS laboratory for analysis.

Analysis was performed by the ethyl acetate method: extraction by ethyl acetate followed by partitioning steps using cyclohexane (Fernandez-Alba *et al.*, 1994). Residue levels were determined by gas chromatography (GC) with electron-capture detection (ECD), nitrogen-phosphorus detector (NPD) and ultraviolet (UV). Residue identities were confirmed by GC coupled with mass spectrometry in the selected ion-monitoring mode. Pesticides were identified according to their retention times, target and qualifier ions and qualifier-to-target abundance ratios (Ueno *et al.*, 2003).

Workshop and expert interviews

A stakeholder workshop was organized at the end of the data collection to validate, extrapolate and substantiate the findings from the farmer survey, identify the main drivers perceived to encourage the use of pesticides and examine the factors that could reduce the indirect and external effects of pesticides. Workshop participants had been identified during the initial secondary data collection. The workshop alternated plenary and working group sessions, lasting for 2 days.

Each participating expert was asked to complete a questionnaire survey about his/her own observations and opinions on the indirect and external effects of pesticide use and their severity. A modified Delphi approach was then used, i.e. the results from the first round were presented and the participants were allowed to revise their responses in the second round. The workshop was hosted at *icipe*, with 25 experts attending. Follow-up interviews for the expert surveys were conducted with those invitees who had been unable to attend the workshop.

Results and Discussion: Indirect and External Effect Cost Estimates

Human health impairments

An average of 28% of the farmers reported having suffered at least one of a variety of acute illness symptoms of pesticide poisoning during and within 24 h of spraying pesticides in 2005 and in 2008. In order of most frequently reported, the most common symptoms were sneezing, headaches, stomach pains, dizziness, burning skin/rash, eye irritation, shortage of breath, backache, vomiting, blurred vision and coughing. These symptoms have all been associated with acute pesticide poisoning (Extension Toxicology Network, 2004), and confirm the results from other studies conducted in developing countries, such as Indonesia, the Philippines and Vietnam, which show that 58–99% of the farmers exposed to pesticides had at least one reported symptom (Antle and Pingali, 1994; Pingali *et al.*, 1994). The present survey showed that, for treatment of minor poisoning, many farmers used home remedies such as milk, lemon juice, honey or herbs. Painkillers from the local pharmacy were bought in cases where the symptoms of illness were mild, and farmers visited the health clinic if the symptoms either persisted or became serious, i.e. the victim was unable to talk, walk, see or vomited continuously.

Applying the cost of illness approach, the indirect costs of pesticide use on farmers' health were estimated at an average of US$6.55/farmer/season for the 28% of the farmers who reported the problem. Considering the total sample, this translates to a mean of US$1.77/farmer/season and, assuming two crop seasons per year, these costs amount to US$3.54/farmer/year. However, health costs are likely to be much higher than that because chronic diseases resulting from long-term pesticide exposure, non-monetary costs (i.e. suffering) and income lost by family members assisting the farmer in seeking treatment were not included.

In Tanzania, farmers' expenses on health due to pesticide exposure ranged between US$0.018 and US$116 per year (Ngowi *et al.*, 2007). In West Africa, the economic value of pesticide-related health costs equalled US$3.92/household/season in the case of cotton–rice systems (Ajayi *et al.*, 2002). The study by Atreya (2005) in Nepal estimated health care costs due to pesticide use at US$16.8/household/annum. In Zimbabwe, cotton growers incurred an annual mean of US$4.73 in Sanyati and US$8.31 in Chipinge for pesticide-related direct and indirect acute health effects (Maumbe and Swinton, 2003).[2]

Residues in vegetables

Of the 208 samples screened for pesticide residues, no sample was found contaminated with organochlorines, organophosphates or pyrethroids above the limit of detection (LOD). However, nearly all samples (92%) had a value higher than the LOD for dithiocarbamates, of which 9% exceeded the set maximum residue levels (MRLs)[3] (Table 15.3). The MRLs expressed as carbon disulfide (CS_2) arise from different dithiocarbamates, which include maneb, mancozeb, metiram, propineb, thiram and ziram. These are the most frequently detected pesticides in monitoring programmes worldwide (Dogheim *et al.*, 2002; EU, 2003), and are suspected of having reproductive and mutagenic effects in human cells (Puzy-Mino *et al.*, 2002). The mean for domestic crops (0.22 ± 0.02 mg CS_2/kg) was statistically different from the mean of export crops (0.14 ± 0.01 mg CS_2/kg) (t-test = 3.47, $P < 0.01$),[4] suggesting two different standards of pesticide handling patterns, depending on the target market. Those vegetables with the highest number of samples deviating from established dithiocarbamate MRLs were spinach (94%) and kale (5%). However, spinach MRLs are set at the lower limit of analytical determination since no more detailed information exists. Spinach and kale are produced for the domestic market only. None of the export crops exceeded the established MRLs.

Considering all active ingredients, tomatoes had the highest average residue levels (0.26 mg CS_2/kg), more than twofold higher than the lowest (baby corn). Vegetables with the lowest average levels were baby corn, French beans, courgettes, peas, cabbage and kale, in order of increasing levels. However, these low levels do not define safety and 'safe' daily intake, since MRLs are not toxicological limits and do not account for acceptable daily intake.

The direct toxic effects of pesticides on animals and humans are easily recognized, but the effects that result from either long-term exposure to low doses of a regular intake of pesticide residues in vegetables or food or multiple pesticide residue combinations are hard to detect and quantify. Assuming that the fraction of each vegetable crop with residues above the established MRLs represents the yield of that vegetable crop produced in 2008, exceeding MRLs results in the total volume that, in theory, should be taken away or abandoned as it is

Table 15.3. Dithiocarbamate residue data (mg CS_2/kg) in vegetable crops.

Vegetable	*N*	Mean	SE	Median	Range	MRL[a]	Samples > MRL (%)[c]
Kale	44	0.25	0.05	0.15	< LOD–1.62	0.5	5
Tomatoes	17	0.26	0.11	0.12	0.02–0.59	3	0
Cabbages	76	0.20	0.01	0.18	< LOD–0.71	3	0
French beans	27	0.12	0.01	0.11	0.09–0.26	1	0
Peas	25	0.16	0.01	0.16	0.10–0.27	1	0
Spinach	17	0.20	0.07	0.13	0.01–1.18	0.05[b]	94
Courgettes	1	0.12	0.00	0.12	0.12	2	X
Baby corn	1	0.11	0.00	0.11	0.11	0.05[b]	X
Total	208	0.20[c]	0.00	0.15	2.59		9

LOD, limit of detection; MRL, maximum residue level.
[a] Maximum residue level (EU, 2008, also cross-referenced with the Codex Alimentarius, 2009).
[b] Indicated as the lower limit of analytical determination (pesticides not authorized for use).
[c] Weighted average; X, not considered due to the low number of samples.

not suitable for human consumption. Multiplying this volume by the farm gate prices from the farmer survey gives the total loss of vegetables due to pesticide residues. Overall, the cost of external effects due to pesticide residues was estimated at US$9.64 million (Table 15.4).

Livestock poisoning

Livestock such as goats, cattle, chickens and pigs were commonly found in the study areas, and nearly all the farmers (97%) kept livestock between 2002 and 2007. Livestock are an important asset providing regular income and can be sold in times of hardship, providing a safety net. For most farmers, cattle are the most important type of livestock. Next to milk and manure, they provide draught power, especially in Kirinyaga and Meru districts. According to the responses of 13% of the farmers who reported having experienced livestock poisoning due to pesticides used in vegetable production, approximately 166 cattle, two oxen and two sheep died because of pesticide poisoning between 2002 and 2007. Direct attribution is difficult without clinical analysis; however, many farmers in the study site knew the main symptoms of pesticide poisoning in animals, which included difficulty in breathing, excessive foamy salivation, vomiting, bloat, abdominal cramps and collapsing after feeding or after passing through the sprayed fields.

Applying the direct market approach, consisting of treatment costs and the value of lost livestock, the average livestock loss was estimated at US$118.39/farmer/year for the 13% of farmers (Table 15.5) affected, and at a mean of US$13.45/farmer/year considering all farmers within the survey who kept livestock between 2002 and 2007. Additional losses not considered in this study occurred when milking cows were not milked or the milk from the poisoned animals was disposed of because it was not fit for consumption.

Veterinarians in the study sites confirmed having treated livestock poisoning and advising farmers not to take the milk from the sick animals. Unfortunately, they neither kept records of the animals they treated nor updated the names of the farmers. According to average estimates from veterinarians, about 28 cases of poisoned livestock were treated per year in some divisions,[5] of which five animals (18%) died. The poisoning of chickens and cats was common, although via secondary poisoning, when they ate dead or damaged insects and rats subjected to pesticide spray.

Collection and disposal of empty pesticide containers

Empty pesticide containers are considered hazardous waste unless they are drained, rinsed and disposed of appropriately. The best way to dispose of empty pesticide containers is to take them to a pesticide container collection site, where they can later be recycled. However, this method of disposal does not exist in Kenya and many small-scale farmers throw them in latrines, crop fields or water bodies (wells, dams,

Table 15.4. Pesticide use externality for vegetable production in the sub-sector, 2008.

Vegetable	Production (million t)[a]	Samples > MRL (%)	Volumes > MRL (t)[b]	Price (US$/kg)	Loss (million US$)
Kale	0.423	5	19,261	0.13	2.50
Spinach	0.048	94	44,615	0.16	7.14
Total	1.774	9	19,261	0.19	9.64

[a] MoA (2007) and HCDA (2008).
[b] Quantities required to be discarded as unfit for human consumption.

Table 15.5. Livestock poisoning due to pesticides used in vegetable production.[a]

Average livestock poisoned (n/farmer/year)	2
Consulted a veterinarian (%)	45
Treatment expenses (US$/head/farmer/year)	9
Mortality rate of affected animals (%)	11
Average cost of livestock loss (US$/head/farmer/year)	118

[a] All calculations based on the 13% of the farmers who experienced livestock poisoning from pesticide used on vegetables.

rivers and ponds), bury them or reuse them. The shallow burial of empty containers, which is what most farmers did, might lead to pesticides leaching into the soil and the build-up of pesticide waste underground. This is both a health and a safety risk, because it is unknown whether such a site is close to underground water sources, as groundwater levels in Kenya change between the rainy and dry seasons. In 2007 Crop Life Kenya, with the assistance of Crop Life International, conducted a container survey, the objective of which was to assess empty pesticide container handling, disposal and management in Kenya. It found 24,783,062 empty containers across the country's major horticultural production areas, a figure projected to increase to 32,509,953 in the year 2010 (AAK, 2005–2008). The study recommended development of a project on pesticide container management.

The field observation module of the farmer survey investigated the mode of empty pesticide container disposal and showed approximately 536 freshly used pesticide containers lying in either the vegetable fields or the farm compound of the 425 farmers surveyed. With a conservative assumption that this represents the practices of vegetable farmers in general, and with two crops per year, this adds up to 2.5 empty containers per farmer per year. The cost of collection of these containers was estimated at US$0.34/farmer/year, based on the deposit–refund system of US$0.13/container,[6] which is currently being applied to soda bottles in Kenya. Since this deposit is collected at the point of sale for soda bottles, it would imply that farmers themselves transport the empty pesticide containers to the agrochemical dealers. Fullerton and Kinnaman (1995) concluded that fees for waste collection should be priced as if disposal and recycling were the only two main disposal options. In a deposit–refund system, the user normally pays a deposit on potentially polluting products, and, when pollution is avoided by returning the waste product and its residues, the deposit is refunded.

Damage prevention costs (government regulation and research)

There are two main bodies empowered by Kenyan law to register and control the use of pesticides in Kenya: the Agrochemical Association of Kenya (AAK) and the Pest Control Products Board (PCPB). Their budget costs can be regarded as damage prevention costs, since their main goal is to minimize the risks associated with pesticide use.

It is difficult to segregate the budgetary allocation for vegetable protection measures involving AAK and PCPB. However, a rough estimation was made to indicate what proportional expenses were allocated to regulate pesticide use, through the proportion of the pesticides estimated as being used in the vegetable sub-sector (570 t) relative to the average import level of 2004–2006 (6999 t) (PCPB, 2004–2006).

Table 15.6 shows the approximate budget for the vegetable sub-sector allocated between AAK and PCPB.[7] PCPB allocates about 30% of its budget to training in the proper use of pesticides. Assuming that this relative share of expenditure on such training and control constitutes an indicator for external costs at the regulatory level, external costs amounted to US$18,115.

In addition to these bodies, local extension officers at sub-location levels give advice and information concerning pesticide

Table 15.6. Government budgets to prevent pesticide risks (from PCPB, 2004–2006; AAK, 2005–2008).

Organization	Budget (US$) 2006	2005	Average	Vegetable sub-sector[a]
AAK	263,680	274,719	269,200	21,924
PCPB	499,405	445,048	472,227	38,458
Total	763,085	719,767	741,426	60,382

[a] Authors' calculation.

use. Pesticide companies allocate considerable resources to launch media campaigns, farm-based demonstrations, regular farmer visits, workshops and farmers' meetings. They promote research combined with demonstrations to highlight pesticide efficacy aspects, mainly to prove the superiority of their products. Among the 4805 active NGOs working in the country (Directory of NGOs in Kenya, 2005), 74 were involved in agricultural extension activities. The expenditures incurred by pesticide companies are not included as an externality or indirect effect, because companies do so to promote pesticide use and the impact of their activities was already estimated in terms of health damage, livestock loss and residues in vegetables. The NGOs and extension officers have small budgetary allocations to impart training on the use of pesticides and were, therefore, not accounted for.

Externalities Not Valued

The effects of pesticides on beneficial arthropods and birds

From the farmer survey, approximately 21%, 37%, 35% and 6% of the interviewed farmers, respectively, had observed dead bees, natural enemies, soil biota and birds in their vegetable fields during or within 24 h of spraying pesticides in the 5 years preceding the time of the survey (May 2003–May 2008). Excluding those farmers from the total sample who had not checked the vegetable plot for dead beneficial insects or birds after spraying and those who did not re-enter the field within 24 hours almost doubles the frequencies reported (bees 37%, natural enemies 60%, soil biota 61% and birds 11%).

When asked which pesticides they had sprayed prior to observation, dimethoate (Dimeton 40EC) was the main pesticide, followed by cyhalothrin (Karate 25C), cyfluthrin (Bull dock 25EC), fenpyroximate (Ogor 40EC) and carbofuran (Furadan). In general, all the pesticides listed by farmers were reported by Mineau *et al.* (1999) to be very toxic to beneficial arthropods and birds. Some farmers even explained how they observed earthworms dying as they emerged from the contaminated field, and that the birds which ate the earthworms also died.

Over 64% of the farmers interviewed perceived that the presence of bees, natural enemies, soil biota and birds in their vegetable fields had decreased, but on the other hand 17% of farmers perceived an increase (Fig. 15.1). The majority of farmers attributed this decrease mainly to the increased use of pesticides; however, the decreased number of birds observed was least associated with increased use of pesticides (Fig. 15.2).

Pesticide resistance in pests

Over 50% of the farmers reported that they had, at one time or another, experienced a problem with pests that could not be controlled even when using the right level (the recommended rate) and quality (not expired) of pesticide, although this regime had previously been effective. When asked what they had done, 37% reported that they had increased the spraying frequency, 28% had switched to different products, 20%

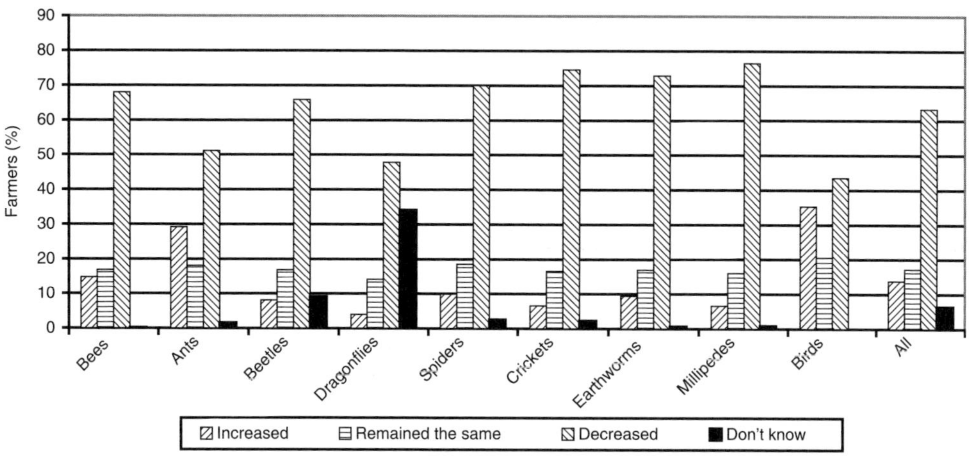

Fig. 15.1. A comparison of farmers' perceptions of the presence of beneficial insects and birds in vegetable fields, May 2003 to May 2008.

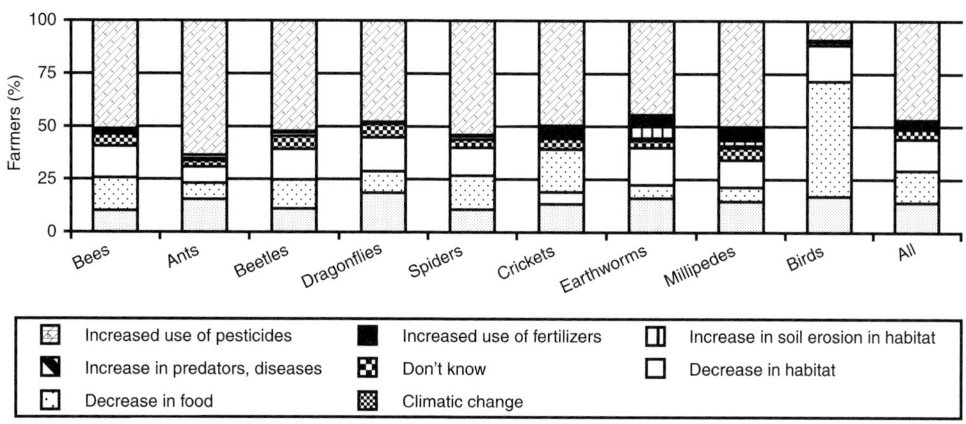

Fig. 15.2. A comparison of farmers' perceptions of the factors associated with the decline in beneficial insects and birds in vegetable fields, May 2003 to May 2008.

had used increased concentrations and 15% had combined pesticides. Answering the question as to how they perceived the pest control efficacy of the current market pesticides, over half (52%) responded that they were weak.

Globally, about 520 insect species, 150 plant-pathogenic species and about 273 weed species are now resistant to pesticides (Stuart, 2003). In the Matale district of Sri Lanka, land was abandoned because pesticides became ineffective in protecting crops (Wilson and Tisdell, 2001).

Expert Assessment of Externalities

Consultations during the expert workshop confirmed the results presented in the preceding section as realistic and, if anything, rather conservative. Experts ranked natural enemies as the major category being negatively affected by pesticide use in the vegetable sub-sector, followed by humans (through direct human health impairment), livestock, farmland birds, aquatic organisms and bees; water quality and residues in vegetables were the least frequently mentioned external and indirect effects. Seventy-four per cent of the experts identified pesticide runoff as the main cause of water pollution in the current vegetable production systems, and 66% expected negative effects on their own health due to pesticide residues on the vegetables they consumed.

Factors that result in pesticides being the dominant form of pest management include the high susceptibility of vegetables to insect attack, the lack of advice on alternative methods, the change from subsistence farming to commercial farming and the effectiveness of pesticides in terms of high knock-down effects, the last of which was perceived by the farmers as 'easier' pest control. The experts unanimously stressed the urgent need for an integrated approach to effectivenly raise awareness among farmers, to introduce effective policies for control of pesticide use and to enforce stiff penalties for those who misuse pesticides as major reforms to minimize the indirect and external effects of pesticide use. They emphasized that integrated pest management could be a solution to reducing reliance on pesticides.

Summary of the Externalities in the Sub-sector

Table 15.7 summarizes the indirect and external costs associated with pesticide use in the vegetable sub-sector. Extrapolation of the human health costs to the entire vegetable sub-sector, with the total number of vegetable farmers estimated at 183,021, gives a welfare loss of US$0.65 million/year. Similarly, livestock losses of US$13.45/farmer/year and the costs of avoiding pollution through a deposit–refund system for empty pesticide containers of US$0.34/farmer/year, extrapolated to include all vegetable farmers, raised the costs significantly.

The overall indirect costs borne by farmers themselves add up to US$3.11 million/year; dividing this cost by the direct costs of pesticides used in the vegetable sub-sector of US$5.06 million[8] gives a ratio of 1.0:0.6, indicating that when a farmer spends US$1.0 in private costs, it costs him or her another US$0.6 indirectly; inclusion of other related 'potential' external costs leads to a ratio of 1.0:2.5.

Summary and Conclusions

The study shows that the indirect and external costs associated with pesticide use in vegetable production in Kenya are high, implying that the 'true' cost of pesticides is currently grossly understated. If external, as well as indirect, costs were appropriately incorporated in the market price of pesticide products, a switch to more human and environmental health-friendly pest control methods, such as the use of less toxic products and a switch to integrated pest management (IPM) practices, would be relatively more favourable. However, design and implementation of such a system needs to be evaluated given current pest control conditions.

Lessons from biocontrol-based IPM for cabbage as an alternative to chemical control have already been found to decrease the negative human health effects associated with pesticide use (Jankowski, 2007). Other pest management strategies such as intercropping (Legutowska *et al.*, 2002), tillage type and crop rotation (Hummel *et al.*, 2002) have been shown to reduce pests significantly. These strategies are cost effective and environmentally friendly, and thus there is a need to bring these pest control methods to the attention of farmers.

In addition, a deposit–refund system where a pesticide container charge (the deposit) is left with the pesticide vendors until the container is taken back (refund) is taken as a proxy for the costs of reducing the external effects of inappropriate pesticide container disposal methods.

The findings on pesticide residue testing support recommendations for the routine monitoring of these pollutants in vegetables to minimize the human health risks to consumers.

A mechanism to facilitate the formal documentation of pesticide-related cases of

Table 15.7. Total estimated costs from the use of pesticides in the vegetable sub-sector in Kenya.

Type of costs	US$/farmer/ year	Million US$/ year	Other effects identified, but not valued
Health impact	3.54	0.65	Chronic health effects for farmers, deaths, effects on hired labour and consumers
Livestock	13.45	2.46	Produce loss
Vegetable production loss	52.67	9.64	Higher consumer prices due to low supply
Collection and disposal of empty pesticide containers	0.34	0.06	
Government regulations to prevent damage	0.10	0.02	Costs of administration by pesticide-related research at universities and environmental agencies
Loss of natural enemies	NQ	NQ	Loss of natural enemies
Cost of pesticide resistance	NQ	NQ	Cost of extra pesticides for the control of red spider mite, which farmers claimed was resistant to most of the common pesticides they were using, and they responded by increasing the number of applications and application rates.
Honeybee and pollination losses	NQ	NQ	Loss of honey and honeybees
Bird losses	NQ	NQ	Loss of farmland birds
Indirect costs borne by farmers themselves	16.99	3.11	
Potential external costs	53.11	9.72	
Direct costs	27.66	5.06	
Ratio (indirect farmer: direct)		0.62	
Ratio (external: direct)		2.54	

NQ, effects not quantified.

poisoning in both humans and livestock should be put in place, as this would help greatly in the monitoring of the extent of the hazard. This could be achieved by advocating free medical assistance in medical and veterinary centres for individuals affected, and the costs for these services could be recovered by an appropriate tax imposed on the pesticides that are responsible for the majority of poisoning.

Further research

Further research is recommended to refine the estimates of the indirect and external effects. In particular, effects that could be identified but not quantified and valued in this study should be included. Health costs should cover all individuals exposed to pesticides, e.g. pesticide traders, hired workers and consumers, and incorporate pesticide-induced chronic illnesses and deaths. It would be of value to investigate whether pesticide-related health impairments render farmers more vulnerable to poverty. The feasibility of introducing a deposit–refund system for pesticide containers should be analysed, including an assessment of incentive mechanisms for implementation of such a system, and this analysis should take both private and social perspectives into account.

Acknowledgements

This research was supported by DAAD and an *icipe*-GTZ/BMZ funded project. The authors express appreciation to farmers as well as staff members of the Ministry of Agriculture and other individuals who provided valuable assistance.

Endnotes

[1]GlobalGAP (formerly known as EurepGAP) is a private sector body that sets voluntary standards for the certification of agricultural products around the globe.
[2]These figures are exactly as reported in the studies cited and, therefore, are not in absolute dollars or adjusted by purchasing power parity.
[3]MRLs represent the maximum concentration of pesticide residue (expressed in mg/kg) that is legally permitted in a crop, and are often referred to as the legal trading limit. They are derived from an assessment of the residues found when the crop is treated according to Good Agricultural Practices (GAP). They are not permanent, and they can be either raised or lowered to take account of new information and data. They are normally fixed at the lower limit of analytical determination where there are no authorized uses.
[4]Export crops were taken from a random mix of GlobalGAP-certified and non-certified farms.
[5]The third administrative unit of Kenya.
[6]The pesticide containers were made of glass, metal or plastic, in the range 200–500 ml, which is similar to that of glass soda bottles of 300 or 500 ml. Both types of containers can be recycled, but the cost of recycling varies depending on both the type of material used and the cost of cleaning. Based on existing information, using the deposit rate for the soda bottles is probably the lower bound for a deposit–refund system for pesticide containers. More research is required on appropriate incentives for the implementation of a deposit–refund system for pesticide containers.
[7]These budgets also include the registration and control of less harmful biopesticides as well as biological controls which, due to the unavailability of more differentiated figures, cannot be taken out of the calculations.
[8]The private cost of US$14 was multiplied by 2, assuming two cropping seasons per year, then extrapolated by the total number of farmers estimated.

References

AAK (2005–2008) Agrochemicals Association of Kenya, news reports. Agrochemicals Association of Kenya, Nairobi, Kenya.

Ajayi, O.C. (2000) Pesticide use practices, productivity and farmers' health: The case of cotton–rice systems in Côte d'Ivoire, West Africa. *Pesticide Policy Project Special Issue* 3, Institute of Horticultural Economics, Hannover, Germany.

Ajayi, O.C., Camara, M., Fleischer, G., Haidara, F., Sow, M., Traore, A. *et al.* (2002) Socio-economic assessment of pesticide use in Mali. *Pesticide Policy Project Special Issue* 6, Institute of Horticultural Economics, Hannover, Germany.

Antle, J.M. and Capalbo, S.M. (1995) Measurement and evaluation of the impacts of agricultural chemical use: A framework for analysis. In: Pingali, L.P. and Rogers, P.R. (eds) *Impact of Pesticides on Farmer Health and the Rice Environment.* Kluwer, Norwell, Massachusetts, pp. 23–55.

Antle, J.M. and Pingali, P.L. (1994) Pesticides, productivity, and farmer health: A Philippine case study. *American Journal of Agricultural Economics* 76, 418–430.

Asfaw, S. (2008) Global agrifood supply chain, EU food safety standards, and African small-scale producers: The case of high value horticultural exports from Kenya. PhD. dissertation. Leibniz University of Hannover, Hannover, Germany.

Atreya, K. (2005) Health costs of pesticide use in a vegetable growing area, central mid-hills, Nepal. *Himalayan Journal of Sciences* 3, 81–84.

Azeem, M.K., Soomro, M.H. and Ahmad, I. (2003) Economic evaluation of pesticide use externalities, in the cotton zone of Punjab, Pakistan. *The Pakistan Development Review* 41, 683–698.

Carson, R. (1962) *Silent Spring*. Houghton Mifflin Co., New York.

Codex Alimentarius (2009) Codex Alimentarius. www.codexalimentarius.net/mrls/pestdes/jsp/pest_q-e.jsp (accessed 21 August 2009).

Directory of NGOs in Kenya (2005) Annual Reports, Directory of NGOs in Kenya. Government printer, Nairobi, Kenya.

Dogheim, S.M., El-Marsafy, A.M., Salama, E.Y., Gadalla, S.A. and Nabil, Y.M. (2002) Monitoring of pesticide residues in Egyptian fruits and vegetables during 1997. *Food Additives and Contaminants* 19, 1015–1027.

EU (2003) Monitoring of pesticide residues in products of plant origin in the European Union, Norway, Iceland and Liechtenstein, 2003 Report. www.europa.eu.int/comm/food/fvo/specialreports/pesticide_residues/report_2003_en.pdf (accessed 4 April 2008).

EU (2008) Pesticide MRLs database Reg. (EC) No 839/2008. www.ec.europa.eu/sanco_pesticides/public/index.cfm (accessed 4 September 2009).

Extension Toxicology Network (2004) Cholinesterase Inhibition. Toxicology information brief, Cornell University. www.pmep.cce.cornell.edu/profiles/extoxnet/TIB/cholinesterase. html (accessed 12 December. 2008).

Farber, S.C., Costanza, R. and Wilson, M.A. (2002) Economic and ecological concepts for valuing ecosystem services. *Ecological Economics* 41, 375–392.

Fernandez-Alba, A.R., Valverde, A., Aguera, A. and Contreras, M. (1994) Gas-chromatographic determination of organochlorine and pyrethroid pesticides of horticultural concern. *Journal of Chromatography A* 686, 263–274.

Fullerton, D. and Kinnaman, T.C. (1995) Garbage, recycling, and illicit burning or dumping. *Journal of Environmental Economics and Management* 29, 78–91.

Garming, H. and Waibel, H. (2009) Pesticides and farmer health in Nicaragua – a willingness to pay approach to evaluation. *European Journal of Health Economics* 10, 125–133.

Hanley, N., Shogren, J.F. and White, B. (1997) *Environmental Economics in Theory and Practice*. Oxford University Press, New York.

HCDA (2008) Annual Report 2008, Horticultural Crops Development Authority. Horticultural Crops Development Authority, Nairobi, Kenya

Houndekon, V.A., De Groote, H. and Lomer, C. (2006) Health costs and externalities of pesticide use in the Sahel. *Outlook on Agriculture* 35, 25–31.

Hummel, R.L., Walgenbach, J.F., Hoyt, G.D. and Kennedy, G.G. (2002) Effects of production system on vegetable arthropods and their natural enemies. *Agriculture, Ecosystems and Environment* 93, 165–176.

Jankowski, A. (2007) Economics of biological control of diamondback moth in vegetable production in East Africa. MSc thesis, Leibniz University of Hannover, Hannover, Germany.

Jungbluth, F. (1996) Crop protection policy in Thailand. *Pesticide Policy Project Special Issue* 5, Institute of Horticultural Economics, Hannover, Germany.

Kinyamu, J.K., Kanja, L.W., Skaare, J.U. and Maitho, T.E. (1998) Levels of organochlorine pesticides residues in milk of urban mothers in Kenya. *Bulletin of Environmental Contamination and Toxicology* 60, 732–738.

Leach, A.W., Mullié, W.C., Mumford, J.D. and Waibel, H. (2008) *Spatial and Historical Analysis of Pesticide Externalities in Locust Control in Senegal – First Steps*. FAO, Rome.

Legutowska, H., Kucharczyk, H. and Surowiec, J. (2002) Control of thrips infestation on leek by intercropping with clover, carrot, or bean. In: Paroussi, G., Voyiatzis, D. and Paroussis, E. (eds) *Proceedings of the Second Balkan Symposium on Vegetables and Potatoes*. International Society Horticultural Science, Leuven, Belgium, pp. 571–574.

Maumbe, B.M. and Swinton, S.M. (2003) Hidden health costs of pesticide use in Zimbabwe's smallholder cotton growers. *Social Science and Medicine* 57, 1559–1571.

Mineau, P., Fletcher, M.R., Glaser, L.C., Thomas, N., Brassard, J.C., Wilson, L.K. *et al.* (1999) Poisoning of raptors with organophosphorus and carbamate pesticides with emphasis on Canada, US, and UK. *Journal of Raptor Research* 33, 1–37.

MoA (2007) Annual reports, Ministry of Agriculture. Kenya Government printer, Nairobi, Kenya.

Mugambi, J.M., Kanja, L., Maitho, T.M., Skaare, J.U. and Lökken, P. (1989) Organochlorine pesticide residues in domestic fowl (*Gallus domesticus*) eggs from central Kenya. *Journal of the Science of Food and Agriculture* 48, 165–176.

Ngowi, A.V.F., Mbise, T.J., Millinga, L., Ijani, A.S.M., London, L. and Ajayi, O.C. (2007) Smallholder vegetable farmers in northern Tanzania: Pesticides use practices, perceptions, cost and health effects. *Crop Protection* 26, 1617–1624.

Ohayo-Mitoko, G.J., Kromhout, H., Simwa, J.M., Boleij, J.S. and Heederik, D. (2000) Self-reported symptoms and inhibition of acetylcholinesterase activity among Kenyan agricultural workers. *Occupational and Environmental Medicine* 57, 195–200.

Okello, J.J. (2005) Compliance with international food-safety standards: The case of green bean production in Kenyan family farms. PhD dissertation, Michigan State University, East Lansing, Michigan.

PCPB (2004–2006) Annual reports, Pest Control Products Board. PCPB printer, Nairobi, Kenya.

Pimentel, D. (2005) Environmental and economic costs of the application of pesticides in the United States. *Environment, Development and Sustainability* 7, 229–252.

Pingali, L.P., Marquez, C.B. and Palis, G.F. (1994) Pesticides and Philippine rice farmer health: A medical and economic analysis. *American Journal of Agricultural Economics* 76, 587–592.

Pretty, J., Brett, C., Gee, D., Hine, R., Mason, C.F., Morison, J.I.L. *et al.* (2000) An assessment of the total external costs of UK agriculture. *Agricultural Systems* 65, 113–136.

Puzy-Mino, C., Bustamente, G., Sanchez, M.E. and Leone, P.E. (2002) Cytogenetic monitoring in a population occupationally exposed to pesticides in Ecuador. *Environmental Health Perspectives* 110, 1077-1080.

Salameh, P.R., Baldi, I., Brochard, P. and Abi-Saleh, B. (2004) Pesticides in Lebanon: A knowledge, attitude and practice survey. *Environmental Research* 94, 1–6.

Stuart, S. (2003) Development of resistance in pest populations. www.nd.edu/_chem191/e2.html (accessed 20 January 2003).

Tegtmeier, E.M. and Duff, M.D. (2004) External costs of agricultural production in the United States. *International Journal of Agricultural Sustainability* 2, 1–20.

Ueno, E., Oshima, H., Saito, I. and Matsumoto, H. (2003) Determination of nitrogen- and phosphorus-containing pesticide residues in vegetables by gas chromatography with nitrogen-phosphorus and flame photometric detection after gel permeation chromatography and a two-step mini column cleanup. *Journal of AOAC INTERNATIONAL* 86, 1241–1251.

van Ravensway, E. and Hoehn, J.P. (1991) The impact of health risk on food demand: A case study of alar and apples. In: Caswell, J.A. (ed.) *Economics of Food Safety*. Elsevier Science Publishing Company, New York, pp. 155.

Waibel, H., Fleischer, G. and Becker, H. (1999) The economic benefits of pesticides: A case study from Germany. *Agrarwirtschaft* 48, 219–229.

Waikwa, S.W. (1998) Kenyan farmers on the treadmill – liberalization threatens food security and sustainable farming. www.pan-uk.org/pestnews/Issue/pn43/pn43p10.htm (accessed 8 January 2010).

Williamson, S. (2005) Breaking the barriers to IPM in Africa: Evidence from Benin, Ethiopia, Ghana and Senegal. In: Pretty, J. (ed.) *The Pesticide Detox: Towards a More Sustainable Agriculture*. Earthscan, London, pp. 165–180.

Wilson, C. and Tisdell, C. (2001) Why farmers continue to use pesticides despite environmental, health and sustainability costs. *Ecological Economics* 39, 449–462.

Yassin, M.M., Abu-Mourad, T.A. and Safi, J.M. (2002) Knowledge, attitude, practice, and toxicity symptoms associated with pesticide use among farm workers in Gaza Strip. *Occupational Environmental Journal* 59, 387–394.

16 Integrated Pest Management Training and Information Flow among Smallholder Horticulture Farmers in Kenya

Nigat Bekele, Dagmar Mithöfer, David Amudavi and Gideon Obare

Introduction

The horticultural sector is one of the most important sectors in the Kenyan economy, contributing 13% of the country's economic GDP (FPEAK, 2006; Republic of Kenya, 2006a) and 58% of agricultural GDP (Wasilwa, 2008). The sector includes fruit and vegetables, and has great potential to improve household welfare through the provision of income and satisfying domestic food needs, as well as creating opportunities for improving human health by providing vitamins and minerals (GHA, 2005). The sector, however, faces serious constraints, such as pests that are a major concern, largely due to the over-reliance on pesticides for pest and disease control (Raini *et al.*, 2005). Thus, there is an increasing need for alternative production systems, such as integrated pest management (IPM) to reduce the reliance on pesticides (IHC, 2006). Apparently, such systems are available for a range of horticultural crops produced in eastern Africa (Sithanantham, 2004). What is needed is appropriate training for farmers to utilize such systems.

Various training approaches have been applied in Kenya in order to improve agricultural production through the uptake of new production technologies. The approaches include training and visit (T&V), the Integrated Rural Development Programme (IRDP) and face-to-face extension approaches. According to recent assessments, these approaches were perceived as being too top-down, lacking participatory possibilities and not always tailored to the demands of the clientele (Gautam, 2000; Republic of Kenya, 2005). The approaches utilized had a tendency towards centralized management systems that could not fit diverse conditions and needs (Antholt, 1993; Chambers, 1997). Due to these limitations, the Government of Kenya and other stakeholders have switched to group-based participatory training approaches that emphasize the clientele's needs and demands. This group-based participatory training approach provides the opportunity to increase the bargaining power of members in order to access information (Lapar *et al.*, 2006). Overall, farmer groups are considered crucial to acquiring basic skills for problem solving and enterprise management, as well as serving as an efficient channel for delivery of extension services (FAO, 2001). Two approaches, the farmer field schools (FFS) and the common interest groups (CIGs), have become popular in Kenya, as they are considered effective in disseminating information through farmer-to-farmer interaction, aiding production-making decisions.

The FFS approach is a form of adult education in which farmers learn optimally from field observation and their own experimentation (Simpson and Owens, 2002). It is a participatory method of learning (FAO, 2001), originally designed in Asia to help farmers become better at utilizing integrated pest management (IPM) (Van de Fliert *et al.*, 1995). It has since, in the African context, been extended to cover a broader range of plant production and management issues (Witt *et al.*, 2008). The CIG approach is enterprise-based and aims at empowering farmers to take up agribusiness enterprises that are market oriented and demand driven (Githaiga, 2007). It is a participatory extension approach promoting farmer interaction and involves them fully as partners in determining demand and agendas for response by extension service providers (Republic of Kenya, 2005). Moreover, the formation of CIGs is usually undertaken by farmers themselves and is sometimes facilitated by agricultural extension officers. The group members must have the common goal of promoting the enterprise as a business, which is flagged as a commercial opportunity by using pamphlets, posters and media (Republic of Kenya, 2001). The CIG approach was initially designed in 2000 to target marginalized groups of farmers in Kenya. In this approach, extension officers identified a wide range of opportunities to accommodate various categories of farmers with respect to resource endowments and socio-economic status (Githaiga, 2007).

In general, farmers obtain information through formal sources, such as the mass media (radio and print), and informal sources. For example, in Nigeria, neighbours, organized groups, extension agents, other farmers and opinion leaders are important sources for information sharing (Ekoja, 2003). Similar findings hold for Kenya: for example, the farmers' main source of IPM information is their neighbours (Raini *et al.*, 2005). In addition, NGOs, friends, chiefs' *barazas* (public meetings) and agricultural companies are common sources of agricultural information (Rees *et al.*, 2000). What is not clear, though, is the extent to which the two participatory group-based approaches, FFS and CIG, enhance environmentally and health-friendly production practices, due to an increased IPM knowledge in horticultural production. This study aimed to fill the gap, by assessing the factors linked to the acquisition of IPM knowledge and sharing among the two different group-based farmers, as well as among farmers operating individually (the control group).

Methodology

Study area

The study was conducted in five districts in Kenya, namely Muranga, Thika and Maragua in Central Province, and Makueni and Embu in Eastern Province, which are major horticulture production areas. Central Province covers an area of 13,176 km^2 with a population of 3.7 million. The area has 965,000 ha of potential agricultural land of which 78% is devoted to agricultural activities. The province is characterized by both intensive and extensive agricultural activities involving cash and food crops, including horticulture, dairy, poultry and pig production. The areas receive an average annual rainfall of 2600 mm and have a mean annual temperature of 20°C. Soil characteristics include humic nitisol, eutric nitisol, ando-humic nitisol, and nito-rhodic ferralsol.

Eastern Province covers an area of approximately 3952 km^2 with a projected population of 5,587,781 (Republic of Kenya, 2006b) and receives rainfall ranging from 190 mm to 390 mm (Republic of Kenya, 2010). The mean temperature is 26°C and much of the district is characterized by a loamy sand soil type.

Survey design and data

Data were collected from May to July 2008 focusing on active smallholder vegetable and fruit producers grouped in three categories: FFS members, CIG members and

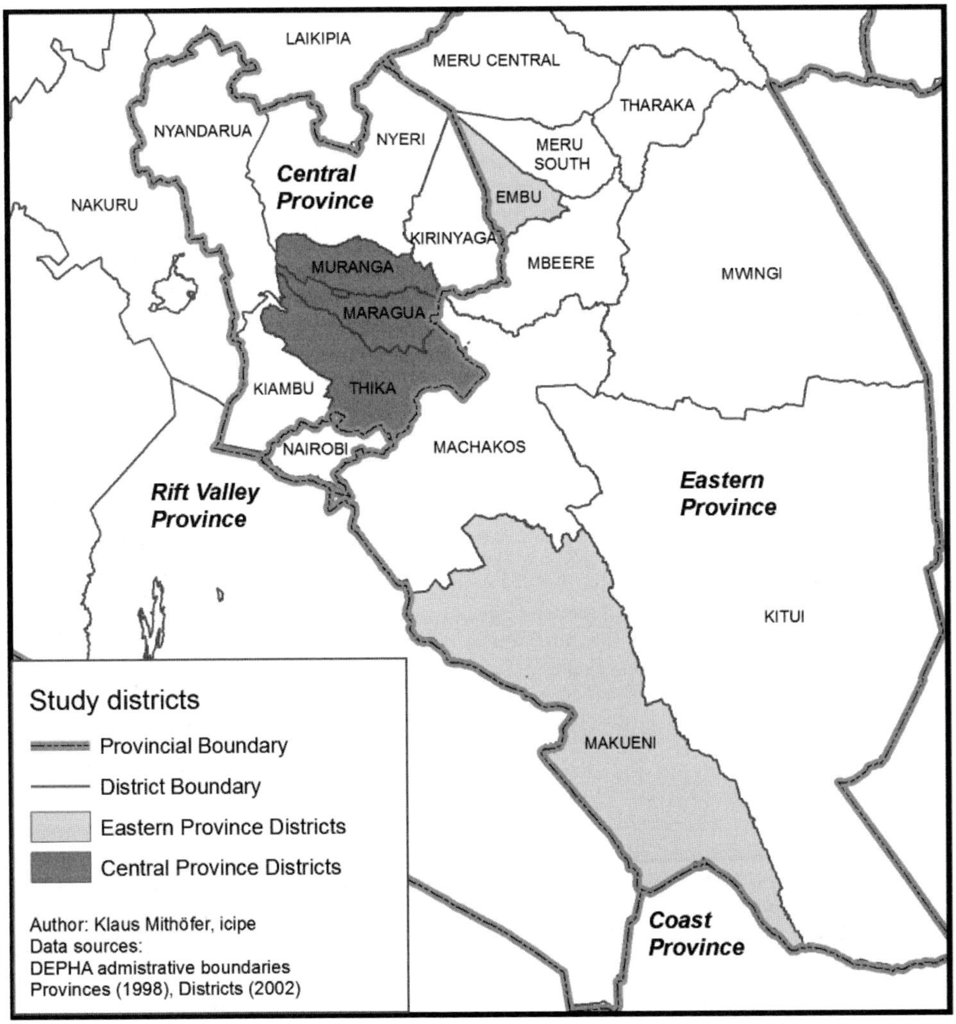

Fig. 16.1. Study area.

control farmers. The control farmers were not members of the two group-based training approaches, but they were sampled from the same villages as the FFS and CIG farmers.

The five study districts were purposely selected. A sampling frame containing all active FFS and CIG horticulture groups and their members was compiled during the formal or informal meetings held in 2007 by district and divisional horticulture extension officers. The sampling units, consisting of small-scale FFS and CIG horticultural producers, were selected from the sampling frame using systematic random sampling. CIG and FFS farmers were selected first, then a sampling frame for control farmers was compiled for the sub-locations selected. From each district, 50% of the sub-locations were randomly selected to acquire the representative sample.

For this study, the sample size was determined following Rea and Parker (1997) as:

$$n = \left(\frac{Z_\alpha \sqrt{p(1-p)}}{Cp} \right)^2 \qquad (16.1)$$

The sample size is given by n; Cp is the confidence interval in terms of proportions and was set at 5%, as this was sufficient to remove 95% bias in sampling. Z_a is the Z score for various levels of confidence (a) and $Z = 1.96$. The proportion of the population containing the major attribute of the population is p, which was assumed at 0.5. Control farmers were selected following the same procedure as for the group farmers. Overall, 33 FFS, 33 CIG and 33 control farmers were selected per district to give a sample of 99 farmers, giving a total sample of 495 horticulture farmers in five survey districts from the two provinces.

To avoid respondent bias, 20 field research assistants who were conversant with the local language of the respondents undertook a 3-day intensive training session on the data collection techniques prior to the survey. During the survey, each of the 20 research assistants completed an average of 25 questionnaires. The survey covered demographic information on farmers such as age, total number of years of schooling, land size, household size and on-farm labour. In addition, information on assets and wealth, main farm and off-farm income sources, pesticide and fertilizer application, horticulture crop portfolio, horticultural training, information access and knowledge of IPM was solicited. Information on group membership, social capital, and the diffusion of information among farmers was also gathered.

Data Analysis

Principal component analysis

Principal component analysis (PCA) is a variable reduction procedure for data sets with a large number of variables that may be correlated with each other. PCA helps to reduce the observed variables to a smaller number of principal components that account for most of the variance in the observed variables. PCA can be applied to IPM as it requires knowledge of different practices. Each of the practices has a knowledge component contributing to the integrated knowledge of IPM practices. However, direct measurement of the knowledge level for each practice by horticulture farmers is difficult, since all the practices simultaneously contribute to integrated knowledge. This study, therefore, chose to use the PCA to obtain an IPM knowledge index that would be a proxy for the knowledge acquired from a range of IPM practices for further analysis.

The PCA method has been used in a number of studies, including that of Amudavi (2005), who investigated the propensity of farmers to adopt integrated natural resource management practices in the Vihiga, Baringo and Embu districts of Kenya. In this study, the PCA was used to generate two components from 12 IPM practices for further analysis.

The PCA procedure begins with a set of K variables: a_{1i}^* to a_{Ki}^* represents knowledge of K IPM practices by the ith farmer, and this is represented by a binary scale: 1 if the farmer knows about each of the practices, measured by checking whether the farmer is able to explain how the practice is to be implemented; 0 if he cannot. The knowledge variable represents the knowledge index of the range of IPM practices. Each variable, a_{1i}^*, is specified by its mean and standard deviation, $a_{1i}i = (a_{1i}^* - a_1^*)/(s_1^*)$, where a_1^* is the mean of a_{1i}^* across all N farmers and s_1^* is the standard deviation. The selected variables are linked with latent components (factors) for each farmer i through the equation:

$$\begin{aligned} a_{1i} &= v_{11} \times A_{1i} + v_{12} \times A_{2i} \ldots\ldots + v_{1K} \\ &\quad \times A_{Ki} \ldots\ldots i = 1, \ldots. N\,(farmer) \\ &\quad k = 1, \ldots\ldots,\ K\,(IPM\ practices) \\ a_{Ni} &= v_{N1} \times A_{1i} + \ldots + v_{KK} \times A_{Ki} \end{aligned} \quad (16.2)$$

where A are the components and v the coefficients on each component for each variable, and these are constant across all households. It is only the left-hand side that is observed, making the solution to the problem indeterminate. The PCA solves this by determining specific linear combinations of the variables with maximum variance accounted for in the first principal component, A_{1i} (Lawley and Maxwell, 1971). The procedure is repeated for each successive component accounting for the maximum of variance remaining. Reversing equation (16.2) yields factor loading from the model that are estimates for each of the K principal components:

$$A_{1i} = f_{11} \times a_{1i} + f_{12} \times a_{2i} ..... + f_{1K} \times a_{Ki}$$
$$A_{K1} = f_{K1} \times a_{1i} + f_{K2} \times a_{2i} ...... + f_{KK} \times a_{Ni} ....i = 1.......N \quad (16.3)$$

where A_{1i} is the first principal component, a_{1i} the normalized variable and f_{11} the factor score coefficient (weight) by which the normalized variable is multiplied to obtain a factor score in the linear combination. Thus, the IPM knowledge index for each farmer is based on the expression

$$A_{1i} = f_{11} \times (a_{1i}^{*} - a_{1}^{*}) / (S_{1}^{*}) + ...f_{1N} \times (a_{ni}^{*} - a_{N}^{*}) / (s_{N}^{*}) \quad (16.4)$$

Furthermore, for the purposes of interpretation, the principal components were rotated using the varimax rotation as proposed by Kaiser (1960). The varimax rotation was used to determine which IPM practices loaded heavily with each of the components. The two principal components with eigenvalue greater than one were extracted and considered as proxy indices for knowledge. They were then used as dependent variables in the recursive binary choice model to assess factors affecting IPM knowledge acquisition and sharing.

Recursive simultaneous binary choice model

Assessing knowledge acquisition and sharing can be analysed using the Heckman two-step sample selection model and the recursive regression model. Heckman's two-stage estimator is the most widely used approach in correcting selection bias that occurs from using non-random selected samples (Heckman, 1978). The first stage involves a selection equation, which is a probit model, to predict the probability. From the probit estimation, the inverse mills ratio (IMR) is generated and then included in the second stage as a regressor to solve the sample selection problem. However, the model requires an exclusion restriction to generate reliable estimates of a variable that appears with a non-zero coefficient in the selection equation, but does not appear in the outcome equation.

Recursive modelling is an analytic tool for studying the relationship between a dependent variable and a collection of predictor variables (Hawkins, 1994), whereby the dependent variable in the first stage becomes an independent variable in the second stage (Heckman, 1978; Maddala, 1983; Greene, 2003). In this study and following Greene (1997), a recursive simultaneous binary choice model is specified to assess the relationship between IPM knowledge acquisition and the sharing of the acquired knowledge as

$$\begin{aligned} y_1 &= x_1'\beta_1 + \varepsilon_1 \qquad y_1 = 1 \text{ if } y_1^{*} > 0,\ 0 \text{ otherwise} \\ y_2 &= x_2'\beta_2 + \delta y_1 + \varepsilon_2 \quad y_2 = 1 \text{ if } y_2^{*} > 0,\ 0 \text{ otherwise} \end{aligned} \quad (16.5)$$

where y_1 is the binary variable (knowledge index) that was generated by using the first two principal components of the 12 IPM practices (see Table 16.7) and y_2 is the dependent variable depicting a binary information-sharing outcome; y_1^* and y_2^* are latent variables for information reception and sharing, respectively; $x_1'\beta_1$ and $x_2'\beta_2$ are index functions of the two equations, and ε_1 and ε_2 are the corresponding disturbance terms assumed not to be normally distributed.

This specification is grounded on the assumption that information acquisition is a necessary condition for sharing the knowledge acquired. This recursive modelling approach imposes less restrictive assumptions than the Heckman approach (Greene, 1997).

Variables in the model

The variables included in this model are shown in Table 16.1. Independent variables that were hypothesized to explain knowledge acquisition included dummies for membership in FFS or CIG with control farmers; number of groups a farmer belongs to (grpnumber); total number of school years of the household head (scholyrs); household members' literacy (hhmbrdwt); gender; age; household size (hhsize); distance to extension services (distextn); total land under horticultural crops (loghectare); number of permanent labourers (permtlbr); number of casual labourers (hwmcaslb); access to horticulture production and pest management information (access); frequency of listening to horticulture production programmes on the radio (freqradio); frequency of reading about horticulture production and pest management in the newspaper (freqnewspaper); and the farmer's locality (district). District is a dummy variable equalling 1 if a farmer lived in a particular district (Muranga, Thika, Maragua or Makueni), and 0 otherwise, with Embu being a reference district.

Participation in FFS and CIG was expected to influence knowledge acquisition and sharing positively and significantly; a higher intensity of training and interaction with trainers enhances knowledge acquisition. Furthermore, participation in FFS and CIG group membership was expected to create social relationships through farmer-to-farmer interaction, leading to information exchange.

It was hypothesized that belonging to many groups could have a significant and positive influence on the acquisition and sharing of IPM knowledge. Farmer groups were expected to facilitate farmer interaction through the creation of social relations, which is a prerequisite for knowledge acquisition and sharing.

Following earlier findings (e.g. Schultz, 1975), we hypothesized that the education of a farmer positively and significantly influenced information acquisition because education enhances the ability of a farmer to acquire, comprehend and process information. Similarly, household member literacy was expected to increase the probability of knowledge acquisition of IPM and was measured as the number of persons who could read and write in the farmers' households.

Gender is a dummy variable that refers to the sex of the farmer, which was hypothesized to influence IPM knowledge acquisition and sharing, since female-headed (compared with male-headed) households were likely to be disadvantaged in information exchanges. For example, a study in Uganda showed that women had a high opportunity cost of time, making them less motivated to participate in information exchanges due to the limited time available for interaction (Katungi *et al.*, 2008). For women to build and maintain a social network was costly in terms of time and other resources (Dasgupta, 2005).

The age of the farm decision maker was expected to influence IPM knowledge acquisition and sharing among smallholder horticulture farmers. Our assumption was that older farmers were not likely to exchange information effectively due to a lack of ability to communicate sufficiently about any new innovation to other farmers. On the other hand, younger farmers were likely to participate in information exchange behaviour since they were capable of interacting with fellow farmers. Thus, the younger the farmer the better she/he could receive or share information on the new technology.

Household size was another important determinant that was expected to affect information flow among farmers. We hypothesized that larger households positively affected information flow activities, since such households had more contacts and a wider social network, which was a prerequisite for information flow. This claim was in agreement with Ketema (2008), who reported the importance of an increased number of household members in creating more contact with different social networks, which meant better access to input and information.

Farm size was expected to influence the farmers' decision whether to acquire IPM knowledge; thus, landholding was

Table 16.1. Description of variables and expected signs.

Variable	Description	Knowledge acquisition	Knowledge sharing
Dependent knowledge index (y_1)	Acquisition of IPM knowledge[a]		
Passinfo (y_2)	Sharing of IPM knowledge[a]		
Independent			
control	Base, not being FFS or CIG member		
FFS	Being FFS member	+	–
CIG	Being CIG member	+	+
Grpnumber	Groups to which farmer belongs, excluding FFS and CIG (*n*)	+	+
Scholyrs	Years schooling (*n*)	+	+
Hhmbrdwt	Household literacy (*n*)	+	+
Gender	Gender of farmer[b]	–	–
Age	Age of farmer	–	–
Hhsize	Household members (*n*)	–	–
Distextn	Distance to extension services (km)		
Loghectare	Total land under horticultural crops (log) (ha)	+	+
Permtlbr	Permanent labourers employed (*n*)	+	+
Hwmcaslb	Casual labourers employed (*n*)	+	+
Access	Whether farmer has adequate (sufficient) access to horticulture production and pest management information[a]	+	
Visitors	Whether farmer receives visitors to share IPM[a]		+
Freqradio	Frequency of listening to radio on horticultural production and pest management information[c]	+	
Freqnewspaper	Frequency of reading newspaper on horticultural production[c]	+	
Embu	Base/reference district		
Maragua	Maragua[d]	+	+
Makueni	Makueni[d]	–	–
Thika	Thika[d]	+	+
Muranga	Muranga[d]	+	+

[a] 1 = yes, 0 = no. [b] 1 = male, 0 = female. [c] 1 = every day, 2 = few times a week, 3 = once a week, 0 = never. [d] 1 = yes, 0 = otherwise.

hypothesized to have a positive influence on acquiring agricultural information such as IPM.

The availability of labour was expected to influence the acquisition and sharing of IPM knowledge positively. A farmer with a large number of permanent labourers was more likely to be in a position to interact with other farmers due to sufficient manpower, which eventually frees up time.

Access to information on horticultural production and pest management information was hypothesized to have a positive influence on IPM knowledge acquisition. Farmers who had adequate access to information could overcome information scarcity through exposure, which could enable them to acquire new agricultural techniques. In contrast, farmers with no access to information would find it difficult to acquire and adopt IPM knowledge, because access to information was expected to increase the probability of acquiring knowledge of IPM. This was measured by assessing whether a farmer had adequate access (farmers receive sufficient horticulture information whenever they need, as determined by the farmers' self-evaluation) to horticulture production and pest management information, with yes = 1 and no = 0.

The distance to extension services was hypothesized as having a negative effect on the probability of IPM knowledge acquisition. This is possibly because farmers who resided closer to extension services were at a greater advantage when it came to acquiring agricultural information. On the other hand, the further a farmer resided from extension services, the less agricultural information he or she acquired. This supposition is in agreement with Ransom *et al.* (2003), who reported that farmers were known to gain access to new information provided through extension services, suggesting the important role that extension services play in disseminating agricultural information.

In addition, receiving visitors into a farmer's home was hypothesized as having a positive and significant influence on IPM knowledge sharing. Farmers who received visitors were likely to share knowledge about new technologies.

Listening to horticulture production and pest management information on the radio was also expected positively to influence IPM knowledge acquisition. This has been shown to create awareness among farmers in East Africa dealing with brassica crops (Nyambo and Löhr, 2005). We expect this variable to enhance the ability of farmers to acquire, synthesize and respond to new information quickly, thereby increasing the probability of IPM knowledge sharing.

In addition, the variable frequency of reading newspaper articles on horticulture production and pest management information was hypothesized as influencing knowledge acquisition on IPM positively. A possible explanation for this is that farmers who are exposed to such articles are likely to acquire new knowledge on a given technology.

Results and Discussion

Descriptive results

The study sample consisted of 487 complete data sets. The descriptive results in Table 16.2 show that the majority (66%) of FFS sample farmers were female. On the other hand, CIG and non-group-based farmers had more male farmers, with only 37% and 24% female farmers, respectively, in the sample. The average age of FFS respondents was 49 years and ranged from 21 to 80 years. The average age of a CIG farmer was 48 years and it ranged from 20 to 82 years. The average number of years of schooling among FFS, CIG and control farmers was 9.2, 10.1 and 9.5, respectively. Land allocated to horticulture among FFS, CIG and non-group-based farmers was around 0.97, 1.03 and 1.01 ha, respectively.

Farmers of the three groups grew a range of horticultural crops including kale, cabbage, tomatoes, bananas, mangoes, passion fruit, French beans, avocados, butternuts, etc. However, a higher share of control farmers (47%) grew kale compared with FFS and CIG farmers, although all farmers ranked kale as their most important crop; kale is a major vegetable consumed by all households in Kenya. FFS farmers grow a smaller variety of crops as compared with CIG and control farmers, which is possibly due to the fact that farmers in FFS rely on key agroecosystem relationships as the basis for making informed management decisions before planting various crops. The total horticulture yield among FFS, CIG and control farmers was recorded as 1804, 2712 and 3255 kg/ha, respectively. These differences in yield among the three groups might be attributed to the variety of crops that farmers grow. On the other hand, the income from the sale of vegetables across FFS, CIG and control farmers was US$405, US$361 and US$322 per household per year, respectively, while income from the sale of fruit generated an average income of US$329, US$383 and US$361 per household per year, respectively. Then again, the negative correlation between crop yield and

Table 16.2. Characteristics of FFS, CIG and control group horticulture farmers.

Variable	FFS (*n* = 157)	CIG (*n* = 159)	Control (*n* = 171)	*p*-value
Gender of household head				
Male	34.39	63.58	76.02	0.000
Female	65.61	36.48	23.98	
Children in household (*n*)	0.8 (0.5)[a]	0.9 (0.5)[a]	0.7 (0.5)[a]	0.638
Adult household members (*n*)	1.3 (0.6)[a]	1.3 (0.6)[a]	1.2 (0.6)[a]	0.403
Horticulture farming practice (years)	20.3 (12.5)[a]	17.3 (12.7)[b]	15.3 (11.5)[b]	0.001
Land under horticultural crops (ha)	0.91 (2.3)[a]	0.87 (2.2)[a]	0.81 (2.4)[a]	0.533
Total horticulture crops output/year (kg/ha)	1804.0 (3.6)[a]	2711.5 (4.2)[a]	3254.6 (3.9)[a]	0.468
Inorganic fertilizer use (kg/ha)	25.55 (5.15)[a]	44.49 (4.40)[a]	52.32 (5.36)[a]	0.453
Organic fertilizer use (kg/ha)	195.90 (8.1)[a]	426.62 (6.1)[b]	654.37 (7.1)[b]	0.001
Pesticide use (kg/ha)	0.27 (6.9)[a]	0.34 (8.3)[a]	0.37 (5.3)[a]	0.288
Total expenditure on inorganic and organic fertilizer (US$)	26.50 (3.2)[a]	36.87 (3.6)[b]	50.90 (3.5)[b]	0.244
Total expenditure on pesticide (US$)	10.21 (2.6)[a]	19.02 (3.3)[b]	17.09 (3.1)[b]	0.001
Distance to extension service (km)	8.21 (8.8)[a]	7.3 (7.3)[a]	6.1 (1.6)[b]	0.045
Distance to Agrovet (km)	6.3 (11.2)[a]	9.1 (40.3)[a]	15.0 (71.5)[a]	0.235
Distance to telephone (km)	3.2 (2.9)[a]	10.5 (51.9)[a]	18.0 (86.3)[a]	0.371
Income from sale of vegetables (US$/year)	405.3 (2.1)[a]	361.4 (2.9)[a]	322.3 (2.6)[a]	0.716
Income from sale of fruit (US$/year)	328.8 (2.9)[a]	383.2 (3.1))[a]	360.7 (2.9)[a]	0.474
Tropical livestock units/total landholding (TLU/ha)	0.4 (2.11)[a]	0.5 (1.9)[a]	0.5 (2.7)[a]	0.083
Permanent labourers engaged (*n*)	0.3 (0.7)[a]	0.3 (0.8)[a]	0.3 (0.7)[a]	0.605
Casual labourers engaged (*n*)	2.4 (3.7)[a]	2.8 (4.9)[a]	2.5 (3.0)[a]	0.608
Family labourers engaged (*n*)	2.6 (2.3)[a]	2.1 (1.7)[a]	2.2 (1.5)[a]	0.097

Figures in parenthesis represent SE; those in a row followed by the same letter are not significantly different at $p < 0.1$ (ANOVA).

income could be attributed to the variety of crops and the different selling price of their produce. The descriptive results further showed that the three most important pesticides used across the three groups were dimethoate, lambda-cyhalothrin and deltamethrin, respectively. Furthermore, the average number of tropical livestock units (TLU) among FFS, CIG and non-group-based farmers was recorded as 0.4, 0.5 and 0.5/ha, respectively. The numbers of casual labourers engaged in farming activities across the three groups averaged 2.4, 2.8 and 2.5, respectively.

Farm accessibility and sources of horticulture production and pest management information

Farmers of the three groups have different access to information, as shown by the results in Table 16.3. Regarding farm accessibility, 55%, 61% and 50% of FFS, CIG and control farmers' farms, respectively, had access throughout the year. On the other hand, 45%, 39% and 46% of FFS, CIG and control farmers' farms, respectively, had access only during the dry season. Around 0.6%, 0.6% and 4.2% of FFS, CIG and control farmers' farms, respectively, had no access all year round. These results were significantly different at 5%, because some farmers live near the main road and have a greater opportunity to acquire agricultural information as compared with others who live in the interior.

On the use of newspaper articles, 7.1%, 5.7% and 1.8% of FFS, CIG and control farmers, respectively, read about horticulture production and pest management in the newspaper daily, while 15%, 20% and 20% of FFS, CIG and control farmers, respectively, read the newspaper a few times a week. In addition, 67%, 58%, and 65% of FFS, CIG and control farmers, respectively, had never read any newspaper articles on horticulture production and pest management.

Forty-four %, 42% and 35% of FFS, CIG and control farmers, respectively, cited Ministry of Agriculture staff as the first and most important source of IPM information, whereas NGOs and friends were second, and neighbours and relatives stood as the third most important source of IPM

Table 16.3. Accessibility of farms and information sources (%) among FFS, CIG and control horticulture farmers.

Variable	FFS (n = 157)	CIG (n = 159)	Control (n = 171)	Chi-square	p-value
Accessibility of farm					
All year	54.80	60.80	50.00	10.258**	0.036
Dry season only	44.60	38.60	45.80		
Never accessible	0.60	0.60	4.20		
Adequate access to horticulture information					
Yes	22.90	17.00	12.30	6.515**	0.038
No	77.10	83.00	87.70		
IPM information sources					
Pamphlet	0.75	0.00	0.00		
Mass media (radio)	0.00	0.83	6.60		
Newspaper	0.75	0.00	1.64		
Field days	5.97	2.50	2.46		
MoA	44.00	41.70	35.30		
Seminars	1.49	0.00	3.28		
Books	0.75	0.00	0.00		
NGOs	12.69	22.50	4.10		
Personal experience	11.19	18.33	11.48	105.482***	0.000
FFS	9.70	0.00	0.00		
CIG	0.00	1.67	2.46		
Jomo Kenyatta University	1.49	2.50	4.92		
Organic institutions	0.00	2.50	0.82		
Friends, relatives, neighbours	9.70	6.67	24.59		
Community leaders	0.75	0.00	0.82		
Agrovets	0.00	0.83	0.82		
Chemical companies	0.75	0.00	0.82		
College	1.49	0.00	0.82		
School	0.00	1.67	4.10		
Use of newspaper					
Daily	7.10	5.70	1.80	9.646*	0.140
Few times a week	14.80	19.80	20.00		
Once a week	11.00	16.60	13.50		
Never	67.10	58.00	64.70		

Statistical significance at the 0.01 (***), 0.05 (**) and 0.1 (*) levels of probability.

information among FFS, CIG and control farmers, respectively. Other information sources used included farmer groups, the radio, books, field days, newspapers, organic institutions, universities and schools. However, during informal group discussions held in 2007 with district and division horticulture officers, we observed that the majority of extension officers in the study area had not attended IPM seminars or training, yet these were the most important source of IPM information for farmers. Munyua and Adams (2007) found that extension officers in Kenya had low to moderate confidence in teaching the IPM concept, attributing this to limited access to information or training on the concept. This implies that the appropriate IPM training of extension workers can greatly improve farmers' knowledge of IPM and good farm management skills.

Integrated pest management technology comprises a range of practices, including physical, chemical, biological, mechanical and cultural techniques that bring the pest populations below the economically damaging level. These practices are described further in Table 16.4.

Knowledge of IPM practices among the three groups of farmers is presented in Table 16.5. Overall knowledge of pesticide application (using selected pesticides, including fungicides, acaricides and insecticides), crop rotation, timely planting and ash application were the most widely known practices. Knowledge of hand picking, mixed cropping, timely planting, crop rotation, ash application, mass trapping, planting resistant varieties and the use of plant extract differed significantly ($p = 0.01$) between the three groups. Moreover, knowledge of sanitation was significantly different ($p = 0.05$) between the three groups.

The results of principal component analysis show that the first component explained 30.5% of the variance in knowledge of IPM practices while the second explained 11% (Table 16.6), and thus both explained the overall figure of 42% of total variation in group variables. This means that the first two components would be correlated with at least some of the observed variables. In this study, following Kaiser (1960), eigenvalues greater than 1 were retained for further analysis.

Furthermore, in this study the components were rotated using varimax rotation with the Kaiser normalization procedure for interpretation. Four of the IPM practices loaded heavily on the first component, while five loaded heavily on the second. The first component had the following IPM practices

Table 16.4. Description of integrated pest management (IPM) practices (from Varela *et al.*, 2003).

Practice	Description
Pesticide application	Using selected pesticides including fungicides, insecticides, acaricides, etc.
Mass trapping	Use of traps to catch large proportion of pest population
Hand picking	Manual removal of pests
Deep ploughing	Killing the pupae of caterpillars and thrips by exposing them to the sun and natural enemies
Plant resistance	Plant breeding for increased genetic resistance to pest damage
Mixed cropping	Mixed cropping of plants with other plants less susceptible or even antagonistic to certain pests and diseases
Timely planting	Planting crops during the appropriate season
Sanitation	Destroying sources of infestation such as crop residues, stems and leaves
Crop rotation	Rotation of crops in different seasons (same purpose as for mixed cropping)
Solarization	Covering soil with clear or transparent polyethylene sheets (mulch) for 3 months: exposure to sunshine reduces soil-borne pests and diseases
Application of ash	Ash application
Plant extracts	Using plant extracts such as neem, alovera, chilli

Table 16.5. Knowledge of integrated pest management (IPM) practices among FFS, CIG and control-based farmers (%).

Knowledge of IPM practices	FFS (n = 157)	CIG (n = 159)	Control (n = 171)	Chi-square	p-value
Pesticide application	69.43	71.07	72.51	0.379	0.827
Mass trapping	21.02	17.61	8.77	10.042***	0.007
Hand picking	57.96	38.99	35.09	19.562***	0.000
Deep ploughing	58.60	50.94	50.29	2.744	0.254
Plant resistance	42.04	34.59	29.24	5.913*	0.052
Mixed cropping	82.80	54.09	52.63	39.637***	0.000
Timely planting	61.15	59.12	41.52	15.575***	0.000
Sanitation	52.87	52.20	40.35	6.607**	0.037
Crop rotation	89.81	71.70	71.35	20.367***	0.000
Solarization	9.55	6.29	4.68	3.169	0.205
Application of ash	73.89	62.26	50.88	18.399***	0.000
Plant extracts	37.58	28.93	22.22	9.329***	0.009

Statistical significance at the 0.01 (***), 0.05 (**) and 0.1 (*) levels of probability.

Table 16.6. Principal component analysis of the range of integrated pest management (IPM) practices among horticulture farmers.

IPM practice	Principal component[a] 1	2
Pesticide application	–0.002	0.333
Mass trapping	0.017	0.414
Hand picking	0.259	0.191
Deep ploughing	0.460	–0.041
Resistant plant varieties	0.027	0.399
Mixed cropping	0.484	–0.066
Timely planting	0.320	0.152
Sanitation	0.074	0.309
Crop rotation	0.514	–0.140
Solarization	–0.119	0.453
Application of ash	0.310	0.063
Plant extracts	0.047	0.408
Eigenvalue (4.97)	3.650	1.320
Percentage of variance explained (41.49)	30.480	11.010

[a] Extraction method, principal component analysis; rotation method, varimax with Kaiser normalization.

loading heavily: crop rotation, mixed cropping, deep ploughing and timely planting. On the other hand, ash application, mass trapping, plant extracts, resistant plant varieties and sanitation heavily loaded on the second component. This suggests that the above-mentioned variables are major contributors to components one and two.

Marginal effect results on IPM knowledge acquisition and sharing

Marginal effect results on factors affecting IPM knowledge acquisition and sharing of that acquired knowledge are presented in Table 16.7. Our findings show that participation in FFS increased the marginal

Table 16.7. Marginal effect results on determinants of integrated pest management (IPM) knowledge acquisition and sharing by horticulture farmers.

Variable	Knowledge acquisition[a]	Knowledge sharing[b]
FFS	0.278***	–0.242***
CIG	0.198***	0.075
No. groups to which farmer belongs, excluding FFS and CIG	0.068***	–0.036
No. school years	–0.007	0.011
Household literacy	0.025**	0.016
Age	0.002	0.002
Gender	–0.041	–0.125**
Household size	–0.022**	0.001
Land under horticulture (log ha)	–0.058**	–0.012
Permanent labourers employed	–0.060**	–0.021
Casual labourers employed	–0.001	0.017
Access to horticultural production information	–0.206***	
IPM knowledge acquisition		0.863***
Distance to extension services	–0.012***	
Farmers' visitors		0.573***
Frequency of listening to radio on horticulture production and pest management information	–0.041**	
Frequency of reading newspaper on horticulture production and pest management information	–0.037**	
Maragua	0.205**	–0.243***
Makueni	0.206**	–0.220**
Thika	0.043	0.181**
Muranga	0.107	–0.132
Log pseudo-likelihood	–225.155	–208.373
Number of observations	487.0	487.0
Wald χ^2 (19)	97.92	166.17
p > chi^2	0.0000	0.0000
Pseudo-R^2	0.2194	0.3526

[a] Index: 1 = yes, 0 = no. [b] Index: yes = 1, 0 = otherwise. Statistical significance at the 0.01 (***), 0.05 (**) and 0.1 (*) levels of probability. The marginal effects were computed by the means of the parameters.

probability of IPM knowledge acquisition by 0.278; these results were significant at 1%, proving the importance of FFS membership in obtaining IPM knowledge. This finding corroborates those of Pontius *et al.* (2002), Godtland *et al.* (2004) and Yang *et al.* (2008), who reported that FFS plays a major role in extending IPM knowledge to farmers. For example, the strategy of the FFS approach was not to train all farmers in the community, but rather to catalyse the spread of knowledge about technologies and practices through farmer-to-farmer diffusion (Feder *et al.*, 2004b). Our results indicate that participation in CIG significantly (p <0.05) increased the marginal probability of IPM knowledge acquisition by 0.198, which is lower than that for FFS membership.

However, the probability of IPM knowledge sharing among FFS farmers was significantly (p <0.05) negative, with a marginal probability of 0.242. The significant and negative influence of FFS participation in knowledge sharing might have reflected the lack of motivation among FFS farmers to disseminate their knowledge of IPM to other farmers. The lack of information sharing among the FFS farmers could also explain why the majority of control farmers lacked information on IPM. The findings of this study agree with those of

Feder *et al.* (2004a), who reported the lack of significant diffusion of knowledge from FFS participants to farmers residing in the same village. Similarly, being a member of a CIG group did not increase the marginal probability of sharing IPM knowledge, possibly since CIG farmers are more interested in market and marketing information.

According to the marginal effect result, FFS and CIG membership, the number of groups to which farmers belonged (excluding FFS and CIG), farmer household members' literacy and locality positively and significantly affected IPM knowledge acquisition, while household size, land size, permanent labour, casual labour, access to horticulture production information, distance to extension services, farmer visitors, frequency of listening to horticulture production information on the radio and frequency of reading newspaper articles on horticulture production negatively and significantly affected IPM knowledge acquisition. Knowledge sharing was significantly and positively linked to the number of casual labourers employed, IPM knowledge acquisition and the number of visitors received, while membership in FFS, gender and locality significantly and negatively affected IPM knowledge sharing.

From our results, belonging to more than one association appeared to have a positive effect on farmers' ability to acquire IPM knowledge, with the marginal probability of 0.068. The significant and positive influence of group membership plays a major role in enhancing farmer interaction which, in turn, enhances IPM knowledge acquisition. Katungi *et al.* (2008) reported that belonging to more associations in rural Uganda appeared to have a strong effect on two-way informal information exchanges, increasing the likelihood that both men and women would engage in information pooling with others.

Literacy among household members provides external effects in terms of benefits to illiterate household members (Basu *et al.*, 2000). In this study, literacy among members of a household increased the probability of acquiring IPM knowledge by 0.022; households with literate members were more likely to acquire IPM information that could subsequently benefit the farmer.

Results on gender indicated that IPM knowledge sharing among female farmers decreased by 0.125, possibly because female farmers were likely to be engaged in household responsibilities that could limit their interaction with other farmers.

The findings on household size showed that the decrease in the number of members negatively influenced the probability of acquiring IPM knowledge by 0.022. Farmers with fewer household members might have more responsibilities that could affect the time left for interaction, as well as for attending agricultural practice enrichment programmes.

Less land under horticulture crops decreased the probability of acquiring IPM information by 0.058; the results were significant at 5%, but with a negative influence on the marginal probability of acquiring IPM knowledge. This finding could imply that farmers with small amounts of land could be risk averse to new technologies. Feder *et al.* (1985), in their adoption studies, demonstrated that less land is a constraint to the rapid adoption of new technologies.

The findings for permanent labour showed that farmers who had employed fewer permanent labourers were less likely to acquire IPM knowledge as compared with those with more permanent labourers. The results were significant at 5%, but with a negative influence on the marginal probability of acquiring IPM knowledge. A possible explanation for these results is that farmers with more permanent labourers were able to afford more time for interaction, resulting in a wider social network compared with farmers having fewer permanent labourers.

On access to horticulture production and pest management information, the findings showed that farmers with insufficient access were less likely to acquire knowledge of IPM compared with those who had adequate access. The results were significant at 1% but with a negative influence on the marginal probability of 0.206, suggesting that the acquisition of information is influenced by accessibility. This means that for a better acquisition of knowledge there is a need to ensure that the sources of information for farmers are easily available and accessible.

The distance to extension services decreased the probability of IPM knowledge acquisition by 0.012, being significant at 1%. This negative influence of distance to the extension service confirms the importance of accessibility of the public extension services, as indicated earlier by the descriptive results. This finding is in agreement with that of Chilot (1994), who reported that the increase in distance to extension services negatively affects agricultural information reception among farmers. This may be due to an increase in the transaction costs associated with distance.

The findings for farmers who received visitors and shared IPM information with them increased the probability of IPM knowledge acquisition by 0.559, the results being significant at 1%. The significance and positive influence of visitors illustrates the importance of farmer-to-farmer interaction in exchanging ideas on new technologies such as IPM.

As expected, the findings showed that farmers who had acquired IPM knowledge were likely to share it. The magnitude of the marginal influence was 0.881, emphasizing the need for training and encouraging farmer-to-farmer extension.

Locality was found to be one of the factors that influenced IPM knowledge acquisition and sharing. In this study, Embu district was taken as a base for the purposes of comparison. It has a diversity of agroecological conditions, ranging from high altitude or temperate vegetable zone (UM1) to very dry low land (Ouma *et al.*, 2002), which serve as a representative for each district. The results showed that farmers in Maragua district were more likely to acquire IPM knowledge compared with those in Embu. The results were significant at 1%, with a marginal probability of 0.205. The significant and positive influence of locality reflected the important role played by proximity to the capital city and good infrastructure. Maragua district has a high farmer density, is nearer to Nairobi, has good infrastructure, has available extension staff, and has many farmer groups such as FFS and CIG, these factors making access to information sources easier and providing it with a large market. The result showed farmers' residence in Makueni district to be statistically significant, in terms of IPM knowledge acquisition, at 5%. Farmers' residence in Makueni, compared with residence in Embu, was more associated with IPM acquisition. This suggests that there are more information-receiving and -sharing networks in Makueni than in Embu.

On the one hand, the probability of sharing IPM knowledge among farmers who resided in Maragua and Makueni as compared with Embu decreased by 0.243 and 0.220, respectively; on the other, the probability of sharing IPM knowledge among farmers who resided in Thika, as compared with Embu, increased by 0.181, the results being significant at 5%. This suggests that the proximity of Thika district to the outskirts of Nairobi and the availability of information-sharing networks could be instrumental here.

The marginal effect results show that the decrease in frequency of listening to the radio and reading newspapers among farmers decreased the probability of acquiring IPM knowledge, the magnitude of the marginal influence being 0.041 and 0.037, respectively. This is possibly because farmers who spend less time listening to the radio and reading newspaper articles on horticulture production are less likely to acquire IPM knowledge.

Summary and Conclusions

Horticulture is an important sector of the Kenyan economy, with a range of pest and disease problems that are mostly dealt with by chemical control. Recently, group-based training approaches have been promoted in Kenya to accelerate the dissemination of new technologies and practices among farmers through various learning and communication networks. Among the most popular examples are the farmer field school (FFS) approach and common interest groups (CIGs). The former approach concentrates more on production and management, whereas the latter centres on commercial opportunities. The aim of this study was to assess the extent to which FFS and CIGs contribute to increasing the knowledge and dissemination of information on integrated pest management. The analysis was based on a random sample of 495 active horticulture farmers in five districts of Kenya. The horticulture farmers were sampled from FFS and CIGs, as well as control farmers who did not belong to either of these two groups. Principal component analysis was used to generate a binary variable (knowledge index), and a recursive regression model was used to analyse the acquisition and sharing of IPM knowledge.

The findings show that the major crops grown by the three categories of farmer were kale, cabbage, tomatoes, bananas, mangoes, passion fruit, French beans, avocados and butternuts. The main sources of income were from the sale of vegetables, fruit, livestock or livestock products and maize, as well as from business and wages. Farmer field schools seemed to be more successful in reaching out to female farmers, as shown by the high proportion of female participants, in contrast to the CIGs and control farmers. In identifying information sources, farmers ranked the Ministry of Agriculture staff, NGOs and other Kenyans (friends, neighbours and relatives) as the three most important sources of horticulture production and pest management information. FFS, CIG and control farmers used different sources of information.

The principal component analysis result showed that the first and second components explained 42% of the total variation in groups. Furthermore, the results indicated that the first component represented latent knowledge gained from crop rotation, mixed cropping, deep ploughing and timely planting while the second was driven by ash application, mass trapping, plant extract, varieties of resistant plants and sanitation.

The findings of the marginal effect indicated that farmer participation in FFS and CIG was more likely to enable the acquisition of IPM knowledge, but it could not enable the sharing of acquired knowledge with other farmers. The study found that the distance to extension services was a major constraint to information flow, and therefore this study recommends that the distance to extension services be reduced by organizing farmers in groups, as well as by increasing the number of well-trained extension staff with an adequate provision of transport facilities to reach out to farmers who live far from the extension service. It is important to consider a large coverage and good road network with an efficient public transport system. The marginal effect result revealed that group membership was of importance in the decision to acquire IPM knowledge, since group interaction enhanced the ability of farmers to acquire information. The study showed that some districts (Maragua and Makueni) facilitated knowledge acquisition, which implies that proximity to Nairobi and IPM information networks plays a major role in knowledge acquisition. In addition, the formation of farmer groups was found to be an important factor in the area of promoting improved horticulture production and IPM, since farmer groups were mentioned as one of the most important sources of IPM information among FFS and CIG groups. A farmer's membership of more than one association plays a major role in the acquisition of IPM knowledge, possibly because group membership enhances farmer interaction, which can be a prerequisite of knowledge acquisition. In addition, group membership has some important characteristics that include better handling of risks, and it is considered a source of resources and inputs.

Policy implication

The findings of this study show that farmers in FFS and CIG groups acquired information on IPM but did not share the information with other farmers, although they were expected to do so. Recently, the aforementioned group-based training approaches have been promoted in Kenya to accelerate dissemination of new technologies among farmers, through various learning and communication networks. It is, therefore, imperative that farmers be encouraged to become part of the two group-based training approaches in order to gain IPM knowledge that is crucial in enhancing their horticultural production. Furthermore, the study suggests that the distance to extension services be reduced by organizing farmers in groups, as well as by providing good road networks with extensive coverage to inaccessible areas.

Further research

From the study, it is not clear why FFS and CIG farmers do not share IPM knowledge, yet they are trained in the IPM concept and have knowledge of IPM practices. In the circumstances, further research is, therefore, recommended to clarify and identify the factors influencing the sharing of information among farmers who participate in group social networks. In addition, further research is needed to investigate the cost-effectiveness of the extension methods used in the FFS and CIG group-based approaches.

Acknowledgements

The Integrated Pest Management Training and Information Flow among Smallholder Horticulture Farmers in Kenya study is supported by the German Academic Exchange Programme (DAAD) and the Impact Assessment Project of the International Centre of Insect Physiology and Ecology (*icipe*) through a grant from the German Development Cooperation (BMZ-GTZ). Thanks are due also to the district agricultural officers who helped in developing the sample frames during the preliminary surveys.

References

Amudavi, D. (2005) Linking institutional partnerships, community groups and rural livelihood improvement in Kenya. PhD dissertation, Cornell University, Ithaca, New York.

Antholt, C.H. (1993) Getting ready for the twenty-first century: Technical change and institutional modernization in agriculture. World Bank technical paper 217, Asia technical department series, World Bank, Washington, DC.

Basu, K., Foster, E.J. and Subramanian, S. (2000) Isolated and proximate illiteracy: And why these concepts matter in measuring literacy and designing education programmes. *Economic and Political Weekly* 35(1/2), 35–39.

Chambers, R. (1997) Whose reality counts? Putting the first last. Publication 71, Intermediate Technology Publications, London.

Chilot, Y. (1994) Factors Influencing adoption of new wheat technologies in the Wolmera and Addis Alem areas of Ethiopia. MSc thesis presented to the school of graduate studies of Alemaya University of Agriculture, Ethiopia, pp. 108.

Dasgupta, P. (2005) Economics of social capital. *The Economic Record* 81 (special issue), S2–S21.

Ekoja, I. (2003) Farmer's access to agricultural information in Nigeria. *Bulletin of the American Society for Information Science and Technology* August/September, 21–23.

FAO (Food and Agriculture Organization) (2001) Farmer innovation and new technology options for food production, income generation and combating desertification (KEN/99/200). Progress Report – 2001. Food and Agriculture Organization of the United Nations, Nairobi, Kenya.

Feder, G., Just, R. and Zilberman, D. (1985) Adoption of agricultural innovations in developing countries: A survey. *Economic Development and Cultural Change* 28, 119–133.

Feder, G., Murgai, R. and Quizon, J.B. (2004a) Sending farmers back to school: The impact of FFS in Indonesia. *Review of Agricultural Economics* 26(1), 45–62.

Feder, G., Murgai, R. and Quizon, J.B. (2004b) The acquisition and diffusion of knowledge: The case of pest management training in farmer field schools, Indonesia. *Journal of Agricultural Economics* 55(2), 217–239.

FPEAK (2006) Horticultural insight. Signs of the times from loose packs to high-care. *Quarterly Publication of the Fresh Produce Exporters Association of Kenya* January–March, 2.
Gautam, M. (2000) *Agricultural Extension: The Kenya Experience. An Impact Evaluation*. The World Bank, Washington, DC.
GHA (2005) Global Horticulture Assessment Report. USAID, Washington, DC.
Githaiga, W. (2007) Economic and social impacts of the common interest group approach to public agricultural extension in Kenya. *Journal of Developments in Sustainable Agriculture* 2, 159–196.
Godtland, E., Sadoulet, E., De Janvry, A., Murgai, R. and Ortiz, O. (2004) The impact of farmer field schools on knowledge and productivity: A study of potato farmers in the Peruvian Andes. *Economic Development and Cultural Change* 53(1), 63–92.
Greene, W. (1997) *Econometric Analysis*, 3rd edn. Prentice Hall, Upper Saddle River, New Jersey.
Greene, W. (2003) *Econometric Analysis*, 5th edn. Prentice Hall, Upper Saddle River, New Jersey.
Hawkins, D.M. (1994) *FIRM: Formal Inference-based Recursive Modelling*. Department of Applied Statistics, University of Minnesota, Minneapolis, St Paul.
Heckman, J. (1978) Dummy endogenous variables in a simultaneous equation system. *Econometrica* 46(4), 931–959.
IHC (2006) Introduction. In: Janick, J. (ed.) *Proceedings of the XXVII International Horticultural Congress on Global Horticulture: Diversity and Harmony*, 13–19 August, Seoul, Korea.
Kaiser, H.F. (1960) The application of electronic computers to factorial analysis. *Education and Psychological Measurement* 20, 141–151.
Katungi. E., Edmeades, S. and Smale, M. (2008) Gender, social capital and information exchange in rural Uganda. *Journal of International Development* 20(1), 35–52.
Ketema, D. (2008) Social networks and diffusion of agricultural technology: The case of sorghum in Metema Woreda, North Gondar, Ethiopia. MSc Thesis, Haramaya University, Alemaya, Ethiopia.
Lapar, L.A., Binh, V.T., Son, N.T., Toingo, M. and Staal, S. (2006) The role of collective action in overcoming barriers to market access by smallholder producers: Some empirical evidence from Northern Vietnam. *Research Workshop on Collective Action and Market Access for Smallholders*, 2–5 October, Cali, Colombia.
Lawley, D. and Maxwell, A. (1971) *Factor Analysis as a Statistical Method*. Butterworth, London.
Maddala, G.S. (1983) *Limited Dependent and Qualitative Variables in Econometrics*. Cambridge University Press, Cambridge, UK.
Munyua, C.N. and Adams, P.F. (2007) Confidence in teaching integrated pest management (IPM) concepts, access to and preferred sources of information: A case study of agricultural extension officers in Kenya. *South African Journal of Agricultural Extension* 36(1), 203–216.
Nyambo, B. and Löhr, B. (2005) The role and significance of farmer participation in biocontrol-based IPM for Brassica crops in East Africa (http://www.bugwood.org/arthropod2005/vol1/6b.pdf).
Ouma, J., Murithi, F., Mwangi, W., Verkuijl, H., Gethi, M. and De Groote, H. (2002) *Adoption of Maize Seed and Fertilizer Technologies in Embu District, Kenya*. CIMMYT, Texcoco, Mexico.
Pontius, J., Dilts, R. and Bartlett, A. (eds) (2002) *Ten Years of IPM Training in Asia – From Farmer Field School to Community IPM*. FAO, Bangkok.
Raini, R., Hoffmann, V. and Zebitz, C.P.W. (2005) Integrated pest management (IPM) information flow: Case study of tomato stakeholders' practices in Kenya. In: *Conference on International Agricultural Research for Development*, 11–13 October. Hohenheim University, Stuttgart, Germany.
Ransom, J., Paudyal, K. and Adhikari, K. (2003) Adoption of improved maize varieties in the hills of Nepal. *Journal of the International Association of Agricultural Economics* 29(3), 299–305.
Rea, L.M. and Parker, R.M. (1997) *Designing and Conducting Survey Research: a Comprehensive Guide*, 2nd edn. Jossey-Bass, San Francisco, California, pp. 114–126.
Rees, D., Momanyi, M., Wekundah, J., Ndungu, F., Odondi, J., Oyure, A.O., Andima, D., Kamau, M., Ndubi, J., Musembi, E., Mwaura, L. and Joldersma, R. (2000) Agricultural knowledge and information systems in Kenya – Implications for technology dissemination and development. Agricultural Research and Extension Network Paper 107, ODI, London.
Republic of Kenya (2001) National agriculture and livestock extension programme (NALEP). Implementation Framework. Nairobi, Kenya.
Republic of Kenya (2005) Review of the National Agricultural Extension Policy (NEAP) and its implementation. Vol. II: main report and annexes. Ministry of Agriculture and Ministry of Livestock and Fisheries Development report, Nairobi, Kenya.

Republic of Kenya (2006a) National agriculture and livestock extension programme (NALEP) report, Nairobi, Kenya.

Republic of Kenya (2006b) *Farm Management Handbook of Kenya II – Natural Conditions and Farm Management Information*, 2nd edn. Part C: East Kenya; Subpart C1: Eastern Province, Nairobi, Kenya.

Republic of Kenya (2010) Ministry of Environment and Mineral Resources. Kenya Meteorological Department. Press Release. Ref. No. Met/ 7/23, Nairobi, Kenya.

Schultz, T.W. (1975) The value of the ability to deal with disequilibria. *Journal of Economic Literature* 13, 827–846.

Simpson, B.M. and Owens, M. (2002) Farmer field schools and the future of agricultural extension in Africa. SD dimensions. Sustainable Development Department, FAO, Rome, pp. 1–8 (www.fao.org/sd/2002/KN0702a_en.htm).

Sithanantham, S. (2004) *Development and Dissemination of IPM for Vegetables in Eastern Africa. icipe* Science Press, Nairobi, Kenya.

Van de Fliert, E., Pontius, J. and Roling, N. (1995) Searching for strategies to replicate a successful extension approach: Training of IPM trainers in Indonesia. *Journal of Agricultural Extension and Education* 1(4), 41–63.

Varela, A., Seif, A. and Löhr, B. (2003) *A Guide to IPM in Brassicas Production in Eastern and Southern Africa. icipe* Science Press, Nairobi, Kenya.

Wasilwa, L. (2008) Horticulture for food. Report, Kenya Agricultural Research Institute, Nairobi, Kenya.

Witt, R., Pemsl, D.E. and Waibel, H. (2008) The farmer field school in Senegal: Does training intensity affect diffusion of information? *Journal of International Agricultural and Extension Education* 15(2), 47–60.

Yang, P., Liu, W., Shan, X., Zhou, J., Lu, J. and Li, Y. (2008) Effects of training on acquisition of pest management knowledge and skills by small vegetable farmers. *Crop Protection* 27(12), 1504–1510.

Index

Page numbers in **Bold** refer to figures, tables and boxes. Suffix 'n' refers to endnotes.